Aquatic Insects: Biodiversity, Ecology and Conservation Challenges

Aquatic Insects: Biodiversity, Ecology and Conservation Challenges

Editors

Marina Vilenica
Zohar Yanai
Laurent Vuataz

MDPI • Basel • Beijing • Wuhan • Barcelona • Belgrade • Manchester • Tokyo • Cluj • Tianjin

Editors
Marina Vilenica
University of Zagreb
Croatia

Zohar Yanai
Tel Aviv University
Israel

Laurent Vuataz
Museum of Zoology in Lausanne
Switzerland

Editorial Office
MDPI
St. Alban-Anlage 66
4052 Basel, Switzerland

This is a reprint of articles from the Special Issue published online in the open access journal *Diversity* (ISSN 1424-2818) (available at: https://www.mdpi.com/journal/diversity/special_issues/aquat_insects).

For citation purposes, cite each article independently as indicated on the article page online and as indicated below:

LastName, A.A.; LastName, B.B.; LastName, C.C. Article Title. *Journal Name* **Year**, *Volume Number*, Page Range.

ISBN 978-3-0365-4941-5 (Hbk)
ISBN 978-3-0365-4942-2 (PDF)

Cover image courtesy of Zohar Yanai

Contents

About the Editors

Marina Vilenica

Marina Vilenica (Assistant professor at University of Zagreb, Faculty of Teacher Education) is an aquatic insect entomologist with a special focus on mayflies and dragonflies of the Balkan Peninsula (Southeastern Europe). She is interested in their taxonomy, conservation status, and ecological requirements in both natural and anthropogenically impacted habitats.

Zohar Yanai

Zohar Yanai (The Steinhardt Museum of Natural History, Tel Aviv) is an entomologist and freshwater ecologist. He studies taxonomic and ecological aspects of various aquatic insects, with a special focus on mayflies (order Ephemeroptera) in the Middle East. He addresses topical conservation issues in Israeli streams, where diversity and distribution patterns of aquatic invertebrates reveal threats such as biological invasions, water diversion and pollution, and hydrobiological barriers.

Laurent Vuataz

Laurent Vuataz (research fellow at Museum of zoology in Lausanne, Switzerland) is a hydrobiologist focusing on freshwater diversity, with a particular interest in aquatic insects. He has expertise in phylogenetics, systematics, barcoding, species delimitation, and river bioindication, with experience in both the academic and private sectors.

MDPI

Editorial

Introduction to the Special Issue "Aquatic Insects: Biodiversity, Ecology, and Conservation Challenges"

Marina Vilenica [1,*], Laurent Vuataz [2,3,*] and Zohar Yanai [4,*]

1 Faculty of Teacher Education, University of Zagreb, Trg Matice Hrvatske 12, 44250 Petrinja, Croatia
2 Museum of Zoology, Palais de Rumine, Place de la Riponne 6, 1014 Lausanne, Switzerland
3 Department of Ecology and Evolution, University of Lausanne (UNIL), 1015 Lausanne, Switzerland
4 The Steinhardt Museum of Natural History, Tel Aviv University, Tel Aviv 6997801, Israel
* Correspondence: marina.vilenica@ufzg.hr (M.V.); laurent.vuataz@vd.ch (L.V.); yanai.zohar@gmail.com (Z.Y.)

Citation: Vilenica, M.; Vuataz, L.; Yanai, Z. Introduction to the Special Issue "Aquatic Insects: Biodiversity, Ecology, and Conservation Challenges". *Diversity* **2022**, *14*, 573. https://doi.org/10.3390/d14070573

Received: 4 July 2022
Accepted: 15 July 2022
Published: 18 July 2022

Publisher's Note: MDPI stays neutral with regard to jurisdictional claims in published maps and institutional affiliations.

In non-marine environments, insects comprise one of the most species-rich and abundant groups of organisms. They have always been the focus of scientific attention on freshwater habitats, such as streams, rivers, lakes, and ponds. Although such habitats cover only 2.3% of the Earth's surface, they accommodate approximately 10% of all known animal species, and represent "hotspots of endangerment" due to disproportionally high biodiversity and anthropogenic pressures [1]. More than 60% of the freshwater species diversity is represented by aquatic insects, with approximately 130,000 described extant species [2–4]. They spend one or more stages of their life cycle in aquatic habitats, with the majority moving to terrestrial areas as adults. Members of the orders Ephemeroptera, Plecoptera, Trichoptera, Megaloptera and Odonata are exclusively aquatic in their immature stages (i.e., nymphs and larvae). Several other insect orders, such as Diptera, Coleoptera, Neuroptera, and Hemiptera also have many aquatic representatives [3,4].

Aquatic insects have important ecological roles in both aquatic and terrestrial habitats as primary consumers, detritivores, and predators. Moreover, they dominate in terms of biomass and productivity, representing an important food resource for a vast number of aquatic and terrestrial, both invertebrate and vertebrate, predators. Therefore, they represent an important link in food and energy transfer from aquatic to terrestrial ecosystems [4]. The composition and structure of their communities are closely related to habitat type, abiotic parameters (e.g., water temperature, water depth, water velocity, oxygen content, pH), predation, microhabitat (substrate) composition, and available food resources. Many aquatic insects, such as mayflies, stoneflies, and caddisflies, have shown to be highly sensitive to anthropogenic alterations in their habitats and have been widely used as valuable taxonomic groups for biomonitoring programs worldwide [5]. Some aquatic insects, such as mosquitoes, have been well-studied due to their important role as disease vectors [6]. In a more anthropocentric view, many aquatic insects are crucial for the provision and support of various ecosystem services (e.g., [7]). Although the efforts of aquatic entomologists tremendously increased during the 21st century, much is yet undiscovered. Our knowledge about aquatic insects is still far from being complete, both in natural systems, such as springs, rivers, streams, lakes, but also in artificial habitats, such as irrigation canals and man-made reservoirs.

The current Special Issue addresses all aspects of biodiversity of aquatic insects, including taxonomic diversity and phylogeny, distribution patterns, and community ecology. In addition to increasing fundamental knowledge, such data are crucial for understanding the importance of anthropogenic disturbances and mitigating their unknown impacts. In a context of rapid global biodiversity loss, encouraging signals, such as the upward trend in the abundance of freshwater insects in some regions, have recently been detected [8]. Past efforts to improve water quality and restore habitats, which most probably explain part of this positive trend, should motivate the scientific community to become even more involved in describing, understanding, and protecting freshwater ecosystems in the future.

Most of the important insect orders with aquatic representatives are represented in this Special Issue (Figure 1). Geographically speaking, the papers published here represent almost all continents and almost all biogeographical realms (Figure 2). Aquatic entomology is widely investigated in Australia, Europe, and North America. The trends reflected here indicate an increasing interest in certain African and Asian countries. This is mainly due to the efforts of local research groups, highlighting the importance of training and supporting scientists in additional countries.

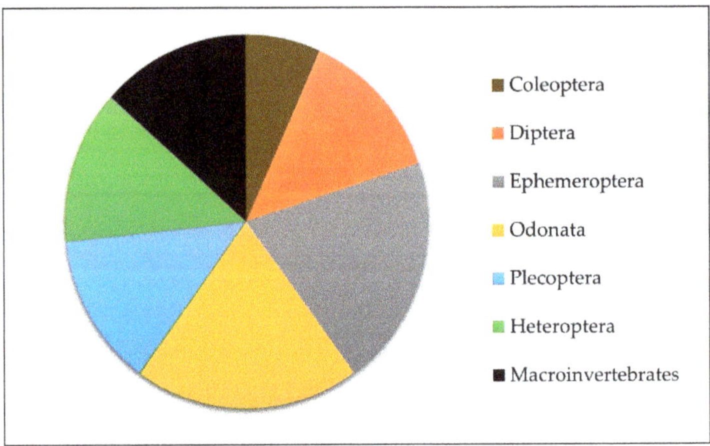

Figure 1. Aquatic insect orders investigated in studies published in the *Diversity* Special Issue: Aquatic Insects: biodiversity, ecology, and conservation challenges.

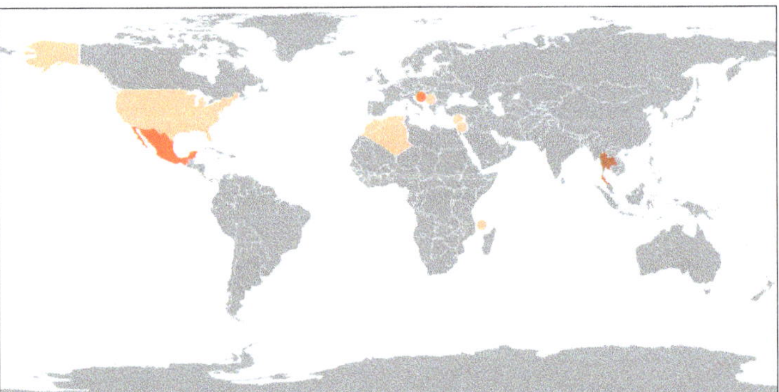

Figure 2. Distribution of contributions to the Special Issue. Colored countries are represented in the Special Issue; color shade represents the number of contributions from a country (light—1, dark—3).

The most speciose aquatic insect orders, Coleoptera and Diptera, are represented with only one [9] and two contributions [10,11], respectively. Studying the aquatic Coleoptera, the authors have reviewed the *Laccophilus alluaudi* species group from the Comoro Islands, and recognized five species, out of which four were newly described [9]. Both publications regarding the aquatic Diptera focused on the very diverse but still relatively poorly investigated family Chironomidae in southeastern Europe. One study investigated the diversity of periphytic Chironomidae in a floodplain aquatic ecosystem and revealed a high dependency of diversity of chironomid assemblages on substrate type, where the highest diversity was recorded on macrophytes [10]. The other study investigated Chironomidae assemblages in urban water bodies, which showed relatively high diversity but also different tolerance levels of chironomid taxa to environmental pressures recorded in such aquatic systems [11].

Relatively small orders of aquatic insects, Ephemeroptera [12–14] and Odonata [15–17], are represented with three papers each. New data about the diversity and distribution of Moroccan mayflies are presented. Currently known Moroccan mayfly fauna consists of 54 species and is characterized by a clear dominance of Mediterranean groups with a strong rate of endemism [12]. Phylogenomic analyses of the family Coloburiscidae, which consists of three extant genera with a Gondwanan distribution (*Coloburiscoides* from Australia, *Coloburiscus* from New Zealand, and *Murphyella* from Chile), confirmed its monophyletic origin [13]. The third publication on mayflies resolved the taxonomy, distribution, and life cycle of the Maghrebian endemic mayfly species, *Rhithrogena sartorii* in Algeria [14]. Taxonomy and distribution were investigated for the gomphid dragonfly species, *Orientogomphus minor* from Thailand, where the nymph was described for the first time, and the male specimen was re-described and illustrated [15]. A checklist of Odonata from Cyprus revealed 37 species, among which some have a very restricted distribution range, such as *Ischnura intermedia* [16]. An ecological study on Odonata of Mediterranean intermittent rivers revealed the importance of aquatic vegetation structure and composition of Odonata assemblages, which were shown to be species-rich in such habitats, with 22 recorded species [17].

Another small aquatic insect order, Plecoptera, was investigated in two publications [18,19]. In the first publication, the stonefly diversity was investigated in Indiana, USA, and revealed 93 species. Plecoptera species richness in the study area was highly influenced by hydrology and glacial history [18]. The Israeli stonefly fauna is extremely species-poor, and historically, only five species were recorded. In the study published in this Special Issue, a strong decrease in stonefly occurrence was observed. The populations of three species have dramatically declined in recent decades (*Protonemura zernyi, Leuctra hippopus*, and *Leuctra kopetdaghi*), whilst the remaining two species (*Brachyptera galeata* and *Marthamea beraudi*) have not been collected at all in over four decades and are considered locally extinct [19].

Aquatic Heteroptera, a species-rich but poorly known insect order, was the focus of two publications, both from the same research group [20,21]. The first publication investigated the aquatic (Nepomorpha) and semiaquatic (Gerromorpha) Heteroptera assemblages in three streams within the Kaeng Krachan National Park in Thailand. The study revealed high species richness, with 60 recorded species [20]. Both mangrove ecosystems, and their biota, are still poorly known. The second publication investigated Gerromorpha assemblages in mangroves located in the central and eastern regions of Thailand and recorded a total of nine species, four of which were new records for the country [21].

In many ecological studies, aquatic insects are sampled and analyzed within the entire macroinvertebrate community. Indeed, two papers adopted a wider perspective and examined the macroinvertebrate community in freshwater habitats of Central America (Mexico) [22,23]. The first publication investigated water quality analysis in a subtropical river using a newly created adapted biomonitoring working party index based on macroinvertebrate communities [22]. The second, and the last publication within this Special Issue, investigated aquatic macroinvertebrates in a biosphere reserve, at sites encompassing different impact and human influence scenarios. The results emphasize the importance of the relationships between the functional macroinvertebrate diversity indices and the physicochemical parameters as well as the environmental indices measured within the study area [23].

The overview of this Special Issue perhaps highlights, in a nutshell, important knowledge gaps in the field of aquatic entomology. Better distribution of taxonomic and geographical foci should be considered in future studies.

Conflicts of Interest: The authors declare no conflict of interest.

References

1. Reid, A.J.; Carlson, A.K.; Creed, I.F.; Eliason, E.J.; Gell, P.A.; Johnson, P.T.; Kidd, K.A.; MacCormack, T.J.; Olden, J.D.; Ormerod, S.J.; et al. Emerging threats and persistent conservation challenges for freshwater biodiversity. *Biol. Rev.* **2019**, *94*, 849–873. [CrossRef] [PubMed]
2. Mayhew, P.J. Why are there so many insect species? Perspectives from fossils and phylogenies. *Biol. Rev.* **2007**, *82*, 425–454. [CrossRef] [PubMed]
3. Balian, E.V.; Lévêque, C.; Segers, H.; Martens, K. Freshwater Animal Diversity Assessment. Developments in Hydrobiology. *Hydrobiologia* **2008**, *595*, 1–637.
4. Dijkstra, K.D.; Monaghan, M.T.; Pauls, S.U. Freshwater biodiversity and aquatic insect diversification. *Annu. Rev. Entomol.* **2014**, *59*, 143–163. [CrossRef] [PubMed]
5. Kitchin, P.L. Measuring the amount of statistical information in the EPT index. *Environmetrics* **2005**, *16*, 51–59. [CrossRef]
6. Roberts, L. Mosquitoes and disease. *Science* **2002**, *298*, 82–83. [CrossRef]
7. Jacobus, L.M.; Macadam, C.R.; Sartori, M. Mayflies (Ephemeroptera) and Their Contributions to Ecosystem Services. *Insects* **2019**, *10*, 170. [CrossRef]
8. Van Klink, R.; Bowler, D.E.; Gongalsky, K.B.; Swengel, A.B.; Gentile, A.; Chase, J.M. Meta-analysis reveals declines in terrestrial but increases in freshwater insect abundances. *Science* **2020**, *368*, 417–420. [CrossRef]
9. Bergsten, J.; Biström, O. Diversification in the Comoros: Review of the *Laccophilus alluaudi* species group with the description of four new species (Coleoptera: Dytiscidae). *Diversity* **2022**, *14*, 81. [CrossRef]
10. Čerba, D.; Koh, M.; Vlaičević, B.; Turković Čakalić, I.; Milošević, D.; Stojković Piperac, M. Diversity of periphytic Chironomidae on different substrate types in a floodplain aquatic ecosystem. *Diversity* **2022**, *14*, 264. [CrossRef]
11. Popović, N.; Marinković, N.; Čerba, D.; Raković, M.; Đuknić, J.; Paunović, M. Diversity patterns and assemblage structure of non-biting midges (Diptera: Chironomidae) in urban waterbodies. *Diversity* **2022**, *14*, 187. [CrossRef]
12. El Alami, M.; El Yaagoubi, S.; Gattolliat, J.-L.; Sartori, M.; Dakki, M. Diversity and Distribution of Mayflies from Morocco (Ephemeroptera, Insecta). *Diversity* **2022**, *14*, 498. [CrossRef]
13. Meecham, J.; Ogden, T.H. Is Coloburiscidae (Ephemeroptera) Monophyletic? A Comparison of Datasets. *Diversity* **2022**, *14*, 505. [CrossRef]
14. Samraoui, B.; Vuataz, L.; Sartori, M.; Gattolliat, J.-L.; Al-Misned, F.A.; El-Serehy, H.A.; Samraoui, F. Taxonomy, distribution and life cycle of the Maghrebian endemic *Rhithrogena sartorii* (Ephemeroptera: Heptageniidae) in Algeria. *Diversity* **2021**, *13*, 547. [CrossRef]
15. Chainthong, D.; Boonsoong, B. Taxonomy and distribution of the gomphid dragonfly *Orientogomphus minor* (Laidlaw, 1931) (Odonata: Gomphidae) in Thailand. *Diversity* **2022**, *14*, 291. [CrossRef]
16. Sparrow, D.J.; De Knijf, G.; Sparrow, R.L. Diversity, status and phenology of the dragonflies and damselflies of Cyprus (Insecta: Odonata). *Diversity* **2021**, *13*, 532. [CrossRef]
17. Vilenica, M.; Rebrina, F.; Matoničkin Kepčija, R.; Šegota, V.; Rumišek, M.; Ružanović, L.; Brigić, A. Aquatic macrophyte vegetation promotes taxonomic and functional diversity of Odonata assemblages in intermittent karst rivers in the Mediterranean. *Diversity* **2022**, *14*, 31. [CrossRef]
18. Newman, E.A.; DeWalt, R.E.; Grubbs, S.A. Plecoptera (Insecta) diversity in Indiana: A watershed-based analysis. *Diversity* **2021**, *13*, 672. [CrossRef]
19. Yanai, Z. The stoneflies (Insecta: Plecoptera) of Israel: Past, present, future...? *Diversity* **2022**, *14*, 80. [CrossRef]
20. Attawanno, S.; Vitheepradit, A. Species composition of aquatic (Nepomorpha) and Semiaquatic (Gerromorpha) Heteroptera (Insecta: Hemiptera) in Kaeng Krachan National Park, Phetchaburi Province, Thailand. *Diversity* **2022**, *14*, 462. [CrossRef]
21. Nakthong, L.-a.; Vitheepradit, A. The Gerromorpha (Heteroptera: Gerridae, Mesoveliidae, Veliidae) of Mangroves of Central and Eastern Regions, Thailand. *Diversity* **2022**, *14*, 466. [CrossRef]
22. Magallón Ortega, G.; Escalera Gallardo, C.; López-López, E.; Sedeño-Díaz, J.E.; López Hernández, M.; Arroyo-Damián, M.; Moncayo-Estrada, R. Water quality analysis in a subtropical river with an adapted biomonitoring working party (BMWP) index. *Diversity* **2021**, *13*, 606. [CrossRef]
23. Rodríguez-Romero, A.J.; Rico-Sánchez, A.E.; Sedeño-Díaz, J.E.; López-López, E. Characterization of the multidimensional functional space of the aquatic macroinvertebrate assemblages in a biosphere reserve (Central México). *Diversity* **2021**, *13*, 546. [CrossRef]

Article

Diversification in the Comoros: Review of the *Laccophilus alluaudi* Species Group with the Description of Four New Species (Coleoptera: Dytiscidae) †

Johannes Bergsten [1] and Olof Biström [2,*,‡]

1 Department of Zoology, Swedish Museum of Natural History, P.O. Box 50007, 10405 Stockholm, Sweden; johannes.bergsten@nrm.se
2 Zoology Unit, Finnish Museum of Natural History, PB 17, University of Helsinki, 00014 Helsinki, Finland
* Correspondence: olof.bistrom@helsinki.fi
† urn:lsid:zoobank.org:act:D1499466-D6B4-44DD-8D0C-8B6D8FC66B42; urn:lsid:zoobank.org:act:EFF85AA5-800C-4DC7-9FC6-781095558A1E; urn:lsid:zoobank.org:act:EF080A54-025D-4BBF-BE67-090FA086E72D; urn:lsid:zoobank.org:act:0D973235-2430-4F83-8BD9-E467BA0CE820.
‡ Contribution to the study of Dytiscidae 93.

Abstract: The *Laccophilus alluaudi* species group is an interesting case of an endemic species radiation of Madagascar and the Comoros. To date, a single species, *Laccophilus tigrinus* Guignot, 1959 (Anjouan), is known from the Comoro Islands, with eight other species known from Madagascar. Here we review the *Laccophilus alluaudi* species group from the Comoro Islands based on partly new material. We recognize five species, out of which four are here described as new: *L. mohelicus* n. sp. (Mohéli), *L. denticulatus* n. sp. (Grande Comore), *L. michaelbalkei* n. sp. (Mayotte) and *L. mayottei* n. sp. (Mayotte). Based on morphology of male genitalia, we hypothesize that the five species form a monophyletic group and originated from a single colonization event from Madagascar. If confirmed, this would constitute one of the few examples of intra-archipelago diversification in the Comoros. The knowledge of species limits in relation to their distribution in the Comoros archipelago is also urgently needed in the face of rapid habitat degradation.

Keywords: island biogeography; new species; taxonomy; biodiversity; colonization; endemism; species radiation; diving beetles; freshwater

Citation: Bergsten, J.; Biström, O. Diversification in the Comoros: Review of the *Laccophilus alluaudi* Species Group with the Description of Four New Species (Coleoptera: Dytiscidae). *Diversity* **2022**, *14*, 81. https://doi.org/10.3390/d14020081

Academic Editors: Luc Legal, Marina Vilenica, Zohar Yanai and Laurent Vuataz

Received: 23 December 2021
Accepted: 18 January 2022
Published: 25 January 2022

1. Introduction

Volcanic oceanic island archipelagos, such as the Galapagos, Hawaii and the Canaries, have long fascinated biologists [1–5]. In contrast to continental islands, oceanic islands have never had land contact with other larger landmasses, they are formed without life to start with, and many have existed for millions of years, forming natural laboratories of evolution [6]. For terrestrial and freshwater organisms, a large expanse of sea constitutes a dispersal barrier. This barrier is a semipermeable filter that has reduced, but over time allowed, arrivals at frequencies dependent on extrinsic factors, such as distance to mainland, ocean currents, trade winds and island size, and intrinsic factors such as dispersal capacity, body size, salinity tolerance and the ability to withstand restricted access to food and water for long periods [2,7]. Except for organism groups of very high dispersal capacity, this filter reduces the geneflow between the source population and a newly formed population after successful colonization, so that over time endemic island species evolve. On especially large and/or more heterogenous islands, or island archipelagos, further in situ or intra-archipelago speciation may occur [8].

The Comoros is one such archipelago of true oceanic islands of volcanic origin that, despite sea-level variation during their lifespan, have never been in land contact with the continent of Africa or with Madagascar [9,10]. The archipelago is situated approximately midway between mainland east Africa and northern Madagascar in the northern part of the Mozambique Channel in the western Indian Ocean. Approximately 300 km of open sea separates the Comoros from either larger landmass. The archipelago consists of the four larger islands: Grande Comore (Ngazidja), Mohéli (Mwali), Anjouan (Ndzuani) and Mayotte (Maore) (Figure 1). Estimated ages of the four islands varies greatly depending on whether the age is inferred from the oldest dated exposed lava rocks, or from the estimated on-start and duration of magmatic activity. This is particularly true for the highest island, Grande Comore, which is dominated by the very active volcano Mount Karthala. The oldest-dated exposed rock is a mere 0.13 Ma, whereas Michon (2016) estimates 9 Ma of magmatic activity [10]. The underestimation of Grande Comore's age from exposed rocks has already been suggested based on dated endemic lineages ([11,12] and references therein). With either estimate, Mayotte, the easternmost island closest to Madagascar, is the oldest island, possibly emerging from the ocean soon after magmatic activity started 20 Ma ago, which considerably expands previous age estimates of the Comoros archipelago as a whole [10]. Uncertainty apart, the Comoros is clearly old enough to attract researchers interested in island colonization and species diversification processes [13–22].

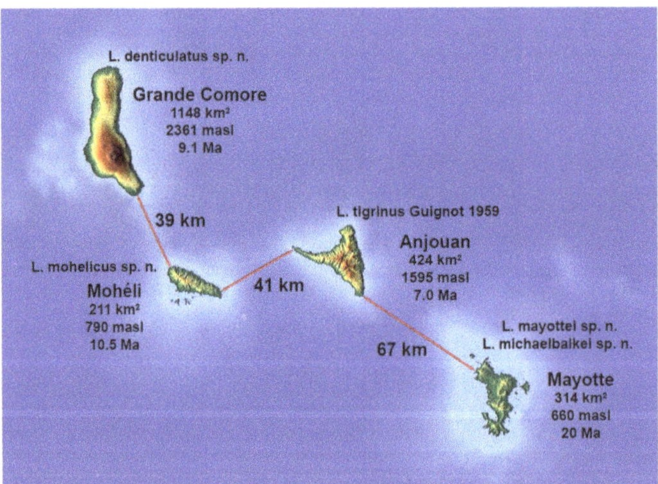

Figure 1. Topographic map of the Comoros archipelago, with indication of *Laccophilus alluaudi* group species occupancy based on our results. Terrestrial surface area in square kilometers and highest point in meters from [9], distance between islands in kilometers (calculated from Google Earth) and estimated beginning of magmatic activity in million years before present following [10] are given for each of the four main islands. Background map from https://maps-for-free.com, (accessed on 4 November 2021) released under Creative Commons CC0.

The Comoros have a maritime tropical climate with a warm rainy season from October to April [23]. From May to September southerly winds dominate, bringing cooler and drier air. Variable topography creates local differences in rainfall and air temperatures. Higher central areas of the islands are generally cooler and wetter than the coastal areas. Vegetation in the Comoro Islands resembles that of Madagascar [9]. Evergreen forest from sea level to approximately 1800 m is the original type of vegetation of the islands. Above the forest area, on Mount Karthala a high-mountain vegetation consists of mountain bushland and thicket [24]. Remaining natural forests constitute a small proportion of what once existed, and are today restricted to higher elevations [23,25]. The islands are partially surrounded by mangrove swamps. The soil of the islands consists of laterite, which is rich in minerals but poor in humus, being subject to erosion when sheltering forests are removed [23].

Maximum altitudes for the islands are the 2631 m peak of Mount Karthala on Grande Comore, a maximum of 1595 m on Anjouan, and below 800 m for the lower islands of Mayotte and Mohéli (Figure 1).

Based on endemism levels and loss of natural vegetation, the Comoros are part of the western Indian Ocean islands biodiversity hotspot [26]. This biodiversity hotspot is dominated by the much larger island of Madagascar, which seems to have been the most important source for many animal and plant groups on the Comoros [14,22,27–29]. Compared to Madagascar, the plant and animal diversity of the Comoros are certainly much lower. This is to be expected based purely on area-richness relationships, in addition to the Comoros' younger age. However, the Comoros are still relatively rich in endemic species. For plants, although poorly documented, as much as 33% are endemic [23], and for vertebrates approximately 20% [24]. The ongoing process of speciation is also demonstrated through many single-island endemic subspecies among the non-endemic birds [24]. While there is substantial species-level endemism, there are fewer examples of in situ diversification in the Comoros [8,13]. In contrast to the Galapagos, Mascarenes and Canaries, where intra-archipelago speciation has thrived, endemism in the Comoros is largely made up of immigrants and anagenesis from the parental stock, at least for vertebrates [8,13]. For invertebrates, diversity patterns, endemism levels and origination processes are still poorly explored in the Comoros.

Laccophilus is the second most diverse genus of diving beetles (Dytiscidae), with a worldwide distribution [30]. Members inhabit both running and standing water bodies and are found over a large altitudinal and habitat range. The *Laccophilus alluaudi* species group was established by Biström et al. while revising the African fauna of the genus [31]. Six species were recognized, five in Madagascar and one in the Comoros. Recently, Manuel and Ramahandrison added three more species to the group [32], and several additional species, all from Madagascar, are known but have yet to be described (Bergsten unpublished). It has become clear that the lineage is one of just a handful of larger (>10 spp.) in situ diversifications of diving beetles in Madagascar [32,33]. One species from the Comoros, *L. tigrinus*, was described by Guignot [34], and this is the only non-Madagascar species of the group. Guignot based his description on material from both Mohéli and Anjouan. While clearly belonging to the *Laccophilus alluaudi* species group, further investigation of this group in the Comoros has not yet attracted any attention from taxonomists. For instance, the multi-island species hypothesis of *Laccophilus tigrinus* has not been questioned, and material from Grande Comore and Mayotte has been lacking. Access to recently collected material from all four Comoro Islands prompted us to revise the group to shed light on both diversity patterns and diversification processes of Comoran fauna.

Here we show that, for one lineage of aquatic insects that has diversified in Madagascar and colonized the Comoros with a single known species [31,32], it actually constitutes a flock of island-endemic species in the Comoros. While not yet tested in a phylogenetic context, morphological evidence points towards a single colonization and intra-archipelago speciation.

2. Material and Methods

2.1. Preparation Technique and Measurements

The study material consisted of both dry-pinned specimens and specimens preserved in ethanol. Dry specimens were softened in hot water for some minutes prior to dissection. The apical part of the abdomen was then detached under a preparation microscope, and the genitalia extracted from surrounding tissue in warm water. Sometimes hardened tissue needed to be treated in a heater-device for about 10 min in 10% KOH solution. After this procedure, the genitalia were cleaned in water and ethanol and glued by the base on a card for photographing. Genitalia from specimens in ethanol were extracted directly without the need of softening in warm water, and both body and genitalia were glue-mounted on a card.

Measurements of body (length and width) were made using a Wild M 11 microscope (Wild Heerbrugg, Heerbrugg, Switzerland). The detailed technique is described in [31].

2.2. Photography

Photographs of specimens and genitalia were captured with a Canon EOS R digital SLR camera (Canon Inc, Tokyo, Japan) mounted on a motorized rail, Stackshot (Cognisys Inc., Traverse City, MI, USA). A long working distance metallurgic objective from Mitutoyo (Mitutoyo Corporation, Kanagawa, Japan) (5× for habitus; 10× for genitalia) was fitted to the camera using a Balpro bellows and customized adaptors (Novoflex, Memmingen, Germany). The stack-photography system was maneuvered with Zerene Stacker (Zerene Systems, Zerene Systems LLC, Richland, WA, USA) in combination with Canon EOS Utility (Canon Inc., Tokyo, Japan). Captured stacks of photos were merged using the Pmax algorithm in Zerene Stacker. Merged photos were prepared for plates in Adobe Photoshop (Adobe Inc., San Jose, CA, USA) and Adobe Illustrator (Adobe Inc., San Jose, CA, USA).

2.3. Depositories

Studied materials were deposited in the following institutions, which are referred to in the text by their abbreviations:

IRSNB—Institut Royal des Sciences Naturelles de Belqique, Brussels, Belgium
MRAC—Musée Royal de l'Afrique centrale, Tervuren, Belgium
MNHN—Muséum National d'Historie Naturelle, Paris, France
MZH—Finnish Museum of Natural History, Helsinki, Finland
NHRS—Swedish Museum of Natural History, Stockholm, Sweden
ZSM—Zoologische Staatsammlung, München, Germany.

3. Results

Based on morphological examination of material from each of the four main islands of the Comoros archipelago, it became clear that each island has at least one endemic species; we found no evidence of any species in common. In addition, examined material from Mayotte showed that multiple species can exist sympatrically. Below we give first a determination key, followed by the species descriptions. The known fauna of this group on the Comoros increases from one to five species as a result.

3.1. Key to Species of *L. alluaudi* Species Group in the Comoros

1. Pronotum anteriorly and head posteriorly darker; ferruginous to dark brown (Figure 2A–D) (rarely dark area indistinct in old, dry specimens) 2

   Pronotum and head unicolored testaceous to pale ferruginous (Figure 2E,F) 4

2. Smaller species (body length 3.36–3.76 mm); dark area on head and pronotum ferruginous; gradual delimitation of markings (rarely dark markings almost absent) (Figure 2D) (Mohéli) . . . . *L. mohelicus*

   Larger species (body length 3.98–4.12 mm); dark area on head and pronotum darker and more distinctly delimited (Figure 2A–C) (Grande Comore, Anjouan) 3

3. Penis narrower and more strongly curved in ventral view (Figure 3C); apically with an asymmetric apical knob (right side in ventral view pre-apically straight to weakly concave) and subbasally with a distinct denticle (Figure 4B) (Grande Comore) . . . *L. denticulatus*

   Penis broader and less curved (Figure 3A,B); apically without an apical offset knob (right side pre-apically convex but apex with protrusion on left side); subbasal "shelf" not as sharply denticulated (Figure 4A in suboptimal condition, but similar to Figure 4C) (Anjouan) . . . . *L. tigrinus*

4. Humeral region of elytra with an extensive pale area which lacks dark spots (Figure 2F); penis quite robust and sinuate (Figures 3F and 4E); (Mayotte) . . . . . . *L. michaelbalkei*

   Humeral region with two vague, dark spots (Figure 2E); penis moderate-sized, non-sinuate but slightly curved (Figures 3E and 4D); (Mayotte) . . . . *L. mayottei*

Figure 2. Habitus images of *Laccophilus alluaudi* species group members from Comoro Islands: (**A**) *L. tigrinus* (male, PT 13265); (**B**) *L. tigrinus* (female); (**C**) *L. denticulatus* (male); (**D**) *L. mohelicus* (male); (**E**) *L. mayottei* (male); (**F**) *L. michaelbalkei* (male). Scale bar 1 mm. Note that (**A**) (Photo: MNHN/Christophe Rivier) is photographed under different circumstances (equipment, angle and resolution).

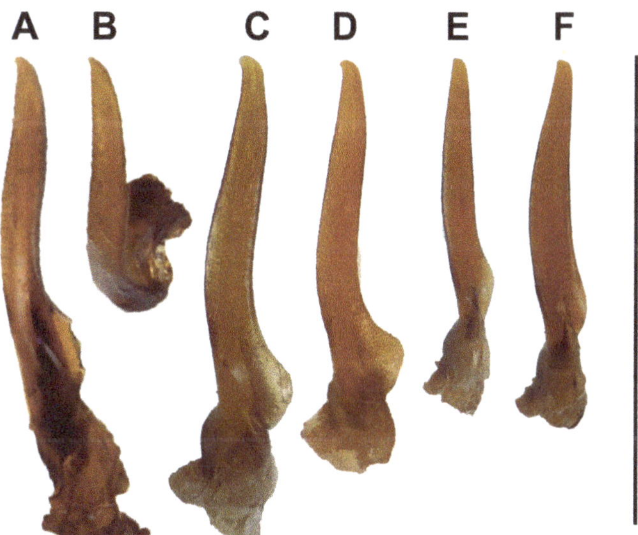

Figure 3. Penis in ventral view of *Laccophilus alluaudi* species group members from Comoro Islands: (**A**) *L. tigrinus* (PT 13265); (**B**) *L. tigrinus* (HT 13263); (**C**) *L. denticulatus*; (**D**) *L. mohelicus*; (**E**) *L. mayottei*; (**F**) *L. michaelbalkei*. Scale bar 1 mm. Note that (**A**,**B**) (Photos: MNHN/Christophe Rivier) are photographed under different circumstances (equipment, angle and resolution).

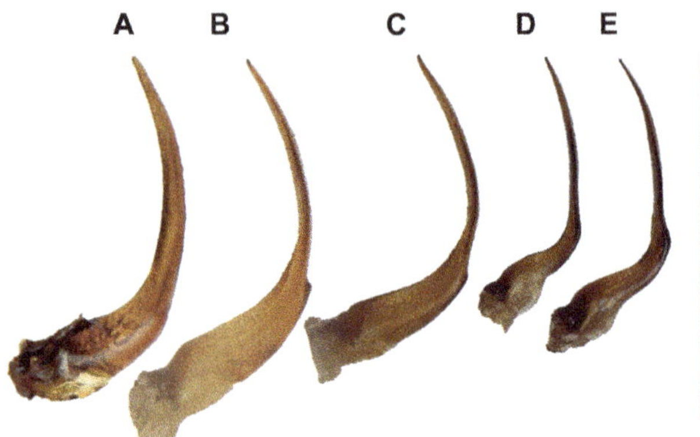

Figure 4. Penis in lateral view of *Laccophilus alluaudi* species group members from Comoro Islands: (**A**) *L. tigrinus* (HT 13263); (**B**) *L. denticulatus;* (**C**) *L. mohelicus;* (**D**) *L. mayottei;* (**E**) *L. michaelbalkei.* Scale bar 1 mm. Note that (**A**) (Photo: MNHN/Christophe Rivier) is photographed under different circumstances (equipment, angle and resolution) and is partly covered with residual glue hiding a subbasal shelf similar to in (**C**).

3.2. *Taxonomic Treatment*

The *Laccophilus alluaudi* species group is characterized by a pale body with elytra exhibiting longitudinal, dark-colored markings. Moreover, members of the species group are moderately sized (approximate length of 3–5 mm), and no stridulation apparatus is exhibited on metacoxal plates ventrally of the body. Male genitalia are asymmetrically shaped, with penis sinuate or curved from the basal part towards apex

Laccophilus tigrinus Guignot, 1959

Guignot, 1959: 76 (original description) [34]; Guignot 1961: 931 (faunistics) [35]; Wewalka 1980: 726 (faunistics) [36]; Nilsson 2001: 251 (catalogue) [37]; Biström et al. 2015: 43 (monograph) [31]; Nilsson & Hajek 2021; 231 (catalogue) [38].

Material studied:

Holotype, male: "Type/Anjouan Foret de M'Remani X-1953 (Millot)/F. Guignot det., 1955 L. tigrinus sp.n. Type [male symbol]/HOLOTYPE/HOLOTYPE Laccophilus tigrinus Guignot/MNHN, Paris EC13263 [QR code]/Data in NHRS JLKB 000030002" (MNHN).

Paratypes: "PARATYPE/F. Guignot det., 19 L. tigrinus sp.n. Paratype [male symbol]/PARATYPE Laccophilus tigrinus Guignot 1959/MNHN, Paris EC13265 [QR code]/Data in NHRS JLKB 000030003" [likely with same collecting data as holotype but not on labels] (1 ex. MNHN). "Paratype/PARATYPE Laccophilus tigrinus Guignot 1959/MNHN, Paris EC13245 [QR code]" (1 ex. MNHN).

Additional material studied: "Anjouan Comoros 12.27483 S, 44.47502 E, 18.3.2010, 798 m asl, Drinking water basin near Ouzini, SOH 0069/Laccophilus tigrinus Guignot, 1959 Det. J. Bergsten 2021" (1 ex. NHRS; 1 ex. MZH; 1ex. ZSM).

Total material studied: 6 specimens (MNHN; MZH; NHRS; ZSM).

Diagnosis: *L. tigrinus* is separated from other species in the group by the combination of a dark area frontally on pronotum and posteriorly on the head, a relatively larger body size (length 3.92-4.08 mm) and the lack of a distinct sharp denticle subbasally on penis.

Description:

Body (Figure 2A,B), length, male 3.92–4.08 mm, width 2.24–2.36 mm; female, length 3.98–4.12 mm, width 2.32–2.40 mm.

Head: Frontal outline of head broadly straight to almost straight; laterally foremargin somewhat curved towards eyes. Testaceous to pale ferrugineous. Posteriorly at pronotum with quite broad, dark ferrugineous area. Rather shiny although with fine reticulation which extensively is irregular. Shape of meshes variable, in part transverse. Extensively impunctate, but at eyes with fine, scattered punctures. Area of punctures extends little towards mid-head. Antenna testaceous to pale ferrugineous, slender. Apical segment slightly longer than adjacent ones; apically pointed.

Pronotum: Testaceous to pale ferrugineous. Anteriorly, posterior to head with broad, brown to dark ferrugineous spot. Delimitation of dark spot sometimes slightly vague. Lateral outline of pronotum evenly curved; non-margined. Rather shiny although distinctly microsculptured. In part with double reticulation but size-classes of meshes indistinct and difficult to discern. Reticulation variable; meshes densest and smallest at margins, on disc slightly larger and irregularly shaped. Frontally and laterally with some irregularly distributed, somewhat indistinct punctures (hidden in dense microsculpture).

Elytra: Testaceous with distinct black to dark ferrugineous, longitudinal markings. Four inner markings reach almost to base of elytra. Three lateral markings end before elytral base and leave a moderately sized pale humeral area. Longitudinal markings in two female specimens broader and in part confluent. Extreme, lateral dark marking in part medially reduced (sometimes broken). Pale humeral area provided with 2–3 dark, vague spots. Slightly matte due to distinct microsculpture. Reticulation dense, meshes moderately sized, almost uniform in shape. Double reticulation reduced; indistinct and rudimentary. A fine and somewhat sparse and irregular row of discal punctures may be discerned in frontal half of elytra. Laterally with fine and sparse punctures which posteriorly form a distinct row which ends clearly before elytral apex. Scattered, very fine punctures discernible in posterior part of elytra.

Ventral aspect: Prosternum testaceous to pale ferrugineous. Metacoxal plates laterally dark, almost black. Towards middle, plates become slightly paler; dark ferrugineous. Metathorax dark ferrugineous; anteriorly pale ferrugineous. Metacoxal process and abdomen testaceous to pale ferrugineous. Sternites laterally sometimes with a vague dark spot. Metathorax and metacoxal plates almost impunctate; with fine, sparse punctures. Plates provided with a few transverse impressions which laterally are strongly bent to short, slightly curved impressions. Abdomen rather shiny, almost impunctate, provided with somewhat curved, sparse striae. Metacoxal plates and abdomen rather shiny; reticulation indistinct, plates with shagrination. Metacoxal lines with a blunt lateral extension. Ventrite in apical half almost keeled, with fine, irregular punctures.

Legs: Testaceous to pale ferrugineous. Pro- and mesotarsus somewhat enlarged, provided with adhesive discs.

Male genitalia: Penis almost evenly broad in medial part; less curved (Figure 3A,B); apically without an offset apical knob (right side pre-apically convex) but with small left-turned apical protrusion (Figure 3A,B); subbasal "shelf" not sharply denticulated (Figures 3A,B and 4A). Earlier dissection and mounting of male genitalia in imaged holotype may have effect on their present configuration except for apex, which is definitely different to shape of apex in *L. denticulatus.* Based on studied paratype material, *L. tigrinus* has subbasal shelf in lateral view, blunt and gently rising as in *L. mohelicus* (Figure 4C; also see illustration in [36]), but not sharply denticulate as in *L. denticulatus* (Figure 4B).

Female: Pro- and mesotarsi slender, lack adhesive discs.

Distribution: Anjouan (Figure 1).

Discussion. *L. tigrinus* was described by Guignot based on material from both Anjouan and Mohéli [34]. A type statement in the introduction qualifies as a holotype designation and the holotype (MNHN EC13263) is from Anjouan. The paratype material from Mohéli (MNHN EC13266) belongs to *L. mohelicus.*

Laccophilus denticulatus n. sp.

urn:lsid:zoobank.org:act:D1499466-D6B4-44DD-8D0C-8B6D8FC66B42.

Material studied:

Holotype, male; "Grande Comore Nioumbadjou R. Joqué 9.8. 1981/Holotype Laccophilus denticulatus n. sp. Bergsten & Biström, 2022" (1 ex. MRAC).

Paratypes: Same data as holotype but with paratype labels "Paratype Laccophilus denticulatus n. sp. Bergsten & Biström, 2022" (1 ex. MRAC; 1 ex. MZH).

Total material studied: 3 specimens (MRAC; MZH).

Etymology: Species name refers to the small basal denticle or tooth (latin noun: denticulus) subbasally on the penis.

Diagnosis: Similar in shape and size to *L. tigrinus* and shares an infuscated area posteriorly on head and anteriorly on pronotum. Separated from *L. tigrinus* based on the distinctly offset (pre-apical sinuation also on right side) asymmetric apical knob and the subbasal sharp denticle of penis. *L. mohelicus* described below has a somewhat similar asymmetric apical knob but not clearly preapically sinuate on right side. Its body size is smaller and it has a more vaguely delimited, less darkened area on head and pronotum.

Description (only differences to description of *L. tigrinus* observed).

Body (Figure 2C), length, male 3.92–4.08 mm, width 2.28–2.36 mm; female, length 3.96 mm, width 2.32mm.

Male genitalia: Penis in ventral aspect sinuate and with left turned and offset asymmetric apical knob (Figure 3C). Subbasally, on left side opposite of anterior portion of subbasal expansion with a distinct denticle (Figure 4B).

Distribution: Grande Comore (Figure 1).

Laccophilus mohelicus n. sp.

urn:lsid:zoobank.org:act:EFF85AA5-800C-4DC7-9FC6-781095558A1E.

Material studied:

Holotype, male: "Moheli Foret de Fomboni 600 m 2eme torrent 6.54/Institut scientifique Madagascar/Type/R. Mouchamps det. Laccophilus mohelicus n.sp. TYPE/Laccophilus tigrinus Guignot O. Biström det./Holotype Laccophilus mohelicus n. sp. Bergsten & Biström, 2022" (IRSNB).

Paratypes: Same data as holotype but labelled as allotype and "Paratype Laccophilus mohelicus n. sp. Bergsten & Biström, 2022" (1 ex. IRSNB); "Moheli Comoros, 1.III.2010 Lat -12.29384 Lon 43.65220, 251 m asl, Chalet St. Antoine, slope, puddle along creek through forest SOH 0031/Paratype Laccophilus mohelicus n. sp. Bergsten & Biström, 2022" (1 ex. dry, 2 exs. in ethanol NHRS; 1 ex. dry in MZH); "Moheli Comoros, 28.II. 2010, 12.30269S, 43.63731E, 23 m asl Miringoni SOH 0026/Paratype Laccophilus mohelicus n. sp. Bergsten & Biström, 2022" (1 ex. dry, 3 exs. in ethanol NHRS, ZSM).

Other material studied: "Paratype/Moheli Foret de Fomboni 600m 2eme torrent 6.54 (J.M.)/PARATYPE Laccophilus tigrinus Guignot, 1959/MNHN, Paris EC13266 [QR code]/[male symbol] (1 ex. MNHN);

Total material studied: 11 specimens (IRSNB, MNHN, MZH, NHRS, ZSM).

Etymology: We use the name "mohelicus" that was suggested, but not published by R. Mouchamps. The name refers to the Comoro island of Mohéli where the species occurs.

Diagnosis: Most similar to *L. tigrinus* and *L. denticulatus* but slightly smaller (maximum length 3.76 mm). Dark frontal marking of pronotum vague and not clearly delimited. In old specimen dark marking indistinct, leached and almost absent. Penis very similar to *L. tigrinus* (body size and colour patterns best diagnostic differences) and also resembles that of *L. denticulatus*, but smaller, ventral denticle blunt and apical knob less clearly offset.

Description:

Body (Figure 2D), male, length 3.36–3.68 mm, width 1.92–2.12 mm; female, length 3.52–3.76 mm, width 2.04–2.16 mm.

Head: Frontal outline straight, laterally towards eyes slightly curved. Testaceous, posteriorly at pronotum slightly darker; with vague, pale brown area. Darker area narrower and with more vague delimitation in comparison with *L. tigrinus*; sometimes dark area almost absent. Rather shiny, although finely to very finely microsculptured. Reticulation variable, in part double. Posteriorly head with fine and dense, almost isodiametric meshes. On disc fine reticulation almost absent and replaced by larger, somewhat irregular shaped meshes. Frontally larger and fine meshes appear mixed but size-classes in part difficult to separate. At eyes with some irregular punctures, which medially extend for a short distance towards mid-head. Antenna testaceous, slender, apical segment slightly longer than preceding segments; apically pointed.

Pronotum: Testaceous to pale ferrugineous. Frontally in middle often with a vague pale brownish to ferrugineous spot (delimitation of spot gradual and diffuse). Lateral outline of pronotum evenly curved; non-margined. Rather shiny, although microsculptured. Reticulation in part double, but size-classes in part difficult to discern. Fine meshes anteriorly and posteriorly distinct; discally fine meshes indistinct and replaced by somewhat irregular-shaped, large meshes. Frontally and laterally with fine, but clearly discernible, irregular punctures.

Elytra: Testaceous with distinct, black to dark ferrugineous, longitudinal markings. Markings sometimes appear "hollow"; slightly paler in middle but outline of separate marking always distinct. Four inner markings reach almost elytral base, the most lateral marking being anteriorly somewhat expanded. Three lateral markings at base shorter, leaving a quite extensive humeral area pale-colored. Vague darker spot may sometimes be discerned in the pale area. Markings also in part confluent, the most lateral being, in part, reduced. Slightly matte due to distinct, dense microsculpture. Double reticulation reduced, indistinct and rudimentary. Fine, sparse but indistinct and irregular discal row of punctures may be discerned. Posteriorly with fine, scattered punctures. Laterally, in posterior part of elytron with fine, clearly discernible row of punctures, which fade away before reaching elytral apex.

Ventral aspect: Prosternum testaceous to pale ferrugineous. Metacoxal plates laterally almost black; towards middle plates become gradually paler; at metathorax and metacoxal process ferrugineous to pale ferrugineous. Abdomen pale ferrugineous to testaceous. Metathorax anteriorly with a few punctures, otherwise impunctate as metacoxal plates. Metathorax shiny without microsculpture or non-shagreened. Metacoxal plates slightly matte due to fine microsculpture, almost shagreened. Metacoxal plates almost impunctate, provided with few indistinct transverse depressions which laterally turn to a few distinct impressions. Metacoxal lines with blunt lateral extension. Abdomen rather shiny and almost impunctate; apical ventrite apically with some fine punctures and shagreened. Abdominal segments with sparse, curved striae.

Legs: Testaceous to pale ferrugineous. Hindlegs somewhat darker, ferrugineous. Pro- and mesotarsus somewhat enlarged and provided with adhesive discs.

Male genitalia: Penis in ventral view slightly sinuate and provided with apical knob, less distinctly offset compared with *L. denticulatus* (Figure 3D). In lateral view with modest smooth denticle or subbasal "shelf" which is blunt but clearly discernible (Figure 4C).

Female: Pro- and mesotarsus slender, no adhesive discs.

Distribution: Mohéli (Figure 1).

Laccophilus mayottei n. sp.

urn:lsid:zoobank.org:act:EF080A54-025D-4BBF-BE67-090FA086E72D.

Material studied:

Holotype, male; "Mayotte Hajangoua REF04 23 VIII.2013 Nathalie Mary/Holotype Laccophilus mayottei n. sp. Bergsten & Biström, 2022" (1 ex. MZH).

Paratypes: same data as holotype except "Paratype Laccophilus mayottei n. sp. Bergsten & Biström, 2022"(2exs. females MZH, NHRS); "Mayotte Djalimou REF09 22.VIII.2013 Nathalie Mary/Paratype Laccophilus mayottei n. sp. Bergsten & Biström, 2022" (1 ex. male MZH); "Mayotte Bouyouni Amont 30.VII. 2012 Nathalie Mary/Paratype Laccophilus mayottei n. sp. Bergsten & Biström, 2022" (2 exs. females MZH, ZSM).

Total material studied: 6 specimens (MZH, NHRS, ZSM).

Etymology: The name refers to the Comoro island Mayotte where the species occurs.

Diagnosis: *L. mayottei* has a unicolorous testaceous head and pronotum and in this respect resembles *L. michaelbalkei*. From the latter, *L. mayottei* can be distinguished based on a narrower and less sinuate penis in ventral view, and a relatively shorter and less strongly developed basal region in lateral view. *Laccophilus mayottei* is also smaller in body size and has two diffuse darker spots in the pale humeral region of elytra.

Description:

Body (Figure 2E), (male), length 3.40–3.48 mm, width 1.92–2.00 mm; female, length 3.40–3.56 mm, width 1.92–2.04 mm. Body dorsally slightly globular; in species from Madagascar body over elytra somewhat flattened. Body a little smaller than in its sister species *L. michaelbalkei*.

Elytra: Testeceous, with dark longitudinal markings. Four inner markings as in sister species. Three lateral longitudinal markings end quite abruptly and a quite extensive pale area formed in humeral region. Pale humeral area anteriorly with two diffuse darker spots, which in part are united (spots absent in sister species). Three lateral longitudinal markings strongly modified and in part reduced, in part united with each other.

Ventral aspect: Bicolored. Metacoxal plates laterally blackish to dark ferrugineous; become gradually paler towards mid-body; metathorax dark ferrugineous to ferrugineous and metacoxal processes ferrugineous. Abdomen unicolored, distinctly paler, ferrugineous to pale ferrugineous. Lateral impression on metacoxal plates absent; replaced by a few backwards pointing striae. Metacoxal lines simple, lacks minor lateral extension, which is present in species from Madagascar.

Legs: Female pro- and mesotarsi slender, no adhesive discs.

Male genitalia: In lateral view similar to *L. michaelbalkei* but basal part prior to angle relatively shorter and not as strongly developed (Figure 4D). In ventral aspect not clearly sinuate but slightly curved with extreme apex almost straight with left side curved (Figure 3E). Basally moderately enlarged, provided with distinct basal extension (Figure 3E).

Distribution: Mayotte (Figure 1).

Laccophilus michaelbalkei n. sp.

urn:lsid:zoobank.org:act:0D973235-2430-4F83-8BD9-E467BA0CE820.

Material studied:

Holotype, male: "Mayotte Anatana REF08 22.VIII.2013 Nathalie Mary/Holotype Laccophilus michaelbalkei n. sp. Bergsten & Biström, 2022" (1 ex. MZH).

Total material studied:1 specimen (MZH).

Etymology: Named after the distinguished specialist of tropical Dytiscidae, Dr. Michael Balke in Munich, Germany.

Diagnosis: Head and pronotum of *L. michaelbalkei* is unicolored pale ferrugineous to testaceous when *L. tigrinus* and *L. denticulatus* has a dark marking frontally on pronotum. *L. mohelicus* also has a dark marking but it is vague and not distinctly delimited. In old dry specimens the dark marking seems to have been leached and it can therefore be almost absent. From *L. mayottei*, *L. michaelbalkei* is separated by larger body, by quite extensive pale area in humeral region of elytra; *L. mayottei* is smaller, with two vague darker spots in pale humeral area. Penis of *L. michaelbalkei* is in ventral aspect more robust and sinuate while in *L. mayottei* penis-size moderate and penis non sinuate but slightly curved. In lateral aspect the basal portion is relatively longer and more strongly developed prior to and at angle.

Description:

Body (Figure 2F), (male), length 3.60 mm, width 2.16 mm. Body dorsally slightly globular.

Head: Testaceous to pale ferrugineous. Rather shiny although finely microsculptured. Reticulation with two kind of meshes. Large meshes, when discernible contain 2–4 small meshes. Size-categories of meshes sometimes difficult to distinguish. At eyes with fine, irregular punctures, which extend short distance towards middle of head. Antenna testaceous, simple, slender, with quite long segments out of which the apical one is most extensive. No distinct modifications exhibited.

Pronotum: Unicolored, testaceous to pale ferrugineous. Lateral outline of pronotum non-margined, curvature moderate and even. Rather shiny, although finely microsculptured. With three kinds of microsculpture; at margins meshes of reticulation almost isodiametric, dense, fine and one-sized. Discally with double reticulation; two kinds of meshes, smaller and larger meshes, which sometimes are difficult to distinguish, being almost of equal size. Larger meshes, when discernible, may contain 2–4 small meshes. Almost impunctate; laterally and at foremargin with quite dense and irregular punctures.

Elytra: Testaceous, with dark, longitudinal markings. Four inner markings reach almost to anterior margin of elytra. Outside them three lateral markings end frontally distinctly before reaching anterior margin of elytra, forming quite distinct, pale humeral area. The seventh (most lateral) marking of elytra is strongly reduced and only fragments of it are discernible. Rather shiny although finely microsculptured; elytra almost totally covered with fine, dense reticulation of one kind. Anteriorly at base and at suture with double reticulation. Large meshes, when discernible, may contain 2–4 small meshes. A fine and sparse discal row of punctures is discernible on elytron; posteriorly row of punctures mixed with scattered punctures. Elytra laterally with scattered, fine punctures not forming distinct rows.

Ventral aspect: Prosternum testaceous to pale ferrugineous. Metathorax and metacoxal process ferrugineous. Metacoxal plates blackish to dark ferrugineous. Abdomen ferrugineous to pale ferrugineous; little paler apically than at base. Metacoxal plates, metathorax and metacoxal process almost impunctate; sparse, very fine scattered punctures may be discerned. Quite shiny, microsculpture strongly reduced to fine shagrination. Metacoxal plates laterally with few, slightly irregular and shallow striae. Abdomen ferrugineous to pale ferrugineous. Sternites striated being densest at basal sternite and sparsest on apical sternite (3–4 reduced striae at each side). Almost impunctate; apical sternite apically with few fine punctures. Microsculpture reduced to fine shagrination. Metacoxal lines with reduced, blunt, lateral extension.

Legs: Pale ferrugineous to ferrugineous. Protarsus slender, mesotarsus slightly enlarged. Both provided with adhesive discs.

Male genitalia: Penis in lateral view with a quite steep angle in basal half (not evenly curved) (Figure 4E). In ventral aspect penis slightly sinuate, basally quite broad with quite distinct lateral extension (Figure 3F). Penis narrows evenly to quite slender apex, which is slightly turned leftwards (Figure 3F).

Female unknown.

Distribution: Mayotte (Figure 1).

4. Discussion

We found that the *Laccophilus alluaudi* species group, previously only known as one species from Anjouan and Mohéli [34], in fact has colonized all four major islands of the Comoros and likely constitutes an intra-archipelago species radiation. In addition, no species are common for any two islands, and on Mayotte sympatric sister-species occur. This contrasts with the broader patterns found for the Comoros, with very few cases of intra-archipelago cladogenetic events among, for instance, reptiles [8] or birds [13]. The *Laccophilus alluaudi* group has its largest diversity on the island of Madagascar [32], where currently eight species are described but several additional species are known (Bergsten unpublished). Since the species group is not known from the African mainland [31], it is almost certain the Comoros were colonized from Madagascar. This is a common pattern for many faunal and floral elements on the Comoros; for instance, Malagasy vangid birds, Mantellid frogs, bats, chameleons, *Phelsuma* day geckos and plants of the coffee family (Rubiaceae) all seem to have colonized the Comoros from Madagascar [14,22,27–29]. A review of reptile colonization patterns of islands in the western Indian Ocean point towards the direction of ocean currents being overwhelmingly important in the region supporting drift from northern Madagascar towards the Comoros [19]. This likely explains why much fewer colonization events of the Comoros from east Africa have taken place [27]. Reviews are lacking for insect colonization patterns of the Comoros, making it difficult to assess if this is a general pattern also for insects. Since insects may variously arrive both via drifting/rafting and through active or passive flight in the air, we can presume the direction of currents is less of a single, outstanding, explanatory variable. Flight capacity and flight activity varies enormously among different insect groups, and the fact that an intra-archipelago diversification likely has taken place among the *Laccophilus alluaudi* species group suggests tens of kilometers of open sea constitutes a significant dispersal barrier (Figure 1).

All five species in the Comoros share the subbasal expansion on the right side of the penis, which is not present in any of the currently described species from Madagascar. This is therefore likely a synapomorphy carrying evidence of a single colonization event followed by diversification in the Comoros rather than multiple events. Furthermore, three circumfactual conditions point towards Mayotte being the island that was first colonized. Firstly, Mayotte is geographically closest to Madagascar, making it the most probable to be colonized based on island biogeography theory [2]. Second, it is the oldest of the four volcanic islands and hence has been available for colonization for the longest amount of time [10]. Finally, we conclude that among our studied material, Mayotte alone has two sympatric species, also in congruence with the species group having been present on the island the longest, allowing the most time for speciation. An alternative explanation for the two species on Mayotte would be secondary re-colonization from the other islands; however, based on morphology there is no doubt that among known species, the two on Mayotte are sister species, which implies intra-island cladogenesis. Since the subbasal expansion of the penis points towards a single colonization event from Madagascar, the other islands would have been colonized from Mayotte, if Mayotte was first. However, the two species in Mayotte are quite different from the three species on Mohéli, Anjouan and Grande Comore, which are more similar to one another (Figures 2–4). Until the diversity in Madagascar is further explored, and a comprehensive sampling of the group from both islands can be set in a phylogenetic context, some uncertainty will remain as to whether one or perhaps two colonization events most likely occurred.

As in many tropical countries, land use has decimated forests and degraded both terrestrial and freshwater ecosystems on the Comoro Islands. The rate of deforestation for the Union of Comoros is among the highest in the world [25,39]. Lowland forests up to 400 m asl are almost completely gone from the four islands [23]. The last remaining large forest block is on Grande Comore on the Karthala volcano [23,25]. The level of poverty and malnourishment, as well as dependency on forest resources, are extremely high. In concert with rapid population growth, the acute need of land for agriculture has led to an alarming rate of habitat loss [23,39]. Estimates from about a decade ago reported remaining closed forest cover made up somewhere between 4–9% of the Comoros [25,29], a number that has certainly decreased further since. Mayotte also has lost 50% of its native forests in the last 30 years, threatening its diversity (including many endemic species). The most recent surge in deforestation since 2010 has caused several calls of alarm, which in May 2021 led to the protection of 2800 hectares of forests on six of the island's massifs in "Réserve naturelle nationale des forêts de Mayotte" ([40]; Décret no 2021-545). Based on experiences from Madagascar, most of the *Laccophilus alluaudi* group of species are highly dependent on water systems in natural forests and are seldom found in, e.g., agricultural landscapes with eutrophied waters. However, a few species of the group do also occur in western Madagascar in more open landscape. It is likely that colonization of the Comoros stemmed from ancestors living in western Madagascar, which may be more adaptable to open arid landscapes. Such a pattern would mirror two examples of Mantellid frogs of Mayotte. Ecologically unusual species living in arid western Madagascar are the closest relatives, despite the fact that the diverse family Mantellidae is dominated by humid forest dwellers [22]. The protection of a large part of the remaining forests in Mayotte is, however, very positive for the long-term survival of the two endemic *Laccophilus alluaudi* group representatives and for many other freshwater organisms on the island.

Author Contributions: Conceptualization, J.B. and O.B.; investigation and formal analysis, O.B. and J.B.; writing—original draft preparation O.B.; writing—review and editing J.B. All authors have read and agreed to the published version of the manuscript.

Funding: Open access funding provided by University of Helsinki.

Institutional Review Board Statement: Not applicable.

Informed Consent Statement: Not applicable.

Data Availability Statement: See Depositories (Section 2.3).

Acknowledgments: We are in debt to Michael Balke, Munich, Germany, and Oliver Hawlitschek, Hamburg, Germany, for providing us with the more recently collected material from the Comoros, which turned out to be highly interesting. We also thank Antoine Mantilleri (MNHN Paris) for supplying photos of the holotype of *L. tigrinus*, Marc de Meyer and Didier Van den Spiegel (MRAC, Tervuren) and Pol Limburg (IRSNB, Brussels) are sincerely thanked for their loan of specimens.

Conflicts of Interest: The authors declare no conflict of interest.

References

1. Darwin, C. *On the Origin of Species by Means of Natural Selection, or the Preservation of Favoured Races in the Struggle for Life*; John Murray: London, UK, 1859.
2. MacArthur, R.H.; Wilson, E.O. *The Theory of Island Biogeography*; Princeton University Press: Princeton, NJ, USA, 1967.
3. Grant, P.R. *Ecology and Evolution of Darwin's Finches*, 2nd ed.; Princeton University Press: Princeton, NJ, USA, 2009.
4. Emerson, B.C. Evolution on oceanic islands: Molecular phylogenetic approaches to understanding pattern and process. *Mol. Ecol.* **2002**, *11*, 951–966. [CrossRef] [PubMed]
5. Whittaker, R.J.; Triantis, K.A.; Ladle, R.J. A general dynamic theory of oceanic island biogeography. *J. Biogeogr.* **2008**, *35*, 977–994. [CrossRef]
6. Gillespie, R.G. Oceanic islands: Models of diversity. In *Encyclopedia of Biodiversity*; Levin, S.A., Ed.; Academic Press: Waltham, MA, USA, 2001; pp. 590–599.
7. Gillespie, R.G.; Baldwin, B.G.; Waters, J.M.; Fraser, C.I.; Nikula, R.; Roderick, G.K. Long-distance dispersal: A framework for hypothesis testing. *Trends Ecol. Evol.* **2012**, *27*, 47–56. [CrossRef] [PubMed]
8. Ali, J.R.; Meiri, S. Biodiversity growth on the volcanic ocean islands and the roles of in situ cladogenesis and immigration; case with the reptiles. *Ecography* **2019**, *42*, 989–999. [CrossRef]
9. Harris, J.; Rocha, S. Comoros. In *Encyclopedia of Islands*; Gillespie, R., Clague, D., Eds.; University of California Press: Oakland, CA, USA, 2009; pp. 177–180.
10. Michon, L. The volcanism of the Comoros archipelago integrated at a regional scale. In *Active Volcanoes of the Southwest Indian Ocean, Piton de la Fournaise and Karthala*; Bachelery, P., Lénat, J.-F., Di Muro, A., Michon, L., Eds.; Springer: Berlin/Heidelberg, Germany, 2016; pp. 233–244.
11. Warren, B.H.; Bermingham, E.; Bowie, R.C.K.; Prys-Jones, R.P.; Christophe Thébaud, C. Molecular phylogeography reveals island colonization history and diversification of western Indian Ocean sunbirds (Nectarinia: Nectariniidae). *Mol. Phyl. Evol.* **2003**, *29*, 67–85. [CrossRef]
12. Hawlitschek, O.; Toussaint, E.F.A.T.; Gehring, P.S.; Ratsoavina, F.M.; Cole, N.; Crottini, A.; Nopper, J.; Lam, A.W.; Vences, M.; Glaw, F. Gecko phylogeography in the Western Indian Ocean region: The oldest clade of Ebenavia inunguis lives on the youngest island. *J. Biogeogr.* **2017**, *44*, 409–420. [CrossRef]
13. Adler, G.H. Avifaunal diversity and endemism on tropical Indian Ocean islands. *J. Biogeogr.* **1994**, *21*, 85–95. [CrossRef]
14. Rocha, S.; Carretero, M.A.; Harris, J.D. Mitochondrial DNA sequence data suggests two independent colonizations of the Comoros archipelago by Chameleons of the genus Furcifer. *Belg. J. Zool.* **2005**, *135*, 39–42.
15. Rocha, S.; Carretero, M.A.; Harris, J.D. Diversity and phylogenetic relationships of Hemidactylus geckos from the Comoro Islands. *Mol. Phyl. Evol.* **2005**, *35*, 292–299. [CrossRef]
16. Rocha, S.; Carretero, M.A.; Vences, M.; Glaw, F.; Harris, J.D. Decipering patterns of transoceanic dispersal: The evolutionary origin and biogeography of coastal lizards (*Cryptoblepharus*) in the Western Indian Ocean Region. *J. Biogeogr.* **2006**, *33*, 13–22. [CrossRef]
17. Rocha, S.; Posada, D.; Carretero, M.A.; Harris, J.D. Phylogenetic affinities of Comoran and East African day geckos (genus Phelsuma): Multiple natural colonisations, indtroductions and island radiations. *Mol. Phyl. Evol.* **2007**, *43*, 685–692. [CrossRef] [PubMed]
18. Hawlitschek, O., Nagy, Z.T., Glaw, F. Island evolution and systematic revision of Comoran snakes: Why and when subspecies still make sense. *PLoS ONE* **2012**, *7*, e42970. [CrossRef] [PubMed]
19. Hawlitschek, O.; Ramírez Garrido, S.; Glaw, F. How marine currents influenced the widespread natural overseas dispersal of reptiles in the Western Indian Ocean region. *J. Biogeogr.* **2017**, *44*, 1435–1440. [CrossRef]
20. Fuchs, J.; Pons, J.-M.; Goodman, S.M.; Bretagnolle, V.; Melo, M.; Bowie, R.C.K.; Currie, D.; Safford, R.; Virani, M.Z.; Thomsett, S.; et al. Tracing the colonization history of the Indian Ocean scops-owls (Strigiformes: Otus) with further insight into the spatio-temporal origin of the Malagasy avifauna. *BMC Evol. Biol.* **2008**, *8*, 197. [CrossRef] [PubMed]

21. Fuchs, J.; Lemoine, D.; Parra, J.L.; Pons, J.-M.; Raherilalao, M.J.; Prys-Jones, R.; Thebaud, C.; Warren, B.H.; Goodman, S.M. Long-distance dispersal and inter-island colonization across the western Malagasy Region explain diversification in brush-warblers (Passeriformes: Nesillas). *Biol. J. Linn. Soc.* **2016**, *119*, 873–889. [CrossRef]
22. Glaw, F.; Hawlitschek, O.; Glaw, K.; Vences, M. Integrative evidence confirms new endemic island frogs and transmarine dispersal of amphibians between Madagascar and Mayotte (Comoros archipelago). *Sci. Nat.* **2019**, *106*, 19. [CrossRef]
23. Schipper, J. Southern Africa: Island Group between Madagascar. 2021. Available online: https://www.worldwidllife.org/ecoregions/at0105 (accessed on 16 December 2021).
24. Safford, R.J. The Comoros. In *Important Bird Areas in Africa and Associated Islands: Priority Sites for Conservation*; Fishpool, L.D.C., Evans, M.I., Eds.; BirdLife Conservation Series No. 11; Pisces Publications: Newbury, UK; BirdLife International: Cambridge, UK, 2001; pp. 411–464.
25. Hawlitschek, O.; Brückmann, B.; Berger, J.; Green, K.; Glaw, F. Integrating field surveys and remote sensing data to study distribution, habitat use and conservation status of the herpetofauna of the Comoro Islands. *ZooKeys* **2011**, *144*, 21–79. [CrossRef]
26. Myers, N.; Mittermeier, R.A.; Mittermeier, C.G.; da Fonseca, G.A.B.; Kents, J. Biodiversity hotspots for conservation priorities. *Nature* **2000**, *403*, 853–858. [CrossRef]
27. Agnarsson, I.; Kuntner, M. The generation of a biodiversity hotspot: Biogeography and phylogeography of the western Indian ocean islands. In *Current Topics in Phylogenetics and Phylogeography of Terrestrial and Aquatic Systems*; Anamthawat-Jonsson, K., Ed.; InTech Publishers: Rijeka, Croatia, 2012; pp. 33–82.
28. Wikström, N.; Avino, M.; Razafimandimbison, S.G.; Bremer, B. Historical biogeography of the coffee family (Rubiaceae, Gentianales) in Madagascar: Case studies from the tribes Knoxieae, Naucleeae, Paederieae and Vanguerieae. *J. Biogeogr.* **2010**, *37*, 1094–1113. [CrossRef]
29. Goodman, S.M.; Weyeneth, N.; Ibrahim, Y.; Said, I.; Ruedi, M. A review of the bat fauna of the Comoro Archipelago. *Acta Chiropterologica* **2010**, *12*, 117–141. [CrossRef]
30. Miller, K.B.; Bergsten, J. *Diving Beetles of the World. Systematics and Biology of the Dytiscidae*; Johns Hopkins University Press: Baltimore, MD, USA, 2016; 320p.
31. Biström, O.; Nilsson, A.N.; Bergsten, J. Taxonomic revision of Afrotropical *Laccophilus* Leach, 1815 (*Coleoptera, Dytiscidae*). *ZooKeys* **2015**, *542*, 1–379. [CrossRef] [PubMed]
32. Manuel, M.; Ramahandrison, A.T. Four new species of the diving beetle genus *Laccophilus* Leach, 1815 from Madagascar (Coleoptera, Dytiscidae, Laccophilini). *Zootaxa* **2020**, *4822*, zootaxa-4822. [CrossRef] [PubMed]
33. Bergsten, J.; Manuel, M.; Ranarilalatiana, T.; Ramahandrison, A.T.; Hájek, J. Dytiscidae, diving beetles, tsikovoka. In *The New Natural History of Madagascar*; Goodman, S.M., Ed.; Princeton University Press: Princeton, NJ, USA, in press.
34. Guignot, F. Dytiscides et Gyrinides des Comores. *Mem. L'institute Sci. Madag.* **1959**, *10*, 75–79.
35. Guignot, F. Revision des hydrocanthares d'Afrique (Coleoptera Dytiscoidea), 3. In *Annales du Musée Royal du Congo Belge*; Série 8vo Sciences Zoologiques; Musée Royal du Congo Belge: New York, NY, USA, 1961; Volume 90, pp. 653–995.
36. Wewalka, G. Results of the Austrian Hydrobiological Mission, 1974, to the Seychelles-, Comores- and Mascarene Archipelagos. Part VII: Dytiscidae, Gyrinidae (Coleoptera). *Ann. Nat. Mus. Wien* **1980**, *83*, 723–732.
37. Nilsson, A.N. Dytiscidae. In *World Catalogue of Insects*; Apollo Books: Stenstrup, Denmark, 2001; Volume 3, pp. 1–395.
38. Nilsson, A.N.; Hajek, J. A World Catalogue of the Family Dytiscidae, or the Diving Beetles (Coleoptera, Adephaga). Version 1.I.2021. 2021, pp. 1–315. Available online: www.waterbeetles.eu (accessed on 4 November 2021).
39. Ibouroi, M.T.; Dhurham, S.A.O.; Besnard, A.; Lescureux, N. What is the main driver of unsustainable natural resource use in the Comoro Islands? *bioRxiv* **2021**. [CrossRef]
40. Le Ministère de la Transition Écologique 2021. Création d'une Réserve Naturelle Nationale des Forêts à Mayotte. Available online: https://biodiversite.gouv.fr/actualite/creation-dune-reserve-naturelle-nationale-des-forets-mayotte (accessed on 16 December 2021).

Article

Diversity of Periphytic Chironomidae on Different Substrate Types in a Floodplain Aquatic Ecosystem

Dubravka Čerba [1,*], Miran Koh [1], Barbara Vlaičević [1], Ivana Turković Čakalić [1], Djuradj Milošević [2] and Milica Stojković Piperac [2]

[1] Department of Biology, Josip Juraj Strossmayer University of Osijek, Cara Hadrijana 8/A, 31000 Osijek, Croatia; koh.miran@gmail.com (M.K.); barbara.vlaicevic@biologija.unios.hr (B.V.); iturkovic@biologija.unios.hr (I.T.Č.)
[2] Department of Biology and Ecology, Faculty of Science and Mathematics, University of Niš, Višegradska 33, 18000 Niš, Serbia; djuradj@pmf.ni.ac.rs (D.M.); milicas@pmf.ni.ac.rs (M.S.P.)
* Correspondence: dcerba@biologija.unios.hr

Abstract: Different types of water bodies in lowland river floodplains represent vital biodiversity havens and encompass diverse microhabitats, which are essential for structuring different macroinvertebrate communities. Chironomidae larvae (Diptera) are an inseparable part of these communities, with their high richness and abundance. In three water body types within the Danube floodplain Kopački Rit in Croatia, over the course of four sampling campaigns, we recorded 51 chironomid taxa in periphyton on macrophytes, twigs, and glass slides. The most diverse were chironomid communities on macrophytes, whilst month-old periphyton on twigs supported the least taxa. *Cricotopus* gr. *sylvestris*, *Dicrotendipes lobiger*, *Dicrotendipes* spp., *Endochironomus albipennis*, *Glyptotendipes pallens* agg., *Polypedilum sordens* and *Polypedilum* spp. were present in all studied microhabitats. The type of substrate is a very important factor influencing Chironomidae diversity and abundance, which was evident in the presence and dominance of *Corynoneura* gr. *scutellata* and *Monopelopia tenuicalcar* in the dense macrophyte canopy epiphyton. Finding pristine floodplains such as Kopački Rit can be very challenging, as such areas are increasingly altered by human activities. Studies of resident species and the extent to which changes in the parent river influence floodplain communities are important for the protection and restoration of the floodplains.

Keywords: chironomid larvae; taxonomic diversity; substrate preference; Danube; floodplain

Citation: Čerba, D.; Koh, M.; Vlaičević, B.; Turković Čakalić, I.; Milošević, D.; Stojković Piperac, M. Diversity of Periphytic Chironomidae on Different Substrate Types in a Floodplain Aquatic Ecosystem. *Diversity* **2022**, *14*, 264. https://doi.org/10.3390/d14040264

Academic Editors: Marina Vilenica, Zohar Yanai and Laurent Vuataz

Received: 30 January 2022
Accepted: 28 March 2022
Published: 31 March 2022

1. Introduction

Riverine floodplains and different types of wetlands represent very dynamic and diverse habitats, created by prolonged interactions of water inflow from the parent river, ground water and the terrestrial area [1–3]. The hydrological connectivity of adjacent water bodies to the main river channel can be continuous or alternating, depending on the water level, as flooding occurs only during maximum river water level [1,2,4]. The lower reaches of the Danube and its major tributaries are representative of rivers typical for temperate regions of Europe, with wide meandering river channels that have the potential to create floodplains as natural water retention areas, as in the case of Kopački Rit [1,2]. Such floodplains are comprised of both deep and shallow lentic and lotic water bodies, which can be permanent or temporary. Aquatic and semi-aquatic habitats intermingle with forests, dry terrain and water meadows creating habitats suitable for many invertebrate and vertebrate species, providing shelter, food or spawning areas within these ecosystems [4–6]. Negative impacts on the river, e.g., pollution, riverbed destruction, and invasive species introduction, affect the whole watershed [7]. Floodplains around the world have different characteristics, and threats to their ecosystem can be from river regulation, drainage-basin alterations, deforestation, global climate change and extended drought periods [7]. In Europe, floodplain areas or specific segments can be given forms of protective status due to their uniqueness and sensitivity to anthropogenic activities—such as Nature or National Parks—even if they have been modified to some extent [7]. Organisations such as the Danube River Network of Protected Areas, the International Commission for the Protection of the Danube River, and the WWF have an important part in the protection process. Besides being biodiversity hotspots, floodplains provide a very broad spectrum of ecological services, making them even more important for researchers and the general public [7–9].

High biodiversity indicates the high functional diversity of organisms in the floodplain and complex trophic interactions, among others [10]. One of the key elements in the normal functioning of the floodplain food webs is macrozoobenthos, particularly the early life stages of insects. Among them, the prevalent taxonomic group often standing out in its abundance, species and functional diversity is the dipteran family Chironomidae [11–13]. Chironomid larvae are fascinating organisms, distributed across the globe and adapted to an array of different living conditions [11,14–16]. Even though most of the chironomid taxa are euryvalent [11,17], there are some very tolerant species subject to habitat degradation [11,13,17,18], while some have narrow ecological valences and are found only in pristine environments [15,19]. Previous studies have shown that the whole chironomid community responds to changes in their environment [11,13,17,18], which is why they are becoming one of the basic tools in water-quality assessment projects [17,18,20]. However, this is only one aspect of why hydrobiologists find this group interesting. Their ecological traits enable them to fill many niches and serve as different functional groups in aquatic ecosystems. Feeding on algae, detritus, microorganisms or other invertebrates, while at the same time being preyed upon by other aquatic insects, fish or waterfowl, they link different trophic levels [20–22].

Substrate type can influence the structure of Chironomidae communities, since the larvae often exploit the substrate by boring into plant or animal tissue, mining wood, burrowing into the sediment surface, or attaching themselves to a hard substrate [11,23,24]. Chironomid larvae or pupae inhabit not only sediment, but also periphytic communities developed on natural and artificial substrates [25–27]. Woody debris and aquatic macrophytes are common and ecologically important types of natural substrates in floodplain water bodies [28–30]. Aquatic macrophytes represent complex colonisation substrates for all aquatic invertebrates—including Chironomidae—but especially submerged macrophytes, providing refuge from predators, a source of food, and an oxygen-rich environment [31]. Hence, this substrate type is one of the most suitable for chironomids, supporting their high abundance and diversity [30,32]. Woody debris such as twigs, branches and tree trunks also provide important microhabitats for aquatic invertebrates, offering a food source and shelter from predators [28,29]. However, information about the chironomid communities on this substrate type is still limited, especially in aquatic ecosystems such as riverine floodplains.

The main objective of the present study was to assess the diversity of chironomid larvae communities from periphytic communities developed on different substrates with different structural complexities. Some authors found no significant differences between communities on artificial and natural substrates [33], so we additionally aimed to compare species composition and relative abundance of Chironomidae taxa to better understand the differences between the communities that form upon different types of substrate.

2. Materials and Methods

2.1. Study Area

Kopački Rit is one of the largest preserved Danube floodplain landscapes with a total surface area of 231 km^2, situated in the eastern part of Croatia. The Danube borders the east side of the floodplain, from 1383 to 1410 river km, while the river Drava creates the southern border from 0 to 15 river km (Figure 1). Kopački Rit has been protected as a Nature Park since 1999, but it was first declared and protected as an ecologically important area in the 1960s [6]. Furthermore, one part of the park is listed as a Special

Zoological Reserve. The floodplain is on the Important Bird Areas List and is recognized as an important Ramsar and Natura 2000 area; it also conjoins the Mura–Drava–Danube Biosphere Reserve. A multitude of floodplain water bodies within the park (e.g., lakes, ponds, channels) are changing and transforming under the influence of the Danube—its water level and other characteristics [6,34]. Water from the Danube enters Kopački Rit through several channels, but the main avenue is situated in the southern part of the floodplain in the Special Zoological Reserve, through the channels Hulovo and Čonakut, filling along the way Kopačko Lake. Lake Sakadaš, the furthest point from the main river channel, is the deepest lake (6 m in average) and the point of departure for scientific and tourist boats (Figure 1). On the other side of the embankment surrounding the main floodplain area, there is a network of channels and canals, ponds and fisheries that support diverse communities of flora and fauna.

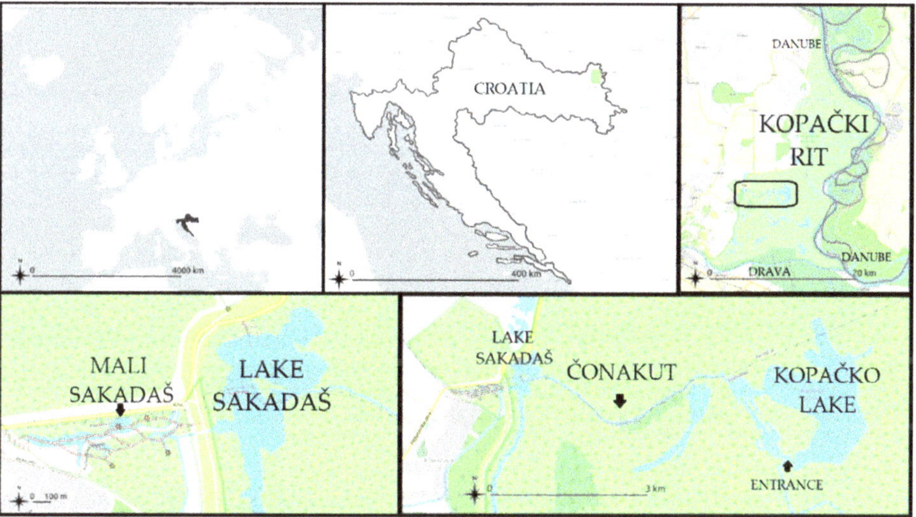

Figure 1. Research locations in the floodplain area of Kopački Rit Nature Park. Top middle, green shape: the geographical position of Kopački Rit in Croatia; top right, green colour: the floodplain area of the Danube; black rectangle: sampling area enlarged on the bottom right; left and right, blue colour: water bodies, green depicts the surrounding semi-aquatic and terrestrial area; water bodies are labelled in black letters.

2.2. Sampling Strategies

Communities of Chironomidae larvae have been studied through various projects, sampling campaigns or *in situ* experiments. Different studies, of which the results are presented here, applied different standard sampling techniques, depending on the substrate and habitat type. Periphytic communities have been studied on an artificial substrate (glass slides in 2008 and 2009) and natural substrates (twigs in 2011 and 2012, and macrophytes in 2013 and 2016).

To collect the data on the community structure and colonisation dynamics of periphytic chironomids on an artificial substrate, glass slides for periphyton development were immersed from April until August in Lake Sakadaš at a depth of 25 cm. The slides were sampled after the first seven days of exposure and afterwards every 14 days. On each sampling date three slides were taken for chironomid analysis and *in situ* placed in bottles with 4% formaldehyde. For a detailed description see Vidaković et al. [35].

Epixylon was studied on willow twigs placed in Lake Sakadaš as part of the *in situ* experiment which included the immersion of twigs (length of 10 cm, diameter 1 cm) to a depth of 20–25 cm, for 5 weeks during three different seasons: summer, late autumn, and spring of the following year. The experiment constructions were placed at three sites in Lake Sakadaš. For invertebrate community analysis, three twigs were sampled and preserved in 4% formaldehyde. For a detailed description see Mihaljević et al. [36].

Epiphytic chironomids were sampled in two different types of macrophyte associations. One was a dense, thick layer of floating macrophytes, *Salvinia natans*, *Spirodela polyrhiza*, *Lemna* sp. and, sporadically, submerged *Ceratophyllum demersum*, formed and sampled alongside three locations: the entrance to Kopačko Lake, the Čonakut channel and the entrance to Lake Sakadaš (listed as epi I). The second association type was sampled in the Mali Sakadaš pond. It represents a typical pond macrophyte association and includes different contributions of *Nymphoides peltata*, *Nymphaea alba*, *S. natans*, *Typha* sp., *Hippuris* sp., *C. demersum* and *Utricularia vulgaris* (listed as epi II). Apart from Lake Sakadaš, during the epi I research there were three sampling sites at each location. Macrophytes were sampled within a surface area of 50 × 50 cm, marked by a wooden frame. Triplicate samples were carefully removed from the water to avoid loss of organisms and preserved in 96% ethanol.

2.3. Periphyton

The starting point of the laboratory work was specific for each type of periphyton. In the experiment with the artificial substrate, periphyton was scraped from both sides of the glass slides and collected in a beaker. A similar procedure was applied for epixylon, namely cleaning the surface of the whole twig. Macrophytes were thoroughly rinsed and cleaned on a sieve over white trays to ensure that all organisms were collected. In all samples, the removed remains were rinsed above a sieve and prepared for the separation and isolation of larvae from the rest of the periphyton under stereoscopic microscopes (Carl Zeiss Jena, Olympus SZX9). Chironomidae larvae were prepared for identification either in the form of temporary native slides—in a drop of ethanol—or as a permanent slide mounted in Berlese medium. A microscope (Olympus BX51) and the following identification keys were used to identify the species and genera of Chironomidae: Schmid [37]; Vallenduuk and Moller Pillot [38]; Bitušík [39]; Bitušík and Hamerlík [40]; Andersen et al. [14]; and Vallenduuk [13].

2.4. Statistical Analysis

After the data on Chironomidae community structure were collected, applying diverse methods to enable comparison between different communities, we calculated, for each sample, relative abundances as the number of individuals of a given taxon divided by the total number of individuals collected in the sample. PRIMER 6 software [41] was applied for multivariate statistical analyses. Non-metric multidimensional scaling (NMDS) was used to present the relations between the chironomid communities from different substrates and analysis of similarity (ANOSIM) was applied to identify the significance of differences between substrates. These methods were applied to the Bray–Curtis similarity matrix based on the square root-transformed relative abundance data. The contribution of Chironomidae taxa to the average dissimilarity between groups was assessed using the SIMPER analysis. For every sample in each substrate type, we calculated the following diversity indices as a biotic metric: species richness (S), Shannon index (H'), and Simpson index (1-lambda). To test whether chironomid communities of different substrate types (epi I, epi II, twigs, and glass slides) differed in S, H' and 1-lambda, Kruskal–Wallis tests followed by Mann–Whitney tests were applied. These analyses were performed using SPSS version 19.0 software.

3. Results

3.1. Diversity of Periphytic Chironomidae

In all sampled communities, 51 Chironomidae taxa were recorded, belonging to three subfamilies: Tanypodinae, Orthocladiinae and Chironominae (Table 1). In the epixylon, only Orthocladiinae and Chironominae larvae were recorded. In the periphyton on the glass slides, Tanypodinae represented less than 1% of the community and these larvae were too young to be identified even to the genus level (Figure 2, Table 1). The most diverse was tribe Chironomini, including 32 different taxa belonging to eight genera. The Tanytarsini tribe was represented with only two genera, *Paratanytarsus* and *Tanytarsus*. Five genera and the *Cricotopus*/*Orthocladius* taxon represented the Orthocladiinae subfamily (Table 1).

Table 1. Relative abundance of Chironomidae taxa in periphyton on all substrate types.

Taxa/Substrate	Epiphyton I		Epiphyton II		Twigs		Glass Slides	
	Range	Average (N = 25)	Range	Average (N = 9)	Range	Average (N = 24)	Range	Average (N = 19)
Tanypodinae								
Ablabesmyia (*Ablabesmyia*) *longistyla* Fittkau, 1962	0–1.96	0.29	0–0.46	0.05				
Ablabesmyia (*Ablabesmyia*) *monilis* agg.	0–1.39	0.06	0–2.16	0.32				
Ablabesmyia spp.	0–0.32	0.01	0–0.46	0.05				
Conchapelopia agg.			0–0.01	0.001				
Monopelopia tenuicalcar (Kieffer, 1918)	27.81–79.49	55.96						
Tanypodinae non det.			0–11.70	1.78			0–5.59	0.29
Orthocladiinae								
Chaetocladius spp.	0–3.97	0.58						
Corynoneura gr. *scutellata*	0–45.77	19.49					0–0.74	0.04
Corynoneura spp.			0–1.49	0.23				
Cricotopus (*Cricotopus*) *bicinctus* (Meigen, 1818)	0–0.20	0.01						
Cricotopus (*Isocladius*) *intersectus* agg.	0–14.81	0.99			0–100	25.05	0–5.90	0.50
Cricotopus (*Isocladius*) gr. *sylvestris*	0–18.17	4.20	0–3.68	1.01	0–100	11.36	0–8.07	1.60
Cricotopus spp.	0–0.65	0.03			0–14.29	1.92	0–0.74	0.08
Cricotopus/*Orthocladius* spp.					0–58.33	4.63		
Nanocladius gr. *dichromus*	0–0.65	0.03						
Psectrocladius (*Psectrocladius*) *limbatellus* (Holmgren, 1869)			0–0.27	0.04				
Psectrocladius (*Psectrocladius*) gr. *sordidellus*					0–1.59	0.07		
Orthocladiinae non det.					0–100	6.39	0–50	3.00
Chironominae								
Chironomus (*Chironomus*) *annularis* agg.			0–0.42	0.05				
Chironomus (*Chironomus*) *luridus* Strenzke, 1959			0–1.60	0.57				
Chironomus (*Chironomus*) *plumosus* agg.			0–2.93	0.44			0–2.63	0.14
Chironomus (*Chironomus*) *tentans* Fabricius, 1805			0–0.91	0.10				
Chironomus (*Lobochironomus*) *dorsalis* Meigen, 1818			0–17.07	3.73				
Chironomus spp.	0–2.47	0.44	0–29.07	13.56				
Dicrotendipes lobiger (Kieffer, 1921)	0–2.30	0.12	3.25–21.64	10.33	0–3.13	0.25	0–10.81	0.57
Dicrotendipes modestus (Say, 1823)			0–7.33	1.27				
Dicrotendipes nervosus (Staeger, 1839)	0–1.85	0.14			0–50	6.07	0–56.72	9.76

Table 1. *Cont.*

Taxa/Substrate	Epiphyton I		Epiphyton II		Twigs		Glass Slides	
	Range	Average (N = 25)	Range	Average (N = 9)	Range	Average (N = 24)	Range	Average (N = 19)
Dicrotendipes notatus (Meigen, 1818)			0–0.21	0.02				
Dicrotendipes pulsus (Walker, 1856)	0–0.95	0.06			0–20	0.97	0–0.57	0.03
Dicrotendipes spp.	0–5.56	0.22	0–2.71	1.33	0–2.38	0.10	0–2.63	0.14
Endochironomus albipennis (Meigen, 1830)	0–15.82	1.33	0–2	0.61	0–4	0.53	0–59.86	6.36
Endochironomus tendens (Fabricius, 1775)	0–5.37	0.77	1.22–9.81	4.73				
Glyptotendipes (Glyptotendipes) barbipes (Staeger, 1839)	0–4.18	0.22						
Glyptotendipes (Glyptotendipes) pallens agg.	0–10.54	2.33	4.27–30.96	15.10	0–43.75	13.82	0–61.20	24.74
Glyptotendipes (Glyptotendipes) paripes (Edwards, 1929)			0–0.54	0.06				
Glyptotendipes spp.	0–11.24	2.47	0–5.26	1.28				
Kiefferulus (Kiefferulus) tendipediformis (Goetghebuer, 1921)	0–10.74	3.54	0–2.92	1.43				
Parachironomus gr. *arcuatus*	0–1.03	0.05	0–3.24	0.36	0–60	15.20		
Parachironomus gr. *frequens*							0–10.69	0.56
Parachironomus varus (Goetghebuer, 1921)							0–100	15.56
Parachironomus spp.			0–31.10	8.92				
Paratendipes nudisquama (Edwards, 1929)	0–0.32	0.01						
Paratendipes spp.	0–0.17	0.01						
Polypedilum (Pentapedilum) sordens (van der Wulp, 1875)	0–7.39	1.98	0–4.50	1.09	0–20	2.96	0–21.63	4.76
Polypedilum (Pentapedilum) uncinatum agg.	0–3.75	0.62	0–0.54	0.06				
Polypedilum (Polypedilum) nubeculosum (Meigen, 1804)	0–2.17	0.16					0–2.63	0.18
Polypedilum (Polypedilum) pedestre (Meigen, 1830)							0–18.07	3.13
Polypedilum (Tripodura) scalaenum (Schrank, 1803)	0–0.26	0.01						
Polypedilum (Uresipedilum) cultellatum Goetghebuer, 1931	0–2.25	0.16						
Polypedilum uncinatum agg./*cultellatum*			0–0.54	0.06				
Polypedilum spp.	0–14.46	2.65	0–3.73	0.70	0–50	2.51	0–50	3.20
Paratanytarsus spp.	0–5.37	0.99	2.26–23.13	8.81			0–50	5.15
Tanytarsus spp.	0–1.05	0.06	0–12.68	5.18				
Chironominae non det.			2.40–36.57	16.71	0–42.37	8.17	0–100	20.21

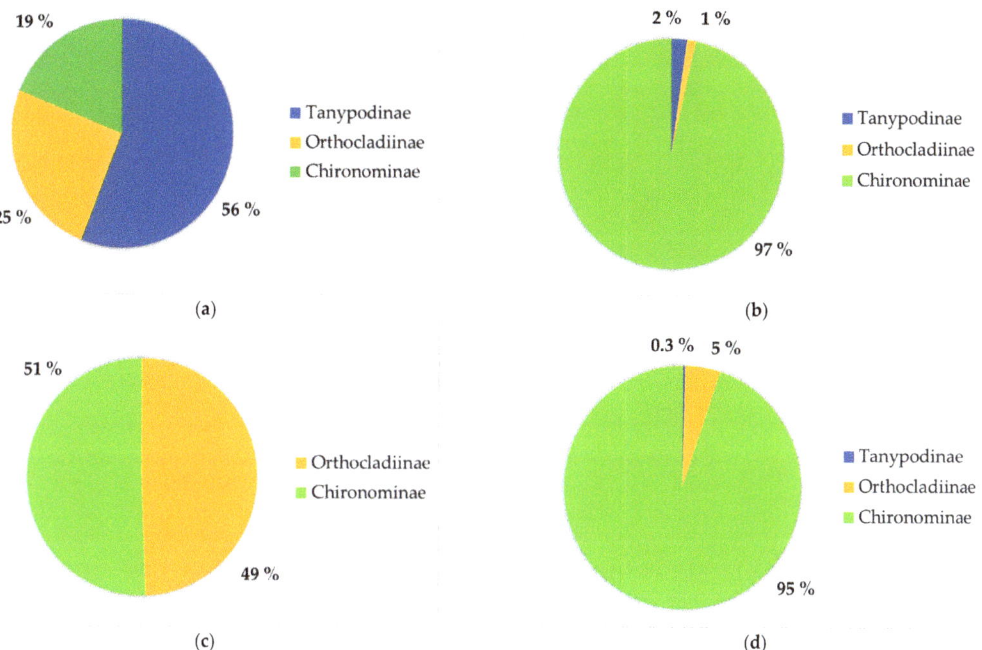

Figure 2. Percentage rate of Chironomidae subfamilies on each substrate type: (**a**) on macrophytes-epi I (epiphyton); (**b**) on macrophytes-epi II (epiphyton); (**c**) on willow twigs (epixylon); (**d**) on glass slides (periphyton).

The richest community type was the epiphyton with 33 chironomid taxa recorded in the first macrophyte study (epi I) and 31 taxa in the second (epi II) (Table 1). One of the important differences among these communities was evident within the subfamily Tanypodinae. *Monopelopia tenuicalcar* was very abundant in all samples and sites in epi I, whereas in epi II (macrophytes in Mali Sakadaš pond) it was not recorded at all, either on glass slides or twigs. In general, macrophytes were the best substrate for Tanypodinae larvae (Figure 2, Table 1). Many taxa were recorded only in the epiphyton, e.g., *Paratendipes* taxa in epi I, or most species of the *Chironomus* genus, which were mainly found in epi II (Table 1). In comparison to the diversity of Chironomidae larvae recorded on macrophytes, periphytic communities on glass slides and twigs were not as rich, comprising 18 and 14 different taxa, respectively. Both communities had high percentages of larvulae that could only be identified to the subfamily level (Table 1). The following species/species groups were recorded on all substrate types: *Cricotopus* gr. *sylvestris*, *Dicrotendipes lobiger*, *Endochironomus albipennis*, *Glyptotendipes pallens* agg., *Polypedilum sordens*, including *Dicrotendipes* spp. and *Polypedilum* spp. Larvae of *Polypedilum pedestre*, *Parachironomus* gr. *frequens* and *Parachironomus varus* were found only on glass slides, whereas *Psectrocladius* gr. *sordidellus* and *Cricotopus/Orthocladius* spp. were only characteristic for epixylon (Table 1).

According to the values of taxonomic diversity indices, the most diverse Chironomidae community was found on macrophytes, especially in epi II, while twigs and glass slides supported the lowest diversity (Figure 3).

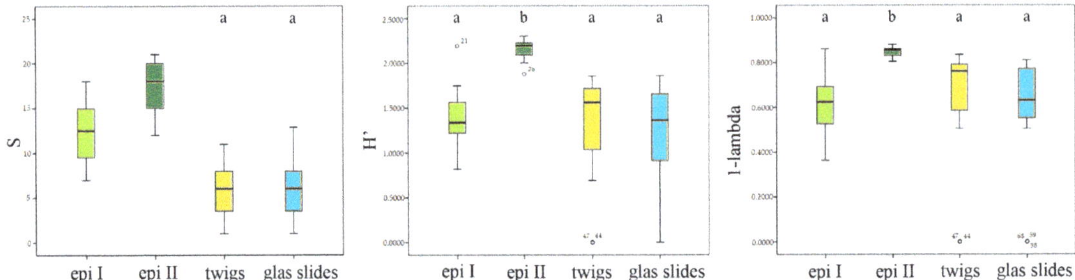

Figure 3. Boxplot representation of diversity indices (species (taxa) richness, S; Shannon index, H′; and Simpson, 1-lambda), across different substrate types. Epi I-macrophyte canopy epiphyton; epi II-pond macrophyte epiphyton; willow twigs; glass slides. Boxes which do not share a common letter are significantly different: a,b at $p < 0.05$ for S, H′, and 1-lambda. If letters are missing, all the boxes are significantly different.

Differences between communities on different substrates were evident in the number of recorded taxa, and the percentage rate of recurrent taxa differed among the substrates. *Corynoneura* gr. *scutellata* displayed a similar trend in appearance and percentage rate as the earlier mentioned *M. tenuicalcar*. *Cricotopus intersectus* agg. and *C.* gr. *sylvestris* from the Orthocladiinae subfamily and *G. pallens* agg. from tribe Chironomini were not only more frequently recorded, but they had a higher relative abundance (Table 1).

3.2. Statistical Analysis

All diversity indices differed between different substrate types (Figure 3). Species richness significantly varied among all substrates except for between twigs and glass slides (Mann–Whitney, $p < 0.05$). Epi II was significantly different in H′ and 1-lambda than all other substrate types (Mann–Whitney, $p < 0.05$).

Differences between the periphytic chironomid communities formed on different substrates were indicated by non-metric multidimensional scaling analysis and ordinated on the NMDS plot (Figure 4). Despite the relatively high stress, the analysis was considered robust by the PRIMER software, i.e., at stress <0.2 the two-dimensional ordination plot can still be considered useful. ANOSIM analysis confirmed the statistical significance of the differences between the communities from different substrate types (Global R = 0.728, $p < 0.001$). Results of the Pairwise tests are given in Table 2. Taxa that contributed the most to the differences among the substrates were indicated using SIMPER analysis (Table 3).

Table 2. Results of the ANOSIM analysis (R statistic values of pairwise tests) showing significant differences between chironomid communities from different substrate types. Results of the pairwise tests are all at $p = 0.001$, with the exception of glass slides vs. epi II at $p = 0.002$. Epi I-macrophyte canopy epiphyton; epi II-pond macrophyte epiphyton.

	Twigs	Glass Slides	epi I	epi II
Twigs				
Glass slides	0.392			
epi I	0.848	0.844		
epi II	0.602	0.360	0.999	

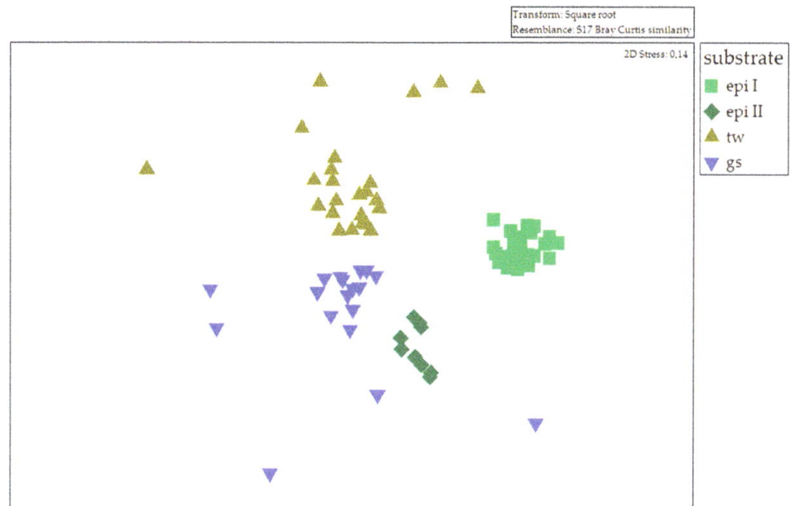

Figure 4. Non-metric multidimensional scaling plot of periphytic chironomid communities from different substrate types based on the relative abundance matrix data. Legend: epi I-macrophyte canopy epiphyton; epi II-pond macrophyte epiphyton; tw-willow twigs; gs-glass slides.

Table 3. Results of the SIMPER analysis showing the contribution of chironomid taxa to dissimilarities between substrate types: epi I-macrophyte canopy epiphyton; epi II-pond macrophyte epiphyton; tw, willow twigs; gs, glass slides.

	Contribution (%)
tw and gs	Average dissimilarity = 78.04
Cricotopus intersectus agg.	12.98
Chironominae non det.	10.37
Glyptotendipes pallens agg.	9.91
Monopelopia tenuicalcar	18.45
Corynoneura gr. *scutellata*	10.14
Cricotopus intersectus agg.	9.83
gs and epi I	Average dissimilarity = 88.53
Monopelopia tenuicalcar	18.79
Corynoneura gr. *scutellata*	10.24
Chironominae non det.	8.96
tw and epi II	Average dissimilarity = 82.64
Cricotopus intersectus agg.	9.39
Chironomus spp.	7.32
Parachironomus gr. *arcuatus*	6.7
gs and epi II	Average dissimilarity = 75.37
Chironomus spp.	8.13
Dicrotendipes lobiger	7.39
Glyptotendipes pallens agg.	7.13
epi I and epi II	Average dissimilarity = 79.28
Monopelopia tenuicalcar	16.03
Corynoneura gr. *scutellata*	8.82
Chironominae non det.	8.37

4. Discussion

The presented results, collected from several studies, allowed us to evaluate the different substrates and the mosaic of diversity in the aquatic communities that they support. Chironomidae larvae, as one of the most abundant, diverse and widely distributed invertebrate groups in aquatic systems of temperate regions, can adequately reflect that diversity [42]. Even though this study represents a "mosaic" of research results, it provides important data on chironomid taxa richness in a floodplain ecosystem.

Kopački Rit, as one of the largest preserved flooding areas of the Danube, harbours great biodiversity and consequently urges us to focus on its protection and conservation [6,32,35]. The park is a part of the Amazon of Europe UNESCO biosphere reserve, and although it is only a fragment of the reserve, it represents a very important component that encompasses a complex network of habitats and hundreds of species, creating a special ecosystem [6]. To better understand it, the study of the biology and ecology of the many communities inhabiting this area, especially aquatic ones, is urgently needed [9]. The main threats to the Kopački Rit floodplain are human activities on the Danube, e.g., pollution, hydromorphological degradation, and embankment. Deepening of the riverbed can lead to a lowering of the groundwater table, which influences at what water level floods enter the floodplain and reduces the overall amount of water available for the entire area [2,7]. Pollution of the Danube has already been detected in the vicinity of urban areas [6,7,18]. At what distance it dwindles and how it affects downstream areas and floodplains can be assessed by monitoring the changes in invertebrate communities, and the presence of tolerant chironomid species in benthic and epiphytic communities [18]. All in all, low biodiversity can indicate degradation of the floodplain ecosystem that motivates protection actions. The described changes and challenges correspond to global problems of floodplain protection and preservation [7]. Finding pristine floodplains such as Kopački Rit can be very challenging in Europe as well as worldwide, as these areas become increasingly altered by human activities. Furthermore, many flooding areas have been detached from the main river channel and have deteriorated over time. In the last decade, there has been much effort to revitalise and restore the already morphologically and hydrologically modified floodplains in Europe, particularly in the Danube and Drava watersheds [7,9]; thus, it is valuable to have data on the biodiversity of preserved ecosystems for comparison in assessment and monitoring projects. Studying the resident species and to what extent the changes in the parent river influence floodplain communities is also important for the protection of remaining intact floodplains.

Our sampling sites were located in the Kopački Rit Nature Park along the main water path of the flood- or flow-pulse from the Danube to the embankment. The connection of the floodplain to the parent river greatly influences all communities, their structure, stability and changes in diversity [43,44]. Concerning this, macrophytes have a varying dynamic of appearance in water bodies of the floodplain area, particularly in the channels leading from the Danube to Lake Sakadaš and in the lake itself. They are constantly present in the floodplain ponds or standing backwater. In the Čonakut channel, after several years, epiphytic communities developed in a dense canopy of floating and submerged macrophytes (epi I). In this community, *Monopelopia tenuicalcar*, which prefers substrates near the surface such as *Lemna* or *Azolla* [13,23], had a high relative abundance, even up to 80% in some samples, but this was not recorded in other communities, not even in epi II. Furthermore, the Orthocladiinae species group *Corynoneura* gr. *scutellata*, which also prefers this type of microhabitat [24] was quite abundant, thereby providing an adequate food source for *M. tenuicalcar*. In 2001 and 2002, in communities developed on macrophytes occurring in the Čonakut channel, Chironomidae larvae were the dominant taxonomic group contributing from 50 to 83% of the total invertebrate abundance [32]. Unfortunately, we do not have any data on the species composition from that research, which hinders a more detailed comparison and evaluation of the overall indicative values of epiphytic chironomids and supports the requirement for a better identification resolution in ecological studies. It also reflects the need to have as precise identification as possible. Epiphyton sampling activities in the Mali Sakadaš pond (epi II) did not meet all of our expectations regarding chironomid diversity on macrophytes, with a low relative abundance of *Cricotopus* species; however, a higher abundance of *Glyptotendipes pallens* agg., *Dicrotendipes*, *Paratanytarsus* and *Chironomus* species was as per other findings for epiphyton in eutrophic water bodies [30]. As Mali Sakadaš is an isolated pond, such differences could have been expected since epiphytic communities and macrophyte diversity and development depend on the connectivity and fluctuations of water level [44,45]. Moreover, Čerba et al. [21,46] previously recorded differences in Chironomidae epiphytic communities on two different submerged macrophytes, indicating the preferences of some taxa (e.g., *Cricotopus* gr. *sylvestris, Endochironomus albipennis, G. pallens* agg.) for specific macrophytes, including their architecture, tissue softness, the ability to hold more detritus, etc. [47]. It could then be anticipated that macrophytes that greatly differ in their leaf and stem architecture, or position related to the water surface described herein, display even greater differences in chironomid community composition. Our research confirmed a previously observed positive interrelationship of Chironomidae larvae abundance and diversity with macrophyte diversity, providing a spectrum of available food, microhabitats to inhabit, and shelter from predators [11].

Despite these differences, macrophytes harbour high chironomid diversity. They also influence other communities and the "health" of the entire floodplain ecosystem. Many fish species come from the main river channel to spawn or to find shelter and food in macrophyte-dominated habitats. Furthermore, firstly reported benthos-feeding fish have been found to feed primarily on the epiphytic Chironomidae larvae [48]. High chironomid taxa richness enables the sufficient colonization of various available microhabitats, depending on their specialties, and in turn caters to the different predatory fish that inhabit them. This does not only highlight the importance of macrophytes, but also the information on taxa diversity, which enables us to better understand the functioning of the relationships among different hydrobiocoenoses in the floodplain.

Another natural basis suited for periphyton development that is often available in floodplain water bodies are branches, tree trunks, or woody debris [49]. Depending on the duration of submergence and size of the wooden surface, epixylon includes various taxa [28] and chironomids can be the dominant invertebrate group [27]. Chironomidae larvae living in such communities can be either xylophagous or feed on algae, fungi or biofilm formed on the surface [11,50,51]. In our research we did not find true xylophagous or wood-boring taxa, which could be the consequence of the five-week immersion period of twigs. Nevertheless, Moller Pillot [24] lists decaying wood as one of the various feeding sources of *G. pallens* agg., as well as the utilisation of the woody microhabitat in self-made mines. Furthermore, *G. pallens* agg. can tolerate winter conditions better than many other species and is ubiquitous in floodplains [24]. During late autumn, besides the mentioned larvae, the chironomid epixylon community mostly included *C. intersectus* and gr. *sylvestris* representatives, previously described as good colonisers, cosmopolitan and pioneer species [11,16,46,52]. In other seasons, the submerged willow twigs represented an additional type of substrate in the lake, providing a temporary feeding and resting place for other chironomid larvae moving from surrounding microhabitats.

Even though glass slides are artificial substrates, they harboured a greater variety of Chironomidae than twigs. One of the reasons could be the presence of other invertebrate taxa in the developed periphyton that created a more suitable and heterogenous microhabitat, with bryozoans, sponges, or *Dreissena polymorpha* clusters [35]. The development of a complex autotrophic component [53] further augmented the colonisation of chironomid larvae as they are the main food source for many species, e.g., *E. albipennis, G. pallens* agg. and *C.* gr. *sylvestris* [54], including *Polypedilum pedestre* and *P. sordens* that feed on detritus, bacteria, diatoms, and other algae [24]. During this research, we found on more than one sampling occasion several larvae with the front part of their body in the mantle cavity of *D. polymorpha*. Since the larvae belonged to different nonparasitic species [55], we cannot state that this is a species-specific relationship, but rather a good example of how Chironomidae larvae successfully exploit available resources [56,57].

One of the important factors influencing the community structure is the life cycle dynamic, i.e., the number of generations per year and the diapause period [11,22,23,37], which can be partly influenced by environmental parameters such as temperature, as they can, in turn, influence the results if the sampling is conducted just after emergence or at the beginning of substrate colonisation. Early Chironomidae larvae stages, also known as larvulae, cannot be accurately identified to species or even genera level to give unambiguous results. Natural seasonal variability of the chironomid community is evident and a very important element to be taken into consideration when studying this group [58,59], and some of the observed differences between communities were undoubtedly a result of seasonal variability. Nevertheless, differences have been observed in the same season on different substrates.

The practical use of the knowledge on Chironomidae diversity in the floodplain would be a construction of biological metrics. Water framework directive (WFD) has regulations for bioassessment and monitoring procedures for lotic and lentic systems [18,60,61]. However, floodplains have different hydrological regimes, and in order to establish a practical chironomid-based assessment protocol it would be necessary to modify standard WFD protocols and biological indices to create specific ones for such ecosystem. Initial research to create a basic dataset would include the sampling of chironomids in all community types in different water bodies, as well as sampling in different seasons and at different water levels for comparison. Simultaneously, biotic and abiotic environmental parameters should be measured to assess the influence of environmental parameters on the community structure [18,60].

To conclude, we showed that the chironomid community's richness and diversity, as well as the relative abundance of Chironomidae taxa, significantly differed depending on the substrate type. As expected, the richest and most diverse community was found on macrophytes. Surprisingly, twigs supported lower taxa richness than an artificial substrate; however, this could be an artifact due to the short immersion period of the twigs. Even though chironomid larvae are considered simple opportunists, many taxa showed preference and adaptation to microhabitats with specific conditions and food availability, such as feeding in a bivalve mantle cavity; the colonisation of clean substrates; and abundant *C.* gr. *scutellata* and *M. tenuicalcar* larvae in dense macrophyte mats. Since floods are important for the accumulation and development of natural substrates such as macrophytes and wood remains in floodplain water bodies that support a high diversity of aquatic organisms (including Chironomidae), the protection of natural hydrological regimes is essential for biodiversity conservation in this unique and endangered aquatic ecosystem. Constant monitoring of diversity within floodplains can help us to better understand the changes of this ecosystem.

Author Contributions: Conceptualization, D.Č., B.V., I.T.Č., D.M., M.S.P.; data curation, D.Č., M.K., B.V., I.T.Č.; formal analysis, D.Č., M.K., B.V., I.T.Č.; funding acquisition, D.Č., D.M., M.S.P.; investigation, D.Č., M.K., B.V., I.T.Č., D.M.; methodology, D.Č., B.V., I.T.Č., D.M., M.S.P.; project administration, D.Č., D.M., M.S.P.; resources, D.Č., M.K., I.T.Č., M.S.P.; supervision, D.Č., D.M., M.S.P.; validation, D.Č., B.V., I.T.Č., D.M.; visualization, D.Č., M.K., B.V., I.T.Č.; writing—original draft, D.Č.; writing—review and editing, D.Č., M.K., B.V., I.T.Č., D.M. All authors have read and agreed to the published version of the manuscript.

Funding: These research projects were funded by the "Institutional financing" of the Department of Biology, Josip Juraj Strossmayer University of Osijek, also as a part of bilateral cooperation scientific project between Croatia and Serbia funded by the Croatian Ministry of Science and Education and Serbian Ministry of Education, Science and Technological Development; and part of the studies were conducted within a national funding projects by Croatian Ministry of Science, Education and Sports, project No. 285-0000000-2674.

Institutional Review Board Statement: Not applicable.

Data Availability Statement: The data presented in this study are available on request from the corresponding author.

Acknowledgments: We are grateful to the students and colleagues who contributed in gathering all these data while working on their master theses or by helping in the field and laboratory work: Tonka Kovačević, Nataša Klasni, Ena Pritišanac, Ivan Balković, Filip Stević and other diligent colleagues who participated in any way. Authors would also like to thank Mayra and Goran Poznanović for language editing.

Conflicts of Interest: The authors declare no conflict of interest. The funders had no role in the design of the study; in the collection, analyses, or interpretation of data; in the writing of the manuscript, or in the decision to publish the results.

References

1. Junk, W.; Bayley, P.; Sparks, R. *The Food Pulse Concept in River-Foodplain Systems;* Canadian Special Publication of Fisheries and Aquatic Science: Ottawa, ON, Canada, 1989; pp. 110–127.
2. Tockner, K.; Malard, F.; Ward, J.V. An extension of the flood pulse concept. *Hydrol. Process.* **2000**, *14*, 2861–2883. [CrossRef]
3. Thomaz, S.M.; Agostinho, A.A.; Hahn, N.S. *The Upper Paraná River and Its Floodplain: Physical Aspects, Ecology and Conservation;* Backhuys Publishers: Leiden, The Netherlands, 2004; p. 393.
4. Amoros, C.; Bornette, G. Connectivity and biocomplexity in waterbodies of riverine floodplains. *Freshw. Biol.* **2002**, *47*, 761–776. [CrossRef]
5. Karaus, U.; Larsen, S.; Guillong, H.; Tockner, K. The contribution of lateral aquatic habitats to insect diversity along river corri-dors in the Alps. *Landsc. Ecol.* **2013**, *28*, 1755–1767. [CrossRef]
6. Sommerwerk, N.; Bloesch, J.; Baumgartner, C.; Bittl, T.; Čerba, D.; Csányi, B.; Davideanu, G.; Dokulil, M.; Frank, G.; Grecu, I.; et al. The Danube River Basin. In *Rivers of Europe*, 2nd ed.; Tockner, K., Zarfl, C., Robinson, C.T., Eds.; Elsevier: Amsterdam, The Netherlands, 2022; pp. 81–180. [CrossRef]
7. Tockner, K.; Stanford, J.A. Riverine flood plains: Present state and future trends. *Environ. Conserv.* **2002**, *29*, 308–330. [CrossRef]
8. Opperman, J.J.; Galloway, G.E.; Duvail, S. The Multiple Benefits of River—Floodplain Connectivity for People and Biodiversity. In *Encyclopedia of Biodiversity*, 2nd ed.; Levin, S.A., Ed.; Academic Press: London, UK, 2013; pp. 144–160. [CrossRef]
9. Jakubínský, J.; Prokopova, M.; Raška, P.; Salvati, L.; Bezak, N.; Cudlín, O.; Cudlín, P.; Purkyt, J.; Vezza, P.; Camporeale, C.; et al. *Managing Floodplains Using Nature-Based Solutions to Support Multiple Ecosystem Functions and Services;* Wiley Interdisciplinary Reviews: Water; Wiley: New York, NY, USA, 2021; Volume 8, p. e1545. [CrossRef]
10. Moss, B. *Ecology of Freshwaters: A View for the Twenty-First Century*, 4th ed.; John Wiley & Sons Ltd.: West Sussex, UK, 2010; p. 470.
11. Armitage, P.D.; Cranston, P.S.; Pinder, L.C.V. *The Chironomidae: Biology and Ecology of Non-Biting Midges;* Chapman & Hall: London, UK, 1995; p. 572.
12. Ferrington, L.C. Global diversity of non-biting midges (Chironomidae; *Insecta-Diptera*) in freshwater. In *Freshwater Animal Diversity Assessment. Developments in Hydrobiology, Volume 198;* Balian, E.V., Lévêque, C., Segers, H., Martens, K., Eds.; Springer: Dordrecht, The Netherlands, 2007; pp. 447–455. [CrossRef]
13. Vallenduuk, H.J. *Chironomini Larvae of Western European Lowland (Diptera: Chironomidae). Keys with Notes to the Species;* Eric Mauch Verlag: Dinkelscherben, Germany, 2017; p. 216.
14. Andersen, T.; Baranov, V.; Hagenlund, L.K.; Ivković, M.; Kvifte, G.M.; Pavlek, M. Blind Flight? A New Troglobiotic Orthoclad (Diptera, Chironomidae) from the Lukina Jama-Trojama Cave in Croatia. *PLoS ONE* **2016**, *11*, e0152884. [CrossRef] [PubMed]
15. Montagna, M.; Urbanelli, S.; Rossaro, B. The species of the genus *Diamesa* (Diptera, Chironomidae) known to occur in Italian Alps and Apennines. *Zootaxa* **2016**, *4193*, 317–331. [CrossRef]
16. Čerba, D.; Hamerlík, L. Fountains—Overlooked Small Water Bodies in the Urban Areas. In *Small Water Bodies of the Western Balkans;* Pešić, V., Milošević, D., Miliša, M., Eds.; Springer International Publishing: Cham, Switzerland, 2021; pp. 73–91. [CrossRef]
17. Rosenberg, D.M. Freshwater biomonitoring and Chironomidae. *Neth. J. Aquat. Ecol.* **1992**, *26*, 101–122. [CrossRef]
18. Milošević, D.; Mančev, D.; Čerba, D.; Stojković Piperac, M.; Popović, N.; Atanacković, A.; Đuknić, J.; Simić, V.; Paunović, M. The potential of chironomid larvae-based metrics in the bioassessment of non-wadeable rivers. *Sci. Total Environ.* **2018**, *616–617*, 472–479. [CrossRef]
19. Lencioni, V.; Rossaro, B. Microdistribution of chironomids (Diptera: Chironomidae) in Alpine streams: An autoecological perspective. *Hydrobiologia* **2005**, *533*, 61–76. [CrossRef]
20. Tarkowska-Kukuryk, M. Environmental Drivers of Macroinvertebrate Assemblages within Peat Pool Habitat-Implication for Bioassessment. *Water* **2021**, *13*, 2369. [CrossRef]
21. Čerba, D.; Mihaljević, Z.; Vidaković, J. Colonisation trends, community and trophic structure of chironomid larvae (Chironomidae: Diptera) in a temporal phytophylous assemblage. *Fundam. Appl. Limnol.* **2011**, *179*, 203–214. [CrossRef]
22. Prejs, A.; Koperski, P.; Prejs, K. Food-web manipulation in a small, eutrophic Lake Wirbel, Poland: The effect of replacement of key predators on epiphytic fauna. *Hydrobiologia* **1997**, *342*, 377–381. [CrossRef]
23. Moller Pillot, H.K.M. *Chironomidae Larvae. Biology and Ecology of the Chironomini;* KNNV Publishing: Zeist, The Netherlands, 2009; p. 270.
24. Moller Pilot, H.K.M. *Chironomidae Larvae. Vol. 3. Orthocladiinae: Biology and Ecology of the Aquatic Orthocladiinae;* KNNV Publishing: Zeist, The Netherlands, 2013; p. 312. [CrossRef]
25. Toth, M.; Móra, A.; Kiss, A.; Dévai, G.Y. *Chironomid Communities in Different Vegetation Types in a Backwater Nagy-Morotva of the Active Floodplain of River Tisza, Hungary;* Boletim Municipal: Margem, Portugal, 2008; pp. 169–175.
26. Keiper, J.B.; Espeland, E.M. Spatial distribution and larval behavior of *Glyptotendipes lobiferus* (Diptera: Chironomidae). *Hydrobiologia* **2000**, *427*, 129–133. [CrossRef]
27. Coe, H.J.; Kiffney, P.M.; Pess, G.R. A Comparison of Methods to Evaluate the Response of Periphyton and Invertebrates to Wood Placement in Large Pacific Coastal Rivers. *Northwest Sci.* **2006**, *80*, 298–307.

28. Braccia, A.; Batzer, D.P. Invertebrates associated with woody debris in a southeastern U.S. forested floodplain wetland. *Wetlands* **2001**, *21*, 18–31.
29. Schneider, K.N.; Winemiller, K.O. Structural complexity of woody debris patches influences fish and macroinvertebrate species richness in a temperate floodplain-river system. *Hydrobiologia* **2008**, *610*, 235–244. [CrossRef]
30. Bazzanti, M.; Coccia, C.; Dowgiallo, M.G. Microdistribution of macroinvertebrates in a temporary pond of Central Italy: Taxonomic and functional analyses. *Limnologica* **2010**, *40*, 291–299. [CrossRef]
31. Dvořák, J. An example of relationships between macrophytes, macroinvertebrates and their food resources in a shallow euthrophic lake. *Hydrobiologia* **1996**, *339*, 27–36. [CrossRef]
32. Bogut, I.; Vidaković, J.; Palijan, G.; Čerba, D. Bentic macroinvertebrates associated with four species of macrophytes. *Biologia* **2007**, *62*, 600–606. [CrossRef]
33. De Camargo, N.S.J.; Lehun, A.L.; Rosa, J.; de Deus Bueno-Krawczyk, A.C. Colonization of benthic invertebrates on artificial and natural substrate in a Neotropical lotic environment in Southern Brazil. *Acta. Sci. Biol. Sci.* **2019**, *41*, e45872. [CrossRef]
34. Preiner, S.; Weigelhofer, G.; Funk, A.; Hohensinner, S.; Reckendorfer, W.; Schiemer, F.; Hein, T. Danube Floodplain Lobau. In *Riverine Ecosystem Management*; Schmutz, S., Sendzimir, J., Eds.; Springer: Cham, Switzerland, 2018; Volume 8, pp. 491–506. [CrossRef]
35. Vidaković, J.; Turković Čakalić, I.; Stević, F.; Čerba, D. The influence of different hydrological conditions on periphytic invertebrate communities in a Danubian floodplain. *Fundam. Appl. Limnol.* **2012**, *181*, 59–72. [CrossRef]
36. Mihaljević, M.; Žuna Pfeiffer, T.; Vidaković, J.; Špoljarić, D.; Stević, F. The importance of microphytic composition on coarse woody debris for nematode colonization: A case study in a fluvial floodplain environment. *Biodivers. Conserv.* **2015**, *24*, 1711–1727. [CrossRef]
37. Schmid, P.E. *A Key to the Larval Chironomidae and Their Instars from Austrian Danube Region Streams and Rivers. Part I: Diamesinae, Prodiamesinae and Orthocladiinae*; Wasser und Abwasser, Suppl. 3; Fed. Inst. for Water Quality: Vienna, Austria, 1993; p. 514.
38. Vallenduuk, H.J.; Moller Pillot, H.K.M. *Chirnomidae Larvae, Vol. 1: Tanypodinae*; KNNV Publishing Company: Zeist, The Netherlands, 2007.
39. Bitušík, P. *Priručka na Určovanie Lariev Pakomarov (Diptera: Chironomidae) Slovenska. Časť I. Buconomyinae, Diamesinae, Prodiamesinae a Orthocladiinae*; Techn. Univ. vo Zvolene, Fak. Ekol. Environm, Katedra Biologie: Zvolen, Slovakia, 2000; pp. 1–133.
40. Bitušik, P.; Hamerlik, L. *Príručka na Určovanie Lariev Pakomárov (Diptera: Chironomidae) Slovenska. Časť 2. Tanypodinae. (Identification Key for Chironomidae of Slovakia. Part 2. Tanypodinae)*; Belianum, Vydavatelstvo Univerzity Matej Bela v Banskej Bystrici: Banská Bystrica, Slovakia, 2014; p. 96.
41. Clarke, K.R.; Gorley, R.N. *PRIMER v6: User Manual/Tutorial*; PRIMER-E Ltd.: Plymouth, UK, 2006; p. 192.
42. Pinder, L.C.V. The habitats of chironomid larvae. In *The Chironomidae: Biology and Ecology of Non-Biting Midges*; Armitage, P.D., Cranston, P.S., Pinder, L.C.V., Eds.; Chapman & Hall: London, UK, 1995; pp. 107–135.
43. Tan, C.; Sheng, T.; Wang, L.; Mbao, E.; Gao, J.; Wang, B. Water-level fluctuations affect the alpha and beta diversity of macroinvertebrates in Poyang Lake, China. *Fundam. Appl. Limnol.* **2021**, *194*, 321–334.
44. Larsen, S.; Karaus, U.; Claret, C.; Sporka, F.; Hamerlík, L.; Tockner, K. Flooding and hydrologic connectivity modulate community assembly in a dynamic river-floodplain ecosystem. *PLoS ONE* **2019**, *14*, e0213227. [CrossRef]
45. Gallardo, B.; Dolédec, S.; Paillex, A.; Arscott, D.B.; Sheldon, F.; Zilli, F.; Mérigoux, S.; Castella, E.; Comín, F.A. Response of benthic macroinvertebrates to gradients in hydrological connectivity: A comparison of temperate, subtropical, Mediterranean and semiarid river floodplains. *Freshw. Biol.* **2014**, *59*, 630–648. [CrossRef]
46. Čerba, D.; Mihaljević, Z.; Vidaković, J. Colonisation of temporary macrophyte substratum by midges (Chironomidae: Diptera). *Ann. Limnol. Int. J. Lim.* **2010**, *46*, 181–190. [CrossRef]
47. Wilson, S.J.; Ricciardi, A. Epiphytic macroinvertebrate communities on Eurasian watermilfoil (*Myriophyllum spicatum*) and native milfoils *Myriophyllum sibericum* and *Myriophyllum alterniflorum* in eastern North America. *Can. J. Fish. Aquat. Sci.* **2009**, *66*, 18–30.
48. Čerba, D.; Miloševič, D.J.; Turković Čakalić, I.; Ergović, V.; Koh, M.; Vuković, A. Functional role of chironomid larvae (Chironomidae, Diptera) within a Danube floodplain. In Proceedings of the 8th Central European Dipterological Conference, Kežmarské Žl'aby, High Tatra Mountains, Slovakia, 28–30 September 2015.
49. Tank, J.L.; Winterbourn, M.J. Microbial activity and invertebrate colonization of wood in a New Zealand forest stream. *N. Z. J. Mar. Freshwat. Res.* **1996**, *30*, 271–280.
50. Moog, O. *Fauna Aquatica Austriaca. Wasserwirtschaftskataster, Bundesministerium fur Land-und Forstwirtschaft, Umwelt und Wasserwirtsdhaft*; FAA: Vienna, Austria, 2002.
51. Spänhoff, B.; Kaschek, N.; Meyer, E.I. Laboratory investigation on community composition, emergence patterns and biomass of wood-inhabiting Chironomidae (Diptera) from a sandy lowland stream in Central Europe (Germany). *Aquat. Ecol.* **2004**, *38*, 547–560.
52. Wotton, R.S.; Armitage, P.D.; Aston, K.; Blackburn, J.H.; Hamburger, M.; Woodward, C.A. Colonization and emergence of midges (Chironomidae: Diptera) in slow sand filter beds. *Neth. J. Aquat. Ecol.* **1992**, *26*, 331–339.
53. Stević, F.; Čerba, D.; Turković Čakalić, I.; Žuna Pfeiffer, T.; Vidaković, J.; Mihaljević, M. Interrelations between *Dreissena polymorpha* colonization and autotrophic periphyton development—A field study in a temperate floodplain lake. *Fundam. Appl. Limnol.* **2013**, *183*, 107–119. [CrossRef]

54. Tarkowska-Kukuryk, M. Periphytic algae as food source for grazing chironomids in a shallow phytoplankton-dominated lake. *Limnologica* **2013**, *43*, 254–264. [CrossRef]
55. Tokeshi, M. On the evolution of commensalism in the Chironomidae. *Freshw. Biol.* **1993**, *29*, 481–489. [CrossRef]
56. Ricciardi, A. Occurrence of chironomid larvae (*Paratanytarsus* sp.) as commensals of dreissenid mussels (*Dreissena polymorpha* and *D. bugensis*). *Can. J. Zool.* **1994**, *72*, 1159–1162. [CrossRef]
57. Coffman, W.P.; Ferrington, L.C. Chironomidae. In *An Introduction of Aquatic Insects of North America*; Merrit, K.W., Cummins, R.W., Eds.; Kendall Hunt Publishing: Dubuque, Iowa, 1996; pp. 551–652.
58. Rossaro, B.; Lencioni, V.; Boggero, A.; Marziali, L. Chironomids from Southern Alpine running waters: Ecology, biogeography. *Hydrobiologia* **2006**, *562*, 231–246. [CrossRef]
59. Milošević, D.; Simić, V.; Stojković, M.; Čerba, D.; Mančev, D.; Petrović, A.; Paunović, M. Spatio-Temporal Pattern of the Chironomidae Community: Toward the Use of Non-Biting Midges in Bioassessment Programs. *Aquat. Ecol.* **2013**, *47*, 37–55. [CrossRef]
60. European Water Framework Directive. *2000/60/EC*; WFD: London, UK, 2000.
61. Karr, J.R.; Chu, E.W. *Restoring Life in Running Waters: Better Biological Monitoring*; Island Press: Washington, DC, USA, 1999; p. 200.

MDPI

Article

Diversity Patterns and Assemblage Structure of Non-Biting Midges (Diptera: Chironomidae) in Urban Waterbodies

Nataša Popović [1,*], Nikola Marinković [1], Dubravka Čerba [2], Maja Raković [1], Jelena Đuknić [1] and Momir Paunović [1]

1 Department of Hydroecology and Water Protection, Institute for Biological Research "Siniša Stanković"—National Institute of the Republic of Serbia, University of Belgrade, Bulevar despota Stefana 142, 11060 Beograd, Serbia; nikola.marinkovic@ibiss.bg.ac.rs (N.M.); rakovic.maja@ibiss.bg.ac.rs (M.R.); jelena.djuknic@ibiss.bg.ac.rs (J.Đ.); mpaunovi@ibiss.bg.ac.rs (M.P.)

2 Department of Biology, Josip Juraj Strossmayer University of Osijek, Cara Hadrijana 8/A, 31000 Osijek, Croatia; dcerba@biologija.unios.hr

* Correspondence: natasa.popovic@ibiss.bg.ac.rs

Abstract: Urban waters are often neglected in biodiversity research; nonetheless, the number of aquatic microhabitats present in a city and the surrounding urban area is impressive. Twenty-two waterbodies in the Belgrade functional urban area (FUA) were investigated for faunistic and diversity patterns and to assess the effects of environmental factors on the differentiation of Chironomidae assemblages. A total of 66 chironomid taxa within four subfamilies was identified. Water quality at the studied sites, expressed by the water pollution index (WPI), varied significantly. K-means clustering gave four homogenous groups of chironomid assemblages, which showed clear preferences to specific habitat conditions and tolerance to anthropogenic pressures. These groups had high values of alpha and beta diversity components. The main component of beta diversity was species turnover. Waterbody type, water temperature, pH, nutrients and overall pollution were the most important factors influencing the distribution and composition of chironomid assemblages, which revealed clear preferences of each assemblage type to the category of waterbody type and tolerances to environmental pressures.

Keywords: chironomid larvae; water pollution index (WPI); alpha and beta diversity; anthropogenic pressure

Citation: Popović, N.; Marinković, N.; Čerba, D.; Raković, M.; Đuknić, J.; Paunović, M. Diversity Patterns and Assemblage Structure of Non-Biting Midges (Diptera: Chironomidae) in Urban Waterbodies. *Diversity* **2022**, *14*, 187. https://doi.org/10.3390/d14030187

Academic Editors: Marina Vilenica, Zohar Yanai, Laurent Vuataz and Michael Wink

Received: 7 February 2022
Accepted: 28 February 2022
Published: 4 March 2022

Publisher's Note: MDPI stays neutral with regard to jurisdictional claims in published maps and institutional affiliations.

1. Introduction

Chironomidae larvae (non-biting midges) are a species-diverse insect group abundant in many freshwater ecosystems, which represent the main food source for many predatory invertebrates, fish and birds [1–3]. This group displays a great capability of adapting to a wide range of environmental conditions, typically occurring at high densities with a key ecological function in lotic freshwater communities [4]. In lotic and lentic waterbodies of temperate regions, chironomids can be one of the most abundant and diverse taxonomic groups within benthos or epiphyton. This is especially true for very productive freshwater ecosystems, where they can represent more than 60% of the macroinvertebrate community [5,6]. The quantitative and qualitative composition of chironomid assemblages can be indicative of changes in water quality. Chironomidae play a crucial ecological role in the cycling of organic matter in rivers, in the export of energy to riparian habitats, and provide a valuable model system for understanding which environmental variables drive species richness [7]. The structure of chironomid assemblages reflect changes in aquatic ecosystems that are strongly correlated with changes in water quality and habitat degradation, clearly pointing to increased saprobity or hydromorphological degradation [8,9]. Chironomids are often used in water quality bioassessment with in situ monitoring programs [9,10] and in laboratory experiments that test their resilience and reactions to heavy metals, nano-

and microplastics or organic pollutants [11,12]. Together with Oligochaeta, non-biting midges are usually dominant and sometimes the only present invertebrate taxa in heavily modified waterbodies regardless of whether they are of artificial or natural origin [13–15]. Urbanisation and landscape conversion in sub-urban areas are a threat and the main cause of aquatic system degradation [16,17]. Some of the known stressors influencing ecological traits and system functioning of urban waterbodies can be from an extrinsic (catchment) source or created within the flow, such as untreated stormwater runoff, various point sources of pollution, septic system leakage, dams or rip-raps [18,19].

In the Belgrade FUA, there are numerous heavily modified and artificial waterbodies (reservoirs, canals and rivers) exposed to different types and intensities of anthropogenic pressures. Having various uses and origins, they differ in hydromorphological characteristics and water quality because of their position and distance from the urban area. Belgrade, with its 1.7 million inhabitants, has no system for treating municipal wastewaters [20]. Some of the running waters flow through industrial zones, agricultural and urban areas and have a role in precipitation drainage, as well as wastewater removal from several rural areas, thus being subjected to combined pressures [21,22]. Reservoirs in the Belgrade sub-urban area are also under high anthropogenic influence, such as wastewater discharge from surrounding settlements, fishing and other recreational activities. Canals and reservoirs are additionally under continual pressure caused by flow regulation. Consequently, they lack a natural water regimen as water is actively drained and pumped. Further, dredging of bed material and removal of aquatic and riparian vegetation contributes to the effect of a complex multi-stressor environment. Many of these reservoirs and canals lie in sanitary protection zones of the water supply system of Belgrade.

Keeping in mind the importance of such aquatic systems, and assessing the intensity of anthropogenic influence and possible degradation, we surveyed modified waterbodies in the Belgrade FUA to investigate whether assemblages of Chironomidae larvae can provide clear signs of environmental changes. We aimed to detect and describe (i) the faunistic patterns of chironomid assemblages and (ii) the patterns of diversity components in analysed chironomid assemblages and to examine (iii) the effects of environmental factors on the differentiation of chironomid assemblages.

2. Materials and Methods

2.1. Study Area

Together with other benthic macroinvertebrates, Chironomidae larvae were collected in 2018 and 2019 during spring and autumn (high and low water level regime), in 22 waterbodies classified according to national legislation [23] to four waterbody types (WBT): (1) non-wadeable (large, non-crossable) rivers: the Danube, Sava and Kolubara rivers; (2) wadeable (small, crossable) rivers: the Peštan, Turija, Beljanica, Topčiderska reka, Barička reka, Barajevska reka, Ralja rivers; (3) canals: the Galovica, Kalovita, Sibnica, Vizelj, Agricultural Plant Belgrade (PKB), Progarska Jarčina, Karaš and Obrenovački kanal canals; (4) reservoirs: the Pariguz, Bela Reka, Duboki potok and Savsko jezero reservoirs (Figure 1) (for the characteristics of sampling sites see Tables S1 and S2).

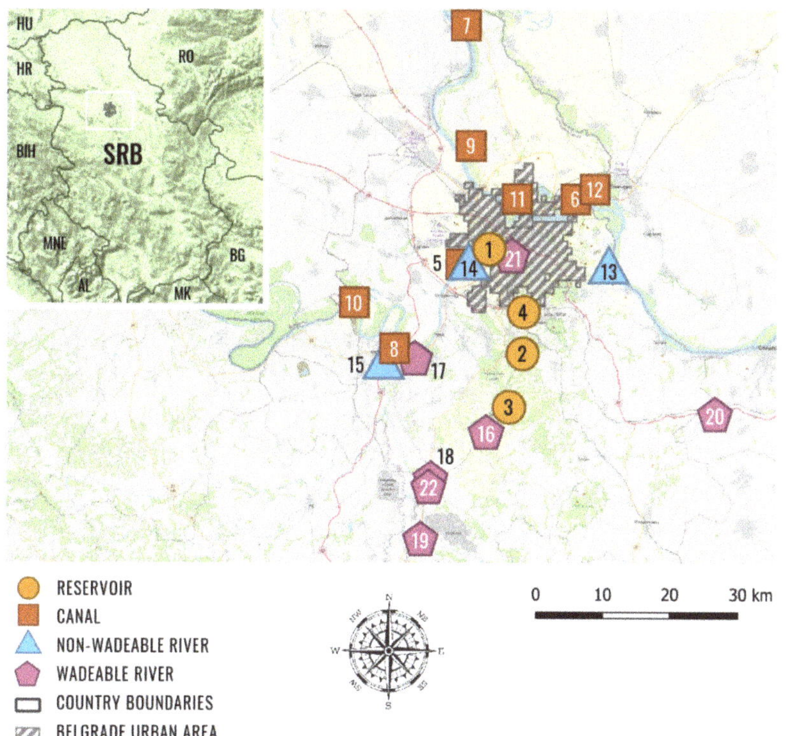

Figure 1. Sampling sites in the Belgrade functional urban area: 1. Savsko jezero, 2. Bela Reka, 3. Duboki potok, 4. Pariguz, 5. Galovica, 6. Kalovita, 7. Karaš, 8. Obrenovački kanal, 9. Agricultural Plant Belgrade (PKB), 10. Progarska Jarčina, 11. Vizelj, 12. Sibnica, 13. Danube, 14. Sava, 15. Kolubara, 16. Barajevska reka, 17. Barička reka, 18. Beljanica, 19. Peštan, 20. Ralja, 21. Topčiderska reka, 22. Turija.

2.2. *Chironomidae*

Samples were collected using a benthological hand net (500 µm mesh size, net frame 25 × 25 cm), following the multihabitat sampling procedure, 20 subsamples pooled into one container per site [24]. Material was collected from the littoral zone down to a depth of 1.5 m from all available microhabitats. The collected material was preserved in 70% ethanol. In the laboratory, Nikon SMZ8000N (magnification 10–120×) and Zeiss Stemi 2000-C (magnification 6.5–50×) stereomicroscopes were used for sorting and identification. Larvae of Chironomidae were identified to the lowest possible taxonomic level (genus, species or species groups and aggregates) using up to date identification keys [25–33].

2.3. *Environmental Variables*

Water temperature (T), pH, conductivity (µS/cm), dissolved oxygen (O_2; mg/L O_2), oxygen saturation (O_2%), nitrite concentration (NO_2; mg/L N; PRI P-V-32/A) and nitrate concentration (NO_3; mg/L N; EPA 300.1) were measured using a Horiba W-23XD multiparametric probe (HORIBA Instruments Corporation, Irvine, CA, USA) in the field. The biochemical oxygen demand (BOD_5; mg/L O_2; SRPS ISO 5813: 1994, SRPS EN 1899–2: 2009), chemical oxygen demand (COD; mg/L O_2; SRPS ISO 6060: 1990), concentrations of total organic carbon (TOC; mg/L C; SRPS ISO 8245:1994), total phosphates (mg/L P; EPA 207. Rev 5, SRPS ENISO 6878: 2008), residue obtained after drying at 105 °C (mg/L; SMEWW 19th method 2540 B), suspended particles (mg/L; SMEWW 19th method 2540 D), ammonium ions (NH_4; mg/L N; PRI P-V-2A) and chlorides (Cl; mg/L Cl; SRPS ISO 9297:

1994) were measured in the laboratory of the Institute of Public Health, Belgrade, Serbia. Microbiological analyses (number of coliforms) were performed in the same laboratory following standard methods (APHA AWWA WEF 1995, SMEWW 2010, SRPS EN ISO 9308–1: 2010).

2.4. Data Analysis

The qualitative composition of chironomids was determined for each site, along with species occurrence frequencies (F = 0–1). The ASTERICS software package Version 3.1.1. [24] was used for the assessment of data and the calculation of metrics.

The modified water pollution index (WPI) [34] was used to estimate water quality classes (Table 1). The WPI is calculated as the sum of the ratio of the measured average value and the standard threshold values for each parameter, divided by the number of used parameters. The standard threshold values for all parameters were specific for each country, given as the national legislative [23], which should minimise bias caused by ecological and geographical differences. The ASTERICS software package, version 3.1.1 [24], was used for the calculation of metrics used for the assessment of the ecological potential for these artificial waterbodies. The ecological analysis of the macroinvertebrate community structure was conducted for each site to calculate WPI. The number of taxa, ASPT (average score per taxon), BMWP (biological monitoring working party score) [35], the Saprobic index (S) [36] using bioindicator valences of each taxon according to [37], α-diversity index (H′) [38] and the percentage of the subfamily Tubificinae (Oligochaeta) in macroinvertebrate communities were calculated and used in the WPI calculation.

Table 1. Water quality classification based on the water pollution index (WPI).

Water Quality Class		WPI
I	very pure	≤0.3
II	pure	0.3–1.0
III	moderately polluted	1.0–2.0
IV	polluted	2.0–4.0
V	impure	4.0–6.0
VI	heavily impure	>6.0

To reveal the variability patterns of analysed chironomid assemblages, we used two powerful non-hierarchical classification methods: K-means clustering [39] and Bayesian classification [40]. Hill et al. [41] emphasised that the number of misclassifications is a key parameter in assessing the analytical power of clustering methods. Contrary to numerous agglomerative and divisive methods, K-means clustering and Bayesian classification enable the allocation of misclassified assemblages to their most similar cluster. Despite their ability to minimise the number of misclassifications, both K-means clustering and Bayesian classification have two serious drawbacks. Both methods allocate assemblages into a pre-specified number of clusters. The main drawback of these methods is subjectivity in the initial selection of the number of clusters. Marinković et al. [42] proposed a simple procedure to avoid this problem. The procedure selects the number of clusters by maximising the variance ratio:

$$VR = \frac{\sigma_B^2}{\sigma_W^2}$$

where σ_B^2 is the between-group variance (i.e., variance of the cluster centroids), and σ_W^2 is the within-group variance (the sum of the variances within each k cluster). Maximising the variance ratio ensures that overlap of homogeneous clusters is minimised. Another drawback of both is that K-means clustering and Bayesian classification are associated with the *local minima* problem. Two closest points in a Euclidean space correspond to the local minimum in one-dimensional space (a linear local minimum). Three closest points in the space represent the local minimum in a two-dimensional space (a planar local minimum). Both planar and linear local minima can disintegrate compact global clusters and replace them with numerous small clusters. To avoid this undesirable effect, we specified a threshold of four points as the minimum size of initial clusters. Selection of initial clusters with at least four points specifies the spatial configuration of the clusters and eliminates the undesirable effects of linear and planar local minima.

MANOVA [43] was used to find a combination of species that maximally discriminates extracted clusters of chironomid assemblages.

Cluster analyses based on Ward's method [44] was used for grouping Chironomidae assemblage types.

For each type of chironomid assemblage, we analysed the components of alpha diversity (species richness, Shannon index and Shannon equitability). The components of beta diversity were detected using the procedures described by Baselga [45] and Podani et al. [46].

The stepwise forward selection (FS) procedure [40] was used to detect environmental variables with statistically significant effects on the chironomid assemblages. At each step of the procedure, we expanded the multiple regression model by adding an environmental variable that explains most of the residual variance (i.e., the variance of faunistic data, not explained by previously selected environmental variables). The statistical significance of the hypothesis that species assemblages are independent of the selected environmental variable was assessed using the non-parametric Monte Carlo permutation test (3000 permutations, $p < 0.05$).

To reduce the weighting of dominant groups, species-abundance data were transformed using the formula:

$$\log(x + 1),$$

where x is the number of recorded individuals. The effects of environmental variables on the faunistic differentiation of assemblages were assessed using canonical correspondence analysis (CCA) [47].

Pearson's correlation test was used to determine the correlation between WBT and WPI and their correlation with measured environmental parameters. For testing data normality, we used the Kolmogorov–Smirnov test.

Statistical analyses were performed using FLORA software [48], updated version.

3. Results

In total, 66 chironomid taxa in four subfamilies (Prodiamesinae, Orthocladiinae, Tanypodinae and Chironominae) were found during the study period. The most diverse and abundant was the Chironominae subfamily (21 genera, 37 taxa). *Polypedilum* was the most diverse genus in the Chironominae subfamily with nine species. *Cricotopus* was the most diverse genus from the Orthocladiinae subfamily. The highest abundance of chironomid species was detected in wadeable rivers, while in reservoirs, canals and non-wadeable rivers, the abundance of chironomid species was comparable. *Cricotopus* gr. *sylvestris* and *Cricotopus bicinctus* (Meigen, 1818) were found in various waterbody types, but they were most abundant in wadeable rivers. In reservoirs, the most abundant was *Ablabesmyia monilis* agg., while in canals, the most numerous were *C.* gr. *sylvestris* and *Parachironomus gracillior* (Kieffer, 1918).

3.1. Chironomid Assemblage Classification

Classification of chironomid assemblages was performed based on the faunistic similarity of the analysed assemblages. The main drawback of hierarchical classification methods is the inability to correct misclassifications. However, non-hierarchical clustering methods allow for the allocation of misclassified species assemblages to their most similar cluster. We, therefore, performed classification of the analysed assemblages using K-means clustering and Bayesian classification. These methods are the most powerful variants of non-hierarchical clustering methods [40]. The calculated variance ratio indicated that K-means clustering produced more acceptable results than Bayesian classification. The main drawback of Bayesian classification is the rigid assumption that all variables must be normally distributed, as concluded in Sekulić et al. [49]. The results of K-means clustering and Bayesian classification are presented in Table 2. To eliminate the undesirable effects of local minima, we used four species assemblages as the minimum size of initial clusters. Since we investigated 22 chironomid assemblages, the greatest number of initial clusters with at least four assemblages was five. The greatest ratio of between-group variance to within-group variance was detected for four clusters. Therefore, we performed K-means clustering using four pre-defined clusters and named these clusters (types) with capital letters A, B, C and D.

Table 2. Dependence of ordering results on the subjectively selected number of clusters. The greatest ratio of between-group variance (BGV) to within-group variance (WGV) $\sigma^2{}_B/\sigma^2{}_W$ assures that the overlap of homogeneous clusters is minimised. In our dataset, the greatest variance ratio (bolded numbers) was obtained for four clusters.

	K-Means Clustering			**Bayesian Classification**		
No. of clusters	WGV σ^2_W	BGV σ^2_B	σ^2_B/σ^2_W	WGV σ^2_W	BGV σ^2_B	σ^2_B/σ^2_W
2	1.1361	0.0522	0.0459	1.1346	0.0591	0.0521
3	1.5603	0.1009	0.0647	1.5768	0.0980	0.0621
4	1.9335	0.1404	**0.0726**	1.9562	0.1341	0.0685
5	2.3112	0.1610	0.0696	2.2934	0.1599	0.0697

MANOVA provided a combination of species that maximally discriminated between four clusters of assemblages (high variance ratio R^2 0.452884).

Species from the *Conchapelopia* aggregate as well as *Procladius* species discriminated assemblage type A from other assemblage types. Type B stands out by the presence of *Microchironomus tener* (Kieffer, 1918). *Polypedilum albicorne* (Meigen, 1838) was the discriminating species for community type C, while *Endochironomus albipennis* (Meigen, 1830) and *Monopelopia tenuicalcar* (Kieffer, 1918) were a distinguishing feature of assemblage type D (Figure 2).

Cluster analyses using Ward's method organised the assemblages into two larger groups based on the similarity of species composition, with each encompassing two assemblage types. The first group incorporates assemblage types A and B, and the second types C and D (Figure 3).

For each group of assemblages, a set of diagnostic species was established based on their frequency of occurrence (Table S3).

Rheocricotopus, *Procladius*, *Conchapelopia* and *Thenemanniella* dominated in type A assemblages and were often associated with *Polypedium convictum* (Walker, 1856), *Nanocladius rectinervis* (Kieffer, 1911), *Microtendipes* gr. *pedellus*, *Chironomus riparius* Meigen, 1804 and *Eukiefferiella claripennis* (Lundbeck, 1898). *Microchironomus tener* dominated in assemblage type B and was usually associated with *Harnischia*, *Cryptotendipes*, *Rheotanytarsus* and *Cladotanytarsus* species. In the type C assemblages, *Glyptotendipes paripes* (Edwards, 1929) was the diagnostic species associated with *Polypedilum albicorne* (Meigen, 1838), *Parachironomus gracillior*, *Dicrotendipes pulsus* (Walker, 1856) and *Kiefferulus tendipediformis* (Goetghebuer, 1921). *Monopelopia tenuicalcar* (Kieffer, 1918) was the diagnostic/main species for assemblage type D and was associated with *Glyptotendipes pallens* agg. and *Xenopelopia* species.

Correspondence between faunistic groups (X-axis) and waterbody types (colours) is presented in a histogram (Figure 4).

Assemblage type A inhabited only wadeable rivers. Assemblage type C occurred in slow-flowing waters such as reservoirs and canals, while assemblages of type D were found mainly in canals and in some wadeable rivers. Assemblage type B was the only type detected in non-wadeable rivers. This type was also found in wadeable rivers and only in one reservoir (Figure 4).

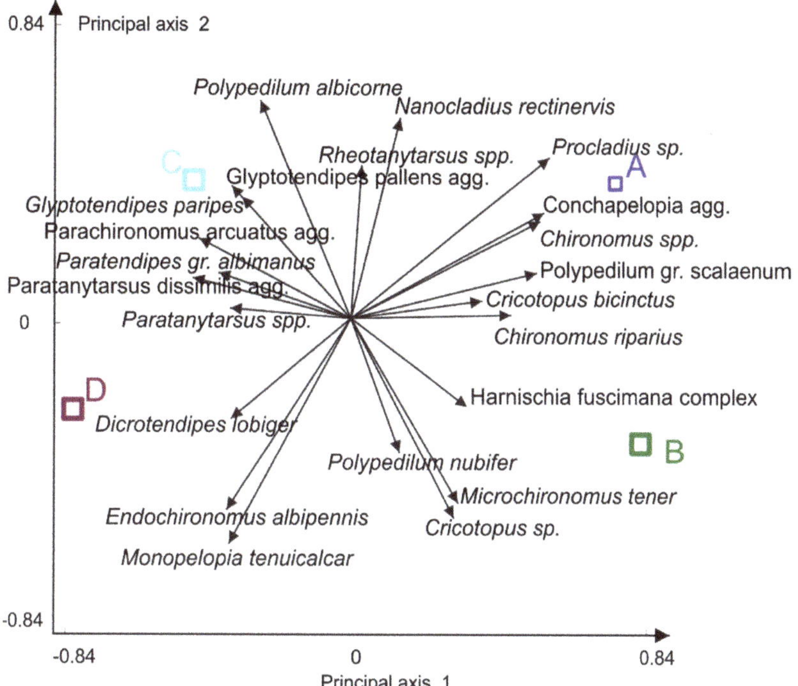

Figure 2. Multivariate analyses of variance (MANOVA) of four Chironomidae assemblage clusters. Four assemblage types (obtained through K-means clustering) are in capital letters A, B, C, D coloured as indicated.

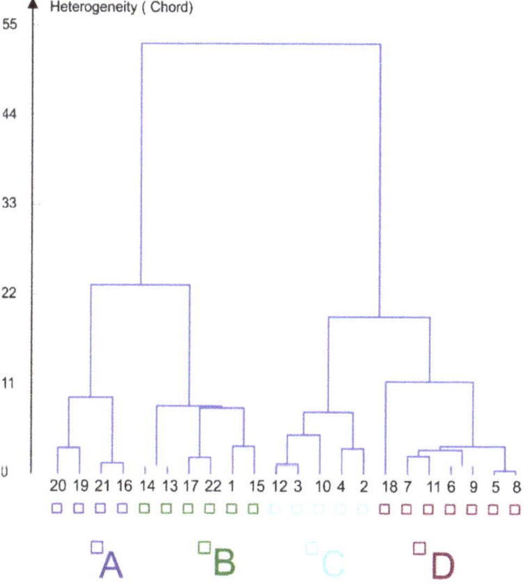

Figure 3. Hierarchical cluster analyses (Ward's method) of the Chironomidae assemblages. Four assemblage types (obtained through K-means clustering) are in capital letters A, B, C, D coloured as indicated. The numbers represent study sites codes (see Figure 1).

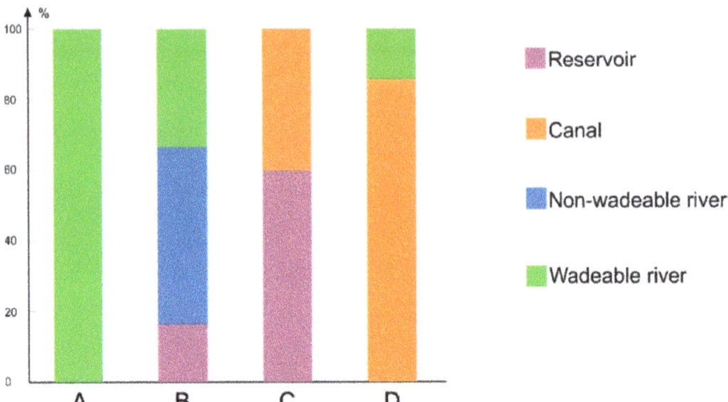

Figure 4. Occurrence (in percentage) of different chironomid assemblage types in four types of waterbodies. Four assemblage types (obtained through K-means clustering) are in capital letters A, B, C, D coloured as indicated.

3.2. Diversity Components

The highest alpha diversity (expressed as Shannon index) was recorded in assemblage types A and B, while in C and D, the values of the Shannon index were lower. The same trend was observed for species richness, while the Shannon equitability index was almost identical in all assemblage types (Figure 5).

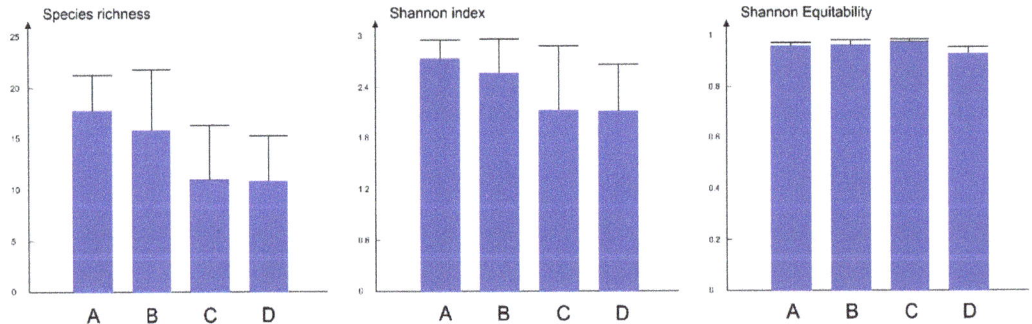

Figure 5. Components of alpha diversity in four types of chironomid assemblages, bars represent average values while lines denote the variance in each group of assemblages. The four community types are in capital letters A, B, C, D coloured as indicated.

High, almost identical, beta diversity values were detected in all analysed types of assemblages using Baselga's method (Figure 6). The dominant component of beta diversity was species turnover, which was high, while nestedness was low. Analyses of beta diversity using the Podani method gave different results between different types of assemblages (Figure 6). While overall beta diversity was similar, its components varied from type to type. Nestedness was lowest in type A and highest in type C, while in types B and D, it was similar. Species turnover was the same in types C and D, which was lower than in types A and B, with type A showing the highest value of this component of beta diversity.

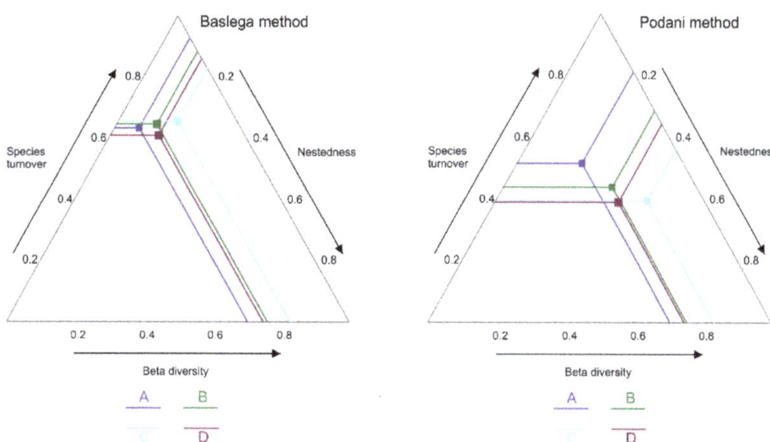

Figure 6. Components of beta diversity in four types of chironomid assemblages (Baselga and Podani methods). Four assemblage types are in capital letters A, B, C, D and coloured as indicated.

3.3. *Patterns in the Chironomidae-Environment Relationship*

Partial forward selection analysis (Table 3) identified ten environmental variables as significant for faunistic differentiation of the analysed urban waterbodies.

Table 3. Results of the forward selection analysis.

Variable	Eigenvalue	F Statistic	Probability
	nutrients		
NO_3	0.383	2.357	**0.000**
NH_4-N	0.244	1.445	**0.004**
NO_2	0.274	1.639	**0.009**
Total nitrogen	0.109	0.622	0.648
Cl	0.133	0.765	0.688
Total phosphate	0.083	0.469	0.804
	oxygen status parameters		
BOD_5	0.279	1.667	**0.001**
COD ($KMnO_4$)	0.294	1.768	**0.005**
O_2%	0.199	1.166	0.052
O_2	0.202	1.180	0.100
TOC	0.226	1.334	0.296
	physical parameters		
WBT	0.380	2.341	**0.000**
pH	0.281	1.681	**0.003**
WPI	0.259	1.539	**0.003**
Suspended solids	0.234	1.381	**0.010**
T	0.175	1.013	**0.034**
Electrical conductivity	0.161	0.929	0.087
Dry residue	0.108	0.618	0.344

Abbreviations: NH_4-N, ammonium concentration (mg/L); NO_3, nitrate concentration (mg/L); NO_2, nitrite concentration (mg/L); Cl, chloride concentration (mg/L); COD, chemical oxygen demand (mg/L); BOD_5, biological oxygen demand (mg/L); TOC, total organic carbon (mg/L); O_2, oxygen concentration (mg/L); O_2%, oxygen saturation; WBT, waterbody types; WPI, water pollution index; T, water temperature (°C); Dry residue, residue obtained after drying at 105 °C (mg/L). Statistically significant values are in bold.

The first two CCA axes explained 34.4% of the variation of the fitted data obtained by multiple regressions. Environmental predictors explained a relatively high portion of the total variability of chironomid distributions (R^2 = 0.585).

CCA showed that the Chironomidae assemblage types were clearly differentiated with respect to nutrients (nitrate, nitrite and ammonium concentration), water temperature, and pH gradients. The importance of other environmental variables was lower. CCA indicated that oxygen demand (chemical and biological oxygen demand) produced effects on faunistic differentiation of the analysed Chironomidae assemblages. WBT also played a role in assemblage differentiation (Figure 7).

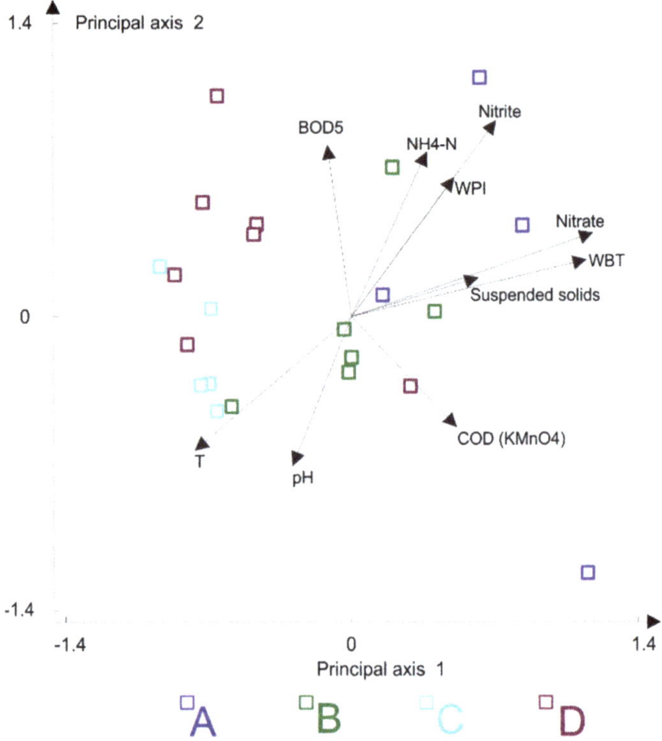

Figure 7. Canonical correspondence analysis of investigated chironomid assemblages. Four assemblage types are in capital letters A, B, C, D coloured as indicated. Nitrate, NO_3 concentration; Nitrite, NO_2 concentration; NH_4-N, ammonium concentration; COD, chemical oxygen demand; BOD_5, biological oxygen demand; WBT, waterbody types; WPI, water pollution index; T, water temperature.

Based on the values of WPI and water quality, the four types of assemblages were different (Figure 8). Assemblage type A was present in sites with various water qualities. Most of these sites were characterised by higher values of WPI and deterioration of water quality (classes IV, V and VI) (Table S2). Assemblage type C was present in less polluted waters (classes II and III of water quality, with lowest WPI values). Assemblage types B and D mainly occurred on sites with pure or moderately polluted water. Extremely high values of WPI (heavily impure water) were recorded at some sites inhabited by assemblage type B (Figure 8).

Since WBT and WPI are considered as derived variables, their correlations with environmental variables were analysed and are presented in Table 4. WPI and WBT were correlated. Nutrients (total phosphate, total nitrogen, nitrate, nitrite and ammonium concentrations) and water temperature showed significant correlation with these two derived variables.

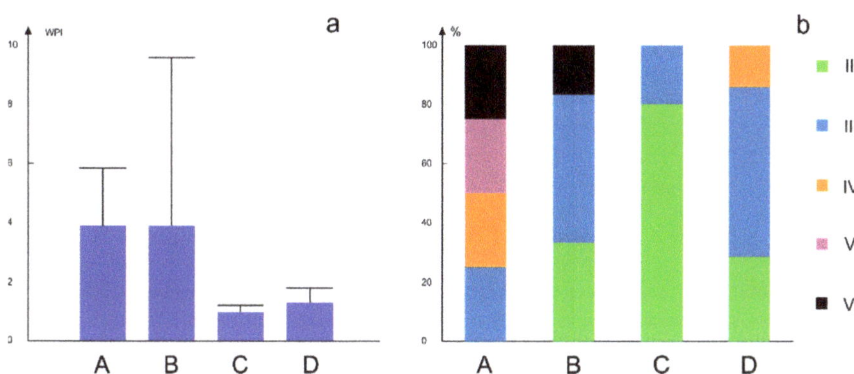

Figure 8. (**a**) Variation of WPI values in four types of chironomid assemblages; bars represent average values while lines denote the variance of WPI values within each group of assemblages. (**b**) Distribution of four chironomid assemblage types in different classes of water quality calculated through WPI; water quality class based on WPI is in Roman numerals II–VI, from very pure to heavily impure (see Table 1); four assemblage types are in capital letters A, B, C, D coloured as indicated.

Table 4. Correlations between WBT and WPI with environmental variables.

	WPI	WBT
Suspended solids	**0.000**	**0.011**
Total nitrogen	**0.000**	0.059
NH_4-N	**0.000**	0.077
BOD_5	**0.000**	0.932
NO_2	**0.013**	**0.017**
Total phosphate	**0.017**	0.315
NO_3	**0.018**	**0.000**
WBT	**0.024**	
WPI		**0.024**
T	0.074	**0.000**
TOC	0.218	0.481
O_2%	0.222	0.648
pH	0.231	0.061
O_2	0.287	0.903
COD ($KMnO_4$)	0.492	0.341
Electrical conductivity	0.601	0.470
Cl	0.741	0.687
Dry residue	0.742	0.417

Abbreviations: NH_4-N, ammonium concentration (mg/L); NO_3, nitrate concentration (mg/L); NO_2, nitrite concentration (mg/L); Cl, chloride concentration (mg/L); COD, chemical oxygen demand (mg/L); BOD_5, biological oxygen demand (mg/L); TOC, total organic carbon (mg/L); O_2, oxygen concentration (mg/L); O_2%, oxygen saturation; WBT, waterbody types; WPI, water pollution index; T, water temperature; Dry residue, residue obtained after drying at 105 °C (mg/L). Statistically significant values are in bold.

4. Discussion

Chironomids, as one of the most diverse aquatic macroinvertebrate groups, have a wide spectrum of biological and ecological preferences [10]; nevertheless, studies of chironomid assemblages mainly focus on the presence/absence of species rather than the abundance and assemblage structure [50]. The diverse fauna of Chironomidae larvae detected during this study enabled us not only to assess biodiversity within them, but also to monitor negative anthropogenic influences and degradation of studied waterbodies. The distribution of the chironomids mainly followed the a priori classification given by the non-chironomid macroinvertebrates [51]. We tried to determine whether assemblages of Chironomidae larvae in urban waters can provide clear indications of environmental changes.

Leszczyńska et al. [50] state that the knowledge of assemblage composition is necessary to assess some diversity components and other (inter-)assemblage parameters, as well as to assess assemblage dependence on environmental factors. High diversity of species characterised all four assemblage types found in Belgrade FUA while exhibiting differences in diversity components. Assemblage types are distinguished by the influence of different components on overall beta diversity. The ratio of components of beta diversity in types A and B pointed to the availability of different microhabitats in each site. A greater availability of habitats and microhabitats, such as sediment composition and the presence of macrophytes and detritus—which provides food, shelter for burrowing, mining and protection from predators—are associated with habitat heterogeneity and can harness the high diversity of chironomids [52–54]. Nevertheless, a huge diversity of the habitats exposed to multi stressors leads to the unclear relationship between biotic and abiotic components [55]. In the other two types (C and D), the nestedness component was slightly higher, pointing to the presence of an ecological gradient decreasing the number of species from site to site and their composition. Aquatic insects display a decrease in alpha diversity as a response to the urbanisation process, where, in sites exposed to substantial influence, only highly tolerant species prevail [16]. However, Chironomidae assemblages in Belgrade urban waters showed relatively high alpha diversity, but components of beta diversity revealed the aftereffects of urbanisation and pollution, high nestedness in canals, reservoirs, and heavily polluted rivers (type C and D) indicated that only tolerable species remained under high anthropogenic pressure.

Assemblage types A and B were similar in species composition, also preferring similar habitats such as wadeable and non-wadeable rivers. Preference for slow and stagnant waters found in canals and reservoirs was exhibited by assemblage types C and D, which were also grouped together based on species composition by cluster analyses.

Diagnostic taxa of assemblage's type A and B *Rheocricotopus*, *Thenemanniella*, *Microchironomus tener* and their associated species are known inhabitants of running waters and reservoirs [28,29,56], which are the types of ecosystems inhabited by the aforementioned assemblage.

Slow-flowing and stagnant waters with ample vegetation are preferred habitats of *Glyptotendipes paripes*, *Monopelopia tenuicalcar*, *Parachironomus gracillior*, *Dicrotendipes pulsus*, *Kiefferulus tendipediformis*, *Glyptotendipes pallens* agg. and *Xenopelopia* [29,32,33,56]. Ecological conditions in sites in which C and D assemblages types were detected corresponded with these ecological preferences. This is also in agreement with the classification of these sites as reservoirs or canals, with the addition of one wadeable river characterised by slow flow and large amounts of vegetation.

Although waterbody type is one of the most important factors determining the distribution of chironomids, the presence of pollution and other anthropogenic pressures (such as habitat degradation) can influence, to a great extent, the abundance and structure of chironomid assemblages [51]. Streams receiving waste effluents are characterised by lower chironomid species richness and the development of more dense populations of *Chironomus riparius* [57,58]. Leszczyńska et al. [50] found that *C. riparius* is the most abundant in low order streams, with low velocity and dense riparian vegetation, preferring stagnant water and soft sediments. It is also known that *C. riparius* may inhabit organically enriched and heavily polluted waterbodies, having efficient oxygen regulation [59]. The results presented herein show the same patterns, e.g., a high abundance of *Chironomus* species in heavily polluted rivers (the Topčiderska, Barička and Barajevska rivers). The Topčiderska river, which flows through the industrial zone and is the recipient of communal wastewaters, was the only site with *Eukiefferiella claripennis*, *Parametriocnemus stylatus* and *Tvetenia clavescens*, species found to be tolerant to habitat degradation [58,60]. Non-wadeable rivers were characterised by the dominance of the subfamily Chironominae (*Dicrotendipes nervosus*, *Polypedilum nubeculosum* and *Chironomus* species). Milošević et al. [9] reported a similar assemblage structure, namely one that was monotonous and driven by frequent and dominant taxa from the Chironominae subfamily, which were also reported in other studies as common dominant taxa in non-wadeable rivers [55,61].

Cricotopus bicinctus, which has already been documented as an indicator species for organic pollution [10], was abundant in wadeable rivers (the Topčiderska, Barajevska and Beljanica rivers) but also present in non-wadeable rivers.

The high values of diversity components observed in assemblage types A and B, despite the high pollution (expressed through WPI and nutrients), were likely supported by the availability of microhabitats and other favourable ecological conditions of the ecosystems they resided in (wadeable and non-wadeable rivers). Canals and reservoirs were inhabited by assemblages (C and D) that exhibited slightly lower alpha and beta diversity components. The limiting factors in these cases could be the uniformity of habitats and the limited availability of resources (food, shelter, substrate), but also the hydro-technical regime and maintenance work on these heavily modified and artificial waterbodies.

5. Conclusions

Urban waters in the Belgrade FUA harbour very diverse chironomid fauna. Based on their preferences for specific waterbody types and tolerance to environmental pressures, Chironomid assemblages can be grouped into several different types characterised by unique species composition.

Our study showed that chironomids could serve as useful indicators of anthropogenic pressures in various waterbody types due to the different sensitivities of the species towards the alteration of environmental conditions in their habitats.

Supplementary Materials: The following supporting information can be downloaded at: https://www.mdpi.com/article/10.3390/d14030187/s1, Table S1: Sampling sites and their characteristics; Table S2: Detected pressures at sampling sites and water quality classes according to the Water Pollution Index (WPI) value; Table S3: Taxa list and frequencies of species in four clusters of species assemblages.

Author Contributions: Conceptualisation, N.P., N.M., M.R. and M.P.; methodology, D.Č., N.P., M.R. and J.Đ.; validation, N.M., M.R. and M.P.; formal analysis, N.P., N.M. and D.Č.; investigation, N.P., N.M., J.Đ., M.R., D.Č. and M.P.; writing—original draft preparation, N.P., N.M., M.R. and D.Č.; writing—review and editing, N.P., N.M., M.R., J.Đ. and M.P.; visualisation, N.M. and N.P.; supervision, N.P. and M.P.; project administration, N.P.; funding acquisition, M.P. All authors have read and agreed to the published version of the manuscript.

Funding: This research was funded by the Ministry of Education, Science and Technological Development of the Republic of Serbia, Contract No. 451-03-9/2021-14/200007.

Institutional Review Board Statement: Not applicable.

Data Availability Statement: The data presented in this study are available on request from the corresponding author.

Acknowledgments: We would like to express our gratitude to the Public Health Institute of Belgrade for providing physical, chemical and microbiological data used in this paper. Special thanks to the Editor and reviewers for their comments and suggestions which greatly helped during shaping of this manuscript.

Conflicts of Interest: The authors declare no conflict of interest.

References

1. Goyke, A.P.; Hershey, A.E. Effects of fish predation on larval chironomid (Diptera: Chironomidae) communities in an arctic ecosystem. *Hydrobiologia* **1992**, *240*, 203–211. [CrossRef]
2. Armitage, P.D.; Cranston, P.S.; Pinder, L.C.V. *The Chironomidae: Biology and Ecology of Non-Biting Midges*; Chapman & Hall: London, UK, 1995; ISBN 10041245260X.
3. Martínez-Sánchez, A.; Allende, A.; Bennett, R.N.; Ferreres, F.; Gil, M.I. Microbial, nutritional and sensory quality of rocket leaves as affected by different sanitizers. *Postharvest Biol. Technol.* **2006**, *42*, 86–97. [CrossRef]

4. Benke, A.C. Production dynamics of riverine chironomids: Extremely high biomass turnover rates of primary consumers. *Ecology* **1998**, *79*, 899–910. [CrossRef]
5. Bogut, I.; Čerba, D.; Vidaković, J.; Gvozdić, V. Interactions of weed-bed invertebrates and Ceratophyllum demersum stands in a floodplain lake. *Biologia* **2010**, *65*, 113–121. [CrossRef]
6. Čerba, D.; Mihaljević, Z.; Vidaković, J. Colonisation of temporary macrophyte substratum by midges (Chironomidae: Diptera). *Ann. Limnol.* **2010**, *46*, 181–190. [CrossRef]
7. Leszczyńska, J.; Grzybkowska, M.; Głowacki, Ł.; Dukowska, M. Environmental Variables Influencing Chironomid Assemblages (Diptera: Chironomidae) in Lowland Rivers of Central Poland. *Environ. Entomol.* **2019**, *48*, 988–997. [CrossRef]
8. Milošević, D.; Čerba, D.; Szekeres, J.; Csányi, B.; Tubić, B.; Simić, V.; Paunović, M. Artificial neural networks as an indicator search engine: The visualization of natural and man-caused taxa variability. *Ecol. Indic.* **2016**, *61*, 777–789. [CrossRef]
9. Milošević, D.; Mančev, D.; Čerba, D.; Piperac, M.S.; Popović, N.; Atanacković, A.; Đuknić, J.; Simić, V.; Paunović, M. The potential of chironomid larvae-based metrics in the bioassessment of non-wadeable rivers. *Sci. Total Environ.* **2018**, *616*, 472–479. [CrossRef]
10. Milošević, D.; Simić, V.; Stojković, M.; Čerba, D.; Mančev, D.; Petrović, A.; Paunović, M. Spatio-temporal pattern of the Chironomidae community: Toward the use of non-biting midges in bioassessment programs. *Aquat. Ecol.* **2013**, *47*, 37–55. [CrossRef]
11. Savić-Zdravković, D.; Jovanović, B.; Đurđević, A.; Stojković-Piperac, M.; Savić, A.; Vidmar, J.; Milošević, D. An environmentally relevant concentration of titanium dioxide (TiO_2) nanoparticles induces morphological changes in the mouthparts of Chironomus tentans. *Chemosphere* **2018**, *211*, 489–499. [CrossRef]
12. Stanković, J.; Milošević, D.; Savić-Zdraković, D.; Yalçın, G.; Yildiz, D.; Beklioğlu, M.; Jovanović, B. Exposure to a microplastic mixture is altering the life traits and is causing deformities in the non-biting midge Chironomus riparius Meigen (1804). *Environ. Pollut.* **2020**, *262*, 114248. [CrossRef] [PubMed]
13. Fusari, L.M.; Fonseca-Gessner, A.A. Environmental assessment of two small reservoirs in southeastern Brazil, using macroinvertebrate community metrics. *Acta Limnol. Bras.* **2006**, *18*, 89–99.
14. Martins, R.T.; Stephan, N.N.C.; Alves, R.G. Tubificidae (Annelida: Oligochaeta) as an indicator of water quality in an urban stream in southeast Brazil. *Acta Limnol. Bras.* **2008**, *20*, 221–226.
15. Rosa, B.J.F.V.; Rodrigues, L.F.T.; de Oliveira, G.S.; da Gama Alves, R. Chironomidae and Oligochaeta for water quality evaluation in an urban river in southeastern Brazil. *Environ. Monit. Assess.* **2014**, *186*, 7771–7779. [CrossRef] [PubMed]
16. Lundquist, M.J.; Zhu, W. Aquatic insect diversity in streams across a rural–urban land-use discontinuum. *Hydrobiologia* **2019**, *837*, 15–30. [CrossRef]
17. Qiu, Z.; Kennen, J.G.; Giri, S.; Walter, T.; Kang, Y.; Zhang, Z. Reassessing the relationship between landscape alteration and aquatic ecosystem degradation from a hydrologically sensitive area perspective. *Sci. Total Environ.* **2019**, *650*, 2850–2862. [CrossRef]
18. Wenger, S.J.; Roy, A.H.; Jackson, C.R.; Bernhardt, E.S.; Carter, T.L.; Filoso, S.; Gibson, C.A.; Hession, W.C.; Kaushal, S.S.; Martí, E. Twenty-six key research questions in urban stream ecology: An assessment of the state of the science. *J. N. Am. Benthol. Soc.* **2009**, *28*, 1080–1098. [CrossRef]
19. Roy, A.H.; Rhea, L.K.; Mayer, A.L.; Shuster, W.D.; Beaulieu, J.J.; Hopton, M.E.; Morrison, M.A.; Amand, A.S. How much is enough? Minimal responses of water quality and stream biota to partial retrofit stormwater management in a suburban neighborhood. *PLoS ONE* **2014**, *9*, e85011. [CrossRef]
20. Živadinović, I.; Ilijević, K.; Gržetić, I.; Popović, A. Long-term changes in the eco-chemical status of the Danube River in the region of Serbia. *J. Serb. Chem. Soc.* **2010**, *75*, 1125–1148. [CrossRef]
21. Kolarević, S.; Knežević-Vukčević, J.; Paunović, M.; Tomović, J.; Gačić, Z.; Vuković-Gačić, B. The anthropogenic impact on water quality of the river Danube in Serbia: Microbiological analysis and genotoxicity monitoring. *Arch. Biol. Sci.* **2011**, *63*, 1209–1217. [CrossRef]
22. Jovanović Marić, J.M.; Kračun-Kolarević, M.J.; Kolarević, S.M.; Đorđević, J.Z.; Paunović, M.M.; Kostić-Vuković, J.M.; Sunjog, K.Z.; Smiljanić, P.B.; Gačić, Z.M.; Vuković-Gačić, B.S. Sensitivity of Bleak (*Alburnus alburnus*) in Detection of the Wastewater Related Pressure in Large Lowland Rivers. *Bull. Environ. Contam. Toxicol.* **2020**, *105*, 224–229. [CrossRef] [PubMed]
23. Official Gazette of the Republic of Serbia. *The Parameters of Ecological and Chemical Status of Surface Waters and Parameters of the Chemical and Quantitative Status of Groundwater*; Off. Gaz. RS No. 74/2011; Official Gazette of the Republic of Serbia: Belgrade, Serbia, 2011.
24. AQEM Consortium. *Manual for the Application of the AQEM System*; Version 1.0; Contract No: EVK1-CT1999-00027; AQEM: Duisburg, Germany, 2002.
25. Schmid, P.-E. A Comprehensive Method to Assess European Streams Using Benthic Macroinvertebrates, Developed for the Purpose of the Water Framework Directive. In *A Key to the Larval Chironomidae and Their Instars from Austrian Danube Region Streams and Rivers with Particular Reference to a Numerical Taxonomic Approach*; Federal Institute for Water Quality of the Ministry of Agriculture and Forestry: Vienna, Austria, 1993; ISBN 390067292X.
26. Vallenduuk, H.J.; Moller Pillot, H.K.M. *Chironomidae Larvae, Vol. 1: Tanypodinae: General Ecology and Tanypodinae*; Brill: Leiden, The Netherlands, 2007; ISBN 9004278036.
27. Maschwitz, D.E.; Cook, E.F. *Revision of the Nearctic Species of the Genus Polypedilum Kieffer (Diptera: Chironomidae) in the Subgenera P. (Polypedilum) Kieffer and P. (Uresipedilum) Oyewo and Sæther*; Ohio Biological Survey, College of Biological Sciences, The Ohio State: Columbus, OH, USA, 2000; ISBN 0867271302.

28. Epler, J.H. *Identification Guide to the Larvae of the Tribe Tanytarsini (Diptera: Chironomidae) in Florida*; Florida Department of Environmental Protection: Tallahassee, FL, USA, 2014.
29. Moller Pillot, H.K.M. *Chironomidae Larvae, Vol. 2: Chironomini: Biology and Ecology of the Chironomini*; Brill: Leiden, The Netherlands, 2009; ISBN 9004278044.
30. Orendt, C.; Spies, M. *Chironomini:(Diptera: Chironomodae: Chironominae)*; Keys to Central European Larvae Using Mainly Macroscopic Characters; Orendt Hydrobiologie: Leipzig, Germany, 2012; ISBN 3000388427.
31. Andersen, T.; Sæther, O.A.; Cranston, P.S.; Epler, J.H. The larvae of Orthocladiinae (Diptera: Chironomidae) of the Holarctic region-Keys and diagnoses. *Insect Syst. Evol.* **2013**, *66*, 189–385.
32. Bitušík, P.; Hamerlík, L. *Príručka na Určovanie Lariev Pakomárov (Diptera: Chironomidae) Slovenska*; Čast'2. Tanypodinae; Belianum Vydav. Univerzity Matej Bela v Banskej Bystrici: Banska Bystrica, Slovakia, 2014.
33. Vallenduuk, H.J. *Chironomini Larvae of Western European Lowlands (Diptera: Chironomidae): Keys with Notes to the Species: With a Redescription of Glyptotendipes (Caulochironomus) Nagorskayae and a First Description of Glyptotendipes (Caulochironomus) Kaluginae New Species*; Erik Mauch: Dinkelscherben, Germany, 2017.
34. Popovic, N.Z.; Duknic, J.A.; Atlagic, J.Z.; Rakovic, M.J.; Marinkovic, N.S.; Tubic, B.P.; Paunovic, M.M. Application of the water pollution index in the assessment of the ecological status of rivers: A case study of the Sava River, Serbia. *Acta Zool. Bulg.* **2016**, *68*, 97–102.
35. Armitage, P.D.; Moss, D.; Wright, J.F.; Furse, M.T. The performance of a new biological water quality score system based on macroinvertebrates over a wide range of unpolluted running-water sites. *Water Res.* **1983**, *17*, 333–347. [CrossRef]
36. Zelinka, M.; Marvan, P. Zur Prazisierung der biologischen Klassifikation des Reinheit fliessender Gewasser. *Arch. Hydrobiol.* **1961**, *57*, 389–407.
37. Moog, O. *Fauna Aquatica Austriaca Katalog zur autökologischen Einstufung aquatischer Organismen Österreichs*; Bundesministerium für Nachhaltigkeit und Tourismus: Vienna, Austria, 2002; p. 670.
38. Shannon, C.E.; Weaver, W. *The Mathematical Theory of Communication*; The University of Illinois Press: Champaign, IL, USA, 1949; pp. 1–117.
39. MacQueen, J. Some methods for classification and analysis of multivariate observations. In *Proceedings of the Fifth Berkeley Symposium on Mathematical Statistics and Probability*; University of California Press: Oakland, CA, USA, 1967; Volume 1, pp. 281–297.
40. James, G.; Witten, D.; Hastie, T.; Tibshirani, R. *An Introduction to Statistical Learning*; Springer: Berlin/Heidelberg, Germany, 2013; Volume 112.
41. Hill, M.O.; Bunce, R.G.H.; Shaw, M.W. Indicator Species Analysis, A Divisive Polythetic Method of Classification, and its Application to a Survey of Native Pinewoods in Scotland. *J. Ecol.* **1975**, *63*, 597–613. [CrossRef]
42. Marinković, N.; Karadžić, B.; Pešić, V.; Gligorović, B.; Grosser, C.; Paunović, M.; Nikolić, V.; Raković, M. Faunistic patterns and diversity components of leech assemblages in karst springs of Montenegro. *Knowl. Manag. Aquat. Ecosyst.* **2019**, *420*, 26. [CrossRef]
43. Bray, J.H.; Maxwell, S.E.; Maxwell, S.E. *Multivariate Analysis of Variance*; SAGE: Thousand Oaks, CA, USA, 1985; ISBN 0803923104.
44. Ward, J.H., Jr.; Hook, M.E. Application of an hierarchical grouping procedure to a problem of grouping profiles. *Educ. Psychol. Meas.* **1963**, *23*, 69–81. [CrossRef]
45. Baselga, A. Partitioning the turnover and nestedness components of beta diversity. *Glob. Ecol. Biogeogr.* **2010**, *19*, 134–143. [CrossRef]
46. Podani, J.; Ricotta, C.; Schmera, D. A general framework for analyzing beta diversity, nestedness and related community-level phenomena based on abundance data. *Ecol. Complex.* **2013**, *15*, 52–61. [CrossRef]
47. ter Braak, C.J.F. Canonical Correspondence Analysis: A New Eigenvector Technique for Multivariate Direct Gradient Analysis. *Ecology* **1986**, *67*, 1167–1179. [CrossRef]
48. Karadžić, B. FLORA: A software package for statistical analysis of ecological data. *Water Res. Manag.* **2013**, *3*, 45–54.
49. Sekulić, D.; Karadžić, B.; Kuzmanović, N.; Jarić, S.; Mitrović, M.; Pavlović, P. Diversity of ostrya carpinifolia forests in ravine habitats of serbia (S-e Europe). *Diversity* **2021**, *13*, 59. [CrossRef]
50. Leszczyńska, J.; Głowacki; Grzybkowska, M.; Przybylski, M. Chironomid riverine assemblages at the regional temperate scale–compositional distance and species diversity. *Eur. Zool. J.* **2021**, *88*, 731–748. [CrossRef]
51. Orendt, C. Results of 10 years sampling of Chironomidae from German lowland running waters differing in degradation. *J. Limnol.* **2018**, *77* (Suppl. S1), 169–176. [CrossRef]
52. Zilli, F.L.; Montalto, L.; Marchese, M.R. Benthic invertebrate assemblages and functional feeding groups in the Paraná River floodplain (Argentina). *Limnologica* **2008**, *38*, 159–171. [CrossRef]
53. de Carvalho, E.M.; Uieda, V.S.; da Motta, R.L. Colonization of rocky and leaf pack substrates by benthic macroinvertebrates in a stream in Southeast Brazil. *Títulos Não-Correntes* **2008**, *22*, 37–44.
54. Vermonden, K.; Brodersen, K.P.; Jacobsen, D.; Van Kleef, H.; Van der Velde, G.; Leuven, R. The influence of environmental factors and dredging on chironomid larval diversity in urban drainage systems in polders strongly influenced by seepage from large rivers. *J. N. Am. Benthol. Soc.* **2011**, *30*, 1074–1092. [CrossRef]
55. Jiang, X.; Xiong, J.; Xie, Z. Longitudinal and seasonal patterns of macroinvertebrate communities in a large undammed river system in Southwest China. *Quat. Int.* **2017**, *440*, 1–12. [CrossRef]

56. Moller Pillot, H.K.M. 2 General Aspects of the Systematics, Biology and Ecology of the Aquatic Orthocladiinae. In *Chironomidae Larvae, Vol. 3: Orthocladiinae*; KNNV: Zeist, The Netherlands, 2013; pp. 7–21. ISBN 9004278052.
57. Hynes, H.B.N. *The Biology of Polluted Waters*, 1st ed.; Liverpool University Press: Liverpool, UK, 1960.
58. Calle-Martínez, D.; Casas, J.J. Chironomid species, stream classification, and water-quality assessment: The case of 2 Iberian Mediterranean mountain regions. *J. N. Am. Benthol. Soc.* **2006**, *25*, 465–476. [CrossRef]
59. De Haas, E.M.; Paumen, M.L.; Koelmans, A.A.; Kraak, M.H.S. Combined effects of copper and food on the midge Chironomus riparius in whole-sediment bioassays. *Environ. Pollut.* **2004**, *127*, 99–107. [CrossRef]
60. Raunio, J.; Paavola, R.; Muotka, T. Effects of emergence phenology, taxa tolerances and taxonomic resolution on the use of the Chironomid Pupal Exuvial Technique in river biomonitoring. *Freshw. Biol.* **2007**, *52*, 165–176. [CrossRef]
61. McCord, S.B.; Kuhl, B.A. Macroinvertebrate community structure and its seasonal variation in the Upper Mississippi River, USA: A case study. *J. Freshw. Ecol.* **2013**, *28*, 63–78. [CrossRef]

Article

Diversity and Distribution of Mayflies from Morocco (Ephemeroptera, Insecta)

Majida El Alami [1,*], Sara El Yaagoubi [1], Jean-Luc Gattolliat [2,3], Michel Sartori [2,3] and Mohamed Dakki [4]

1 Laboratoire Ecologie, Systématique, Conservation de la Biodiversité (LESCB), Unité de Recherche Labellisée CNRST N°18, Département de Biologie, Faculté des Sciences, Université Abdelmalek Essaâdi, P.O. Box 2121, Tétouan 93002, Morocco; sara.elyaagoubi@etu.uae.ac.ma

2 Museum of Zoology, Palais de Rumine, Place Riponne 6, CH-1014 Lausanne, Switzerland; jean-luc.gattolliat@vd.ch (J.-L.G.); michel.sartori@vd.ch (M.S.)

3 Department of Ecology and Evolution, Biophore University of Lausanne, CH-1015 Lausanne, Switzerland

4 Laboratory of Geo-Biodiversity and Natural Patrimony, Scientific Institute, Mohammed V University in Rabat, Av. Ibn Batouta, P.O. Box 703, Rabat 10106, Morocco; dakkiisr@gmail.com

* Correspondence: melalamielmoutaoukil@uae.ac.ma

Abstract: Recent research in various Moroccan areas allowed an update and a revision of the Moroccan Ephemeroptera checklist. In this case, 54 species are now listed, belonging to 10 families and 26 genera. The distribution of all studied species is discussed, as well as their biogeographical affinities. Moroccan Mayflies are characterized by a clear dominance of Mediterranean elements with a strong rate of endemism (33.4%).

Keywords: endemism; distribution; biogeography; Rif; Atlas; Central Plateau; Oriental Morocco

Citation: El Alami, M.; El Yaagoubi, S.; Gattolliat, J.-L.; Sartori, M.; Dakki, M. Diversity and Distribution of Mayflies from Morocco (Ephemeroptera, Insecta). *Diversity* **2022**, *14*, 498. https://doi.org/10.3390/d14060498

Academic Editor: Simon Blanchet

Received: 7 May 2022
Accepted: 8 June 2022
Published: 19 June 2022

1. Introduction

The Mediterranean basin is considered as a world biodiversity hotspots; where aquatic ecosystems are highly threatened by a wide variety of anthropogenic impacts, such as pollution, habitat loss and fragmentation, alien species, and global warming [1–3]. In the southern part of this region, Morocco has the highest wetland diversity [4], due to its situation between two different seas and the Sahara desert, and to the presence of three high mountain chains, with diverse hydrogeological and climate conditions. This provides a complex river network that evolved since the late tertiary in insular conditions, this generated various exceptional ecosystems, which has a high natural heritage value. The conservation of these ecosystems requires an accurate knowledge of their fauna components and their role within these ecosystems. In this sense, aquatic macroinvertebrates constitute ideal indicators of the ecosystem diversity and health [5,6]. Ephemeroptera represent one of the major groups inhabiting lotic ecosystems [7–10] knowing that it constitutes up to 50% of the freshwater total animal biomass.

In Morocco, Ephemeroptera remained practically unknown until the 1970s. Indeed, the first work dedicated to this group was provided by Navás [11] who mentioned two species from the Rif area. Since then, thanks to the work of Lestage [12], Navás [11,13] and Kimmins [14], a first faunal list of 10 species was drawn up. The catalog produced by Dakki and El Agbani [15] was able to complete this list with 16 additional species to raise the number of Ephemeroptera to 26, distributed in the different Moroccan regions. This list has been greatly enriched through following taxonomic studies [15–25] as well as hydrobiological studies carried out on the various Moroccan hydrographic networks (listed below). Despite all these efforts, the knowledge of the Moroccan Ephemeroptera diversity and ecology remains incomplete; in particular, new approaches combining morphological, molecular, ecological and biogeographical evidence challenge the presence of some presumably widely distributed European species.

Thus, the main objective of this study is to summarize the knowledge of this Moroccan fauna with a compilation and an update of available species records of Moroccan mayflies.

2. Materials and Methods

This study included Ephemeroptera collected or identified by the authors in the Rif, Middle and High Atlas (1995–2019) and with material compiled from published works on the hydrographic networks of the main Moroccan domains: Rif: [23,24,26–31], Oriental Morocco: [32–37]; Central Plateau: [38,39]; Middle Atlas: [18,40–45]; High Atlas: [17,46–58].

The mayfly fauna of different Moroccan areas (Figure 1) was reviewed including all hitherto known distribution and ecology records, if some species have migrated to high altitudes to seek a milder temperature for their development, together with references and a few new records.

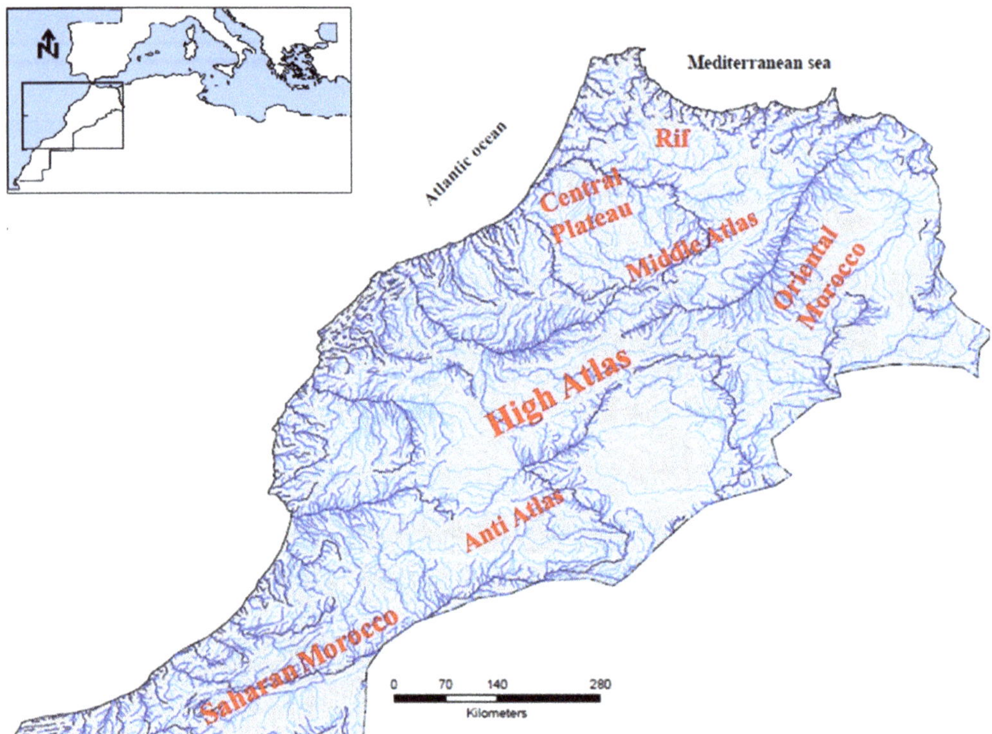

Figure 1. Different hydrographic networks and biogeographic areas of Morocco.

The sampling in the Rif area was performed by Pr. El Alami team (LESCB: Laboratory Ecology, systematics and conservation of biodiversity in the faculty of Science of Tetouan) since 1992. The identified species are conserved in alcohol at 96% or 70% in duly labeled vials and stored in the aquatic macroinvertebrates collection of the LESCB. Other specimens of some species are kept in personal collection of M. Dakki. Specific species identifications sometimes required slides mounting of dissected parts of the nymphs for detailed studies.

3. Results

In case, 54 species of Ephemeroptera representing 26 genera and 10 families have been recorded from Morocco. Among them, 18 (33.4%) are currently considered as endemic to Morocco and 9 (16.7%) endemic to Maghreb.

3.1. Commented Inventory and Distribution of Moroccan Ephemeroptera

3.1.1. Family Leptophlebiidae Banks, 1900

Choroterpes (Choroterpes) atlas **Soldán and Thomas, 1983**

This epipotamal species has a wide distribution and is found in the three Maghreb countries excepting desert areas [22]. In Morocco, this species also has a wide distribution which extends from the Rif basins area to Anti Atlas confines [22–24,26,29,33,45,52]. The streams located between the High Atlas and the Anti Atlas constitute the southern limit for this species. It presents also a wide altitudinal distribution: in the Rif, this species has a clear preference for habitats located between middle and lower courses.

In Morocco, *C. (Choroterpes) atlas* is mainly found along permanent watercourses rich in filamentous algae and occupies habitats bordering the bed in which the current is low with maximum water temperature (30 °C).

Choroterpes (Choroterpes) volubilis **Thomas and Vitte, 1988**

This Moroccan endemic is limited to the northern zones between Rif, Oriental Morocco and Middle Atlas and is absent in High Atlas and Central Plateau [22–24,26,27,29,41,45,59]. This thermophilic species seems to have a more restricted latitudinal distribution than the previous one. It reaches its ecological optimum in facies of rivers at medium and low altitudes where the temperature is high and the current is moderate to low (Table 1).

Table 1. Checklist of the Moroccan Ephemeroptera, with their geographical distribution and habitat preferences.

Species	Authorship	Distribution	Altitudes (m)					Temperature (°C)	Curent	Habitat
			Rif	M. Atlas	H. Atlas	C. Plateau	Oriental			
Choroterpes (Choroterpes) atlas	Soldán and Thomas, 1983	MagE	20–400	470–1300	1100–1630	55–630	245–1300	13.5–31.0	++/+	Pb, Gr
Choroterpes (Choroterpes) volubilis	Thomas and Vitte, 1988	MorE	20–700	210–1340			360–970	7.5–29.0	++/+	Pb, Gr
Choroterpes (Euthraulus) lindrothi	(Peters, 1980)	MagE	5–375	470	175–1600	60–650		13.0–24.0	++	St, Pb
Habroleptoides assefae	Sartori and Thomas, 1986	MorE	800–1600		1700–3000			7.5–20.5	++/+	Pb, Cb
Habrophlebia fusca	(Curtis, 1834)	Wpal		765–1820	765–2600	1000		8.5–29.0	++	Pb, Bld, Bed
Habrophlebia vaillantorum	Thomas, 1986	MorE			1700–3000			5.0–7.0	+++	Pb,Cb
Habrophlebia sp1		MorE	5–1600					7.5–20.5	++/+	Gr, Pb
Habrophlebia sp2		?	750–1550		2480–2610		565–1640	7.5–24.0	++/+	Gr, Pb
Paraleptophlebia cincta	(Retzius, 1783)	Wpal		1470				9.5–24.5	+	Si, Gr, Pb
Thraulus sp1		MagE?					280–435	16.5–20.0	++/+	Si, M
Potamanthus luteus	(Linné, 1767)	Tpal	20–500	210–410	950–1600			13.0–28.0	+++/++	Cb
Ephemera glaucops	Pictet, 1843	Wpal	80	210–1050	450	50	930	12.0–28.0	++/+	Cb
Ephoron virgo	(Olivier, 1791)	Wpal	20–500	210–1425		50–660	3–625	12.0–28.0	+++/++	Gr, Pb
Serratella ignita	(Poda, 1761)	Tpal	5–1600	1340–1500	650–2620	655–1120	931–1616	7.5–20.5	++	Pb
Caenis luctuosa	(Burmeister, 1839)	Wpal	5–1640	210–2050	650–2550	59–1235	3–1820	3.0–30.5	++/+	Gr, Pb
Caenis pusilla	Navás, 1913	Wpal	5–820	215–1425	650–2610		243–1425	12.0–30.5	++/+	Gr, Pb
Brachycercus harrisellus	Curtis 1834	Hol				54		?	+	Si, Pb
Sparbarus cf *kabyliensis*	(Soldán, 1986)	Ib-Mag		720–1600				10.5–29.0	++	Si, Pb
Oligoneuriella skoura	Dakki and Giudicelli, 1980	MorE		215–1450	850–1600		215–470	8.5–29.5	++	Cb
Oligoneuriopsis skhounate	Dakki and Giudicelli, 1980	Ib-Mag	20–500	215–1600	650–1200	115–660	10–1425	13.5–24.5	++	Gr, Pb, Cb
Ecdyonurus ifranensis	Vitte and Thomas, 1988	MorE	5–1640	530–2030	1600		1550 –1820	7.5–25.5	++	Sd, Gr
Ecdyonurus rothschildi	Navás, 1929	MagE	20–410	215–1910	850–2610	115–1120	240–1910	14.5–30.5	++	Gr, Pb
Epeorus cf *torrentium*	Eaton, 1881	Wpal	20–1600		850–2610			5.0–22.5	+++	Pb, Cb
Rhithrogena sp./spp.		?	100–1580	210–1500	650–1650	356	570–1670	7.5–20.5	+++/++	Gr, Pb
Rhithrogena ayadi	Dakki and Thomas, 1986	MorE		1680–2200				6.5–15.0	+++	Cb
Rhithrogena giudicelliorum	Thomas and Bouzidi, 1986	MorE			2400–3000			5.0–13.5	+++	Pb, Cb
Rhithrogena mariae	Vitte, 1991	MorE	140–560					13.5–24.0	++	Pb, Cb
Rhithrogena ourika	Thomas and Mohati, 1985	MorE			850–2620			6.0–19.0	+++/++	Pb, Cb
Rhithrogena ryszardi	Thomas, Vitte and Soldán, 1987	MorE		1260				15.0–18.0	+++	Pb, Cb
Acentrella almohades	Alba-Tercedor and El Alami, 1999	Ib-Mag	20–840	1870			570–930	10.0–27.0	++/+	Pb, Cb
Alainites sp1		MorE	5–1600	755–1820				7.5–27.0	++/+	Pb, Cb, SbV
Alainites oukaimeden	(Thomas and Sartori, 1992)	MorE			950–3200			10.0- 15.0	+++/++	Si, Gr, Pb, AV
Baetis berberus	Thomas, 1986	MorE			2400–3000			5.0–13.5	+++	Pb, Cb
Baetis gr alpinus		MagE?	20–1500				987	10.0–31.5	+++	Pb, Cb
Baetis maurus	Kimmins, 1938	Ib-Mag	20–1500	880–1550	700–2600	659		10.0–31.5	+++	Pb, Cb
Baetis punicus	Thomas, Boumaiza and Soldán, 1983	Ib-Mag	40–1600					7.5–26.0	++/+	Gr, Pb
Baetis gr fuscatus		MorE	20–1600	530–560	730–1600	60–1120		7.5–24.0	++/+	Pb, Cb
Baetis gr lutheri		MorE	20–800					14.5–30.0	+++/++	Pb, Cb
Baetis pavidus	Grandi, 1949	Atl-Med	5–1640	215–1915	50–2600	59–1350	2–1895	7.5–30.5	+++/++	Pb, Cb
Baetis (Rhodobaetis) gr rhodani		?		470–2200	1400–2900	59–1273	10–1670	6.5–27.0	+++/++	Pb, Cb
Baetis atlanticus	Soldán and Godunko. 2006	Atl-Med	5–1640					6.5–27.0	+++/++	Pb, Cb
Centroptilum cf *lutoleum*	(Müller, 1776)	MorE	40–1283					10.5–27.0	++/+	Gr, Pb, FA, Mo
Cheleocloeon dimorphicum	(Soldán and Thomas, 1985)	MagE	5–1400		1433	115–660	3–930	13.5–25.5	+++/++	Gr, Pb
Cloeon gr dipterum		?		210–1820	850	60–1350	5–1670	8.5–30.0	++/+	Gr, Pb

Table 1. *Cont.*

Species	Authorship	Distribution	Altitudes (m)					Temperature (°C)	Curent	Habitat
			Rif	M. Atlas	H. Atlas	C. Plateau	Oriental			
Cloeon peregrinator	Gattolliat and Sartori, 2008	Atl-Med	5–1400					8.0–29.5	++/+	Sd, Gr, Pb
Cloeon simile	Eaton, 1870	Tpal	60–1055				85–870	14.5–22.5	++	Sd, Gr, FA
Labiobaetis neglectus	(Navás, 1913)	Ib-Mag	20–350	210–1425	530	115–660	10–370	13.5–30.5	+++/++	Gr, Pb
Nigrobaetis numidicus	(Soldán and Thomas, 1983)	MagE	20	215–245				19.5–24.0	++	Gr, Pb
Nigrobaetis rhithralis	(Soldán and Thomas, 1983)	MagE	400–950					13.5–18.0	++/+	Gr, Pb
Procloeon cf bifidum	(Bengtsson, 1912)	Tpal	80–1300					14.0–20.5	++	Sd, Gr, SbV
Procloeon cf concinnum	(Eaton, 1885)	Ib-Mag		245–1425	1000–1550	60–1275	10–925	10.0–30.0	++/+	Sd, Gr, Pb
Procloeon stagnicola	Soldán and Thomas, 1983	MagE	20–780					12.0–27.0	++/+	Sd, Gr, Pb
Procloeon pennulatum	(Eaton, 1870)	Hol	60–610	210–1500	1000–1550	290–660	10,0–50	20.5–27.0	+	Sd, Gr
Prosopistoma sp1.		MorE		1650				8.0–26.0	++	Si, M

Table legend: Distribution patterns: Hol: Holarctic. Tpal: Transpalearctic. Wpal: Western Palearctic. Atl-Med: Atlanto-Mediterranean. Ib-Mag: Ibero-Maghrebian. MagE: Maghrebian endemic. MorE: Moroccan endemic. Unknown: ?. Current speed: +++: High. ++: Moderate. +: Low. Habitat types (substrate): Si: Silt. M: Mud. Sd: Sand. Gr: Gravels. Pb: Pebbles. Cb: Cobbles. Bld: Boulders. Bed: Bedrock. SbV: submerged vegetation. AV: Aquatic vegetation. FA: Filamentous algae. Mo: Mosses.

***Choroterpes (Euthraulus) lindrothi* (Peters, 1980)**

This Maghreb endemic is generally less abundant and less frequent than the other species of the subgenus *Choroterpes* [60]. In Morocco, this species has been collected in streams of the Rif [24,29], Middle Atlas [15,42], Central Plateau [15,39] and North High Atlas slopes [48,50,61]. The southernmost locality known for this taxon is Oued Massa located in the Anti Atlas [48,52,57,60]. This species was essentially sampled in the facies of large streams at low (5 m a.s.l) to medium (650 m a.s.l) altitudes characterized by high summer temperatures, rich in filamentous algae and with bed formed of a coarse substrate (Table 1).

***Habroleptoides assefae* Sartori and Thomas, 1986**

This Moroccan endemic presents a discontinuous distribution, and has only been detected in the Rif [24] and the High Atlas.

In the High Atlas, this species is crenobiont and found only in a few limnocrene springs [19,48,52,57]. In the Rif, it is confined in permanent streams with moderate current, stony bed and characterized by submerged macrophytes which provide excellent refuge for larvae when the current is strong.

***Habrophlebia fusca* (Curtis, 1834)**

This Palearctic species has been recorded in Middle Atlas [15,42], Central Plateau [15,38,39,61] and Oriental Morocco [33]. The presence of this species is nevertheless doubtful [24,26,31], since this genus shows a high rate of endemism in the Maghreb countries, particularly in Algeria with the recent description of two new species [62,63]. Therefore, a revision of all Moroccan *Habrophlebia* populations would be necessary as they may represent a complex of species.

***Habrophlebia vaillantorum* Thomas, 1986**

This Moroccan endemic has a restricted distribution area: it has been only located in High Atlas [48,52,53,55], and is alticolous species [64] which prefers biotopes with stony substrate and fast current [55].

***Habrophlebia* sp1**

In North of Morocco, this species is probably new for science. Its description will be carried out soon. It has a large distribution and occupies a wide range of biotopes located between 5 and 1600 m a.s.l [24,26,29]. In the Rif, its ecological optimum is reached in waters characterized by a moderate current and low mineralization.

***Habrophlebia* sp2**

The latest study in Rif rivers revealed the presence of a second species; genetic analysis revealed that it is different from *Habrophlebia* populations found in Algeria. Its identification will be carried out soon.

Paraleptophlebia cincta **(Retzius, 1783)**

This Palearctic species has been reported in Morocco and Algeria [36,40,65], where it seems to be more alticolous than in Europe. This altitudinal shift can be attributed to the water temperature and physico-chemical parameters [42,65]. *Paraleptophlebia cincta* is apparently rheophilous and has a clear preference for permanent streams. This could explain the low number of larvae collected in these two Maghreb countries.

Thraulus **sp1**

Specimens of this species were collected in two permanent stations of Zegzel, one of the Oriental Morocco sub-basin [37].

Its morphological study indicated similarities with its European congener *T. bellus*. However, a genetic study would be necessary to confirm this preliminary identification. This is anyway the first mention of the genus *Thraulus* in Morocco.

3.1.2. Family Potamanthidae Albarda, 1888

Potamanthus luteus **(Linné, 1767)**

This is a Palearctic species found from the British Islands to Korea [66]. It has been also reported in the three Maghreb countries [67,68]. In Morocco, it is recorded from: Middle Atlas [41,42,45], High Atlas [48,49], Central Plateau [38,39] and in the Rif [24,26].

This thermophilic species has a clear preference for the facies of large rivers at low and medium altitudes, characterized by a substrate composed of cobbles and pebbles, with moderate current.

3.1.3. Family Ephemeridae Latreille, 1810

Ephemera glaucops **Pictet, 1843**

This species has a West Palearctic distribution. It is present in the three Maghreb countries. In Morocco, it is found in Middle Atlas [12,13,15], Central Plateau [15], High Atlas [13,51,52], in Oriental Morocco [33]. In the Rif, a single male imago was collected by El Alami [24]. Further studies did not confirm its presence in this part of the country [26,27,29].

In Morocco, *E. glaucops* has a clear preference for running waters with low to moderate current speed, whereas in Europe, it also occupies oligotrophic lacustrine biotopes [65,69,70].

3.1.4. Family Polymitarcyidae Banks, 1900

Ephoron virgo **(Olivier, 1791)**

Ephoron virgo has a West Palearctic distribution. In North Africa, it is known in the three Maghreb countries [62]. In Morocco, the Central Plateau [38,39] and the Middle Atlas [15] seem to constitute the southern limit of its distribution since it is absent in the High Atlas. Its presence in Oriental Morocco has been recently confirmed by Mabrouki et al. [33].

As a hot water stenotherm, *E. virgo* larvae tolerates water temperatures up to 28 °C during the summer period [24]. The larvae mainly inhabit biotopes characterized by a slow to moderate current and a substrate rich in silt and sand in which they dig galleries.

3.1.5. Family Ephemerellidae Klapalek, 1909

Serratella ignita **(Poda, 1761)**

This Palearctic species is distributed in Maghreb only over Algeria and Morocco. In this country, it has a wide latitudinal distribution covering all the Moroccan areas [13,15,24,26,28,29,32,33,39,42,52] reaching its southern limit at the Dr'a wadi [52].

This species essentially favors biotopes which are rich in aquatic plants and detritus. It prefers areas with a fine substrate rich in silt, sand and gravel and with moderate to null flow speed.

3.1.6. Family Caenidae Newman, 1853

***Caenis luctuosa* (Burmeister, 1839)**

This West Palearctic species has a wide distribution and colonizes varied biotopes. In Morocco, its range extends from the northern Rif watershed to the Anti Atlas [15]. Eurytopic and eurythermous, this species abounds in running as well as in stagnant waters.

In the Moroccan hydrographic networks, *C. luctuosa* prefers the lower and middle courses with substrate dominated by a fine grain size; it is only absent in the streams with strong currents and low temperatures. This species can colonize waters with high conductivity (1600 μS/cm) [24] and asserts itself as the most polluo-resistant species [71,72].

***Caenis pusilla* Navás, 1913**

Caenis pusilla is a West-Palearctic species, well known from North Africa [24,29,65]. In Morocco, this species has been found in the Rif [24,26,29]; the Oriental Morocco [33] and in the High Atlas [48,52]. In this area, *C. pusilla* seems to be more alticolous than in the Rif, where it coexists with *C. luctuosa* in some lower watercourses.

Caenis pusilla is less tolerant to thermal variations than *C. luctuosa*, which explains the different and complementary altitudinal distribution of these two species in the High Atlas streams. It adapts better to more rapid flows and does not tolerate low water oxygenation and high salinity values.

***Brachycercus harrisellus* (Curtis 1834)**

In the last decade, this Holarctic species has been found in the Central Plateau [73]. A morphological revision and genetic analyses will be required to confirm its identification.

***Sparbarus* cf. *kabyliensis* (Soldán, 1986)**

In Morocco, the identification of this taxon has not gone beyond the generic level. It was reported by Dakki [40,42] and Dakki and El Agbani [15] in a Middle Atlas stream and in Oriental Morocco [36]. Referring to the work of Gagneur and Thomas [65], these larvae could belong to the species *S. kabyliensis* which was described from NW Algeria [65] and whose presence has been also demonstrated in the Iberian Peninsula [74].

3.1.7. Family Oligoneuriidae Ulmer, 1914

***Oligoneuriella skoura* Dakki and Giudicelli, 1980**

This species is a Maghreb endemic, known only from Algeria and Morocco [24] and it seems to be absent in the Rif [23,24,26,27,29] and Central Plateau [38,39]. In Morocco, it is found between 210 m and 1630 m a.s.l in the Middle and the High Atlas [25,46,52,75] and among 210–460 m a.s.l in Oriental Morocco [33]

Oligoneuriella skoura is rheophilic, and can be found in strong current (rarely in areas with low current) and stony bottom.

***Oligoneuriopsis skhounate* Dakki and Giudicelli, 1980**

The genus *Oligoneuriopsis* is of Afrotropical origin, reaching North Africa and the Iberian Peninsula with the species *O. skhounate* on one side [75–82], and the species *O. orontensis* in the Levant and Iran [83]. In Morocco, this species shows a wide distribution. It is recorded from the Rif [24]; the Oriental Morocco [32,33]; the Middle Atlas [41,42,45] and the High Atlas [52]. This wide latitudinal distribution is associated with a wide altitudinal distribution (Table 1).

In Morocco, this thermophilic and rheophilic taxon prefers large permanent streams with high current speeds and begins its development only when it receives a relatively large thermal sum in summer. In the Rif, this sum is only reached at the beginning of summer, when the majority of rivers are drying up.

3.1.8. Family Heptageniidae Needham, 1901

***Ecdyonurus ifranensis* Vitte and Thomas, 1988**

This Moroccan endemic has been collected in streams of the Middle Atlas [15,41,45], High Atlas [49,52], Oriental Morocco [33] and Rif [24,26–29,83].

This species colonizes the upper courses. It has a clear preference for streams with a stony bottom, strong to moderate current and is replaced downstream by its congener *E. rothschildi*.

***Ecdyonurus rothschildi* Navás, 1929**

This Maghreb endemic is widely distributed in North Africa. In Morocco, its distribution extends from the Tingitane Peninsula to Oriental Morocco and from the Middle Atlas to the Anti Atlas, passing through the Central Plateau. This wide latitudinal distribution is overlaying a wide elevation [32,39,42,53,84,85].

Considered as the most eurythermal and thermophilic species among North African Heptageniidae, *E. rothschildi* colonizes the permanent and temporary streams and only avoids the most upper courses.

***Epeorus* cf. *torrentium* Eaton, 1881**

This taxon was reported under *E. sylvicola* in various Moroccan hydrobiological works which is probably erroneous. Indeed, the morphology of the Moroccan specimens is closer to *E. torrentium* than to *E. sylvicola.* A genetic study would be necessary to confirm the identification of this species.

In the Maghreb, this species is known only in the Moroccan hydrographic networks [61,65,78,86] where it shows a rather discontinuous distribution. It has been recorded in the High Atlas [48,49,52,55,57,58,87] and the Rif [24,26,27,29]. It seems to be absent in the Middle Atlas, the Central Plateau and Oriental Morocco. In the Rif, this rheophilic and cold water stenothermic species is confined to the upper courses (Figure 2). This important rise is probably due to the high warming of these rivers, particularly during the summer period, which becomes a limiting factor for the development of this species.

Figure 2. Oued Maggo a typical habitat of *Epeorus* cf. *torrentium*.

***Rhithrogena* spp.**

This genus presents a high degree of endemism in the Maghreb. Thus, a revision of the *Rhithrogena* collected in the different Moroccan domains must be realized, which could further increase the specific richness of Moroccan Ephemeroptera. Especially, some specimens of this genus are confined to middle and lower streams and tolerate relatively high temperature and mineralization, while others are located in upper streams and have a preference for cold temperatures. Unfortunately, the larval taxonomy of this genus in Morocco is almost unknown, therefore larval identification remains impossible.

***Rhithrogena ayadi* Dakki and Thomas, 1986**

This Moroccan endemic species has only been found in Middle Atlas [18,42,83,88]. It was collected in streams located between 1680 and 2200 m a.s.l.

This rheophilous species has a clear preference for small torrents of high mountains whose current is fast, the substrate is coarse, and the maximum temperature of the water does not exceed 15 °C.

***Rhithrogena giudicelliorum* Thomas and Bouzidi, 1986**

This High Atlas endemic was collected in Assif n'Ouarzane in a restricted altitudinal range varying between 2400 and 3000 m.

R. giudicelliorum is confined to cold torrents fed by permanent snowfields and with a maximum temperature not exceeding 10 °C [48,52,54].

***Rhithrogena mariae* Vitte, 1991**

This Moroccan endemic has a wide distribution in the western Rif. Specimens were found in the middle and lower reaches of streams of this area, whose altitude varies between 140 and 560 m [21,24].

This Rifian species is less rheophilic and more thermophilic than the Atlas ones and has a clear preference for potamal waters. Its development cycle must include specific adaptations since it supports even intermittent watercourses which undergo strong seasonal variations in flow [24].

***Rhithrogena ourika* Thomas and Mohati, 1985**

This High Atlas endemic was sampled in an altitudinal range varying between 850 and 2600 m [17,57].

In a recent extensive survey of the benthic macroinvertebrates of the Ourika watershed, Abessolo et al. [57] failed to collect this species in the different prospected localities, suggesting it may be locally extinct or extremely rare.

R. ourika has a clear preference for the cool waters of the upper courses, but it is less alticolous than its congener *R. giudicelliorum*.

***Rhithrogena ryszardi* Thomas, Vitte and Soldán, 1987**

This is an endemic species of the Middle Atlas belonging to the so-called *germanica group*. Since its discovery in O. Tout and O. Bençmim, at 1260 m [20], this species has not been collecting again, despite numerous samplings realized in waterways of this area. Anthropogenic impacts, such as water diversion by dams and canals, irrigation and organic pollution from villages, could be the cause of its local disappearance.

3.1.9. Family Baetidae Leach, 1815

***Acentrella almohades* Alba-Tercedor and El Alami, 1999**

This Ibero-Moroccan endemic is replaced in the other Maghreb countries by its congener *A. sinaica* [61,89,90]. In Morocco, this species is present in the Middle Atlas, Oriental Morocco and the Rif [24,33,90].

This thermophilic species has a clear preference for the facies of rivers with a clement temperate winter and tolerates wide conductivity variations.

***Alainites* sp1**

This species is preferentially confined to northern streams of Morocco, since it appears to be absent in the High Atlas [52,53], where it is replaced by its congener *A. oukaimeden*. Its absence in Oriental Morocco [32,33] and the Central Plateau [39] is probably linked to the excessive water warming.

In the Rif and the Middle Atlas, this species has been identified as *A. muticus* by previous authors because it has seven pairs of gills [24,26,29].

In the Rif, this species has a wide distribution and a wide altitudinal range. It prefers biotopes with cobbles, pebbles and submerged vegetation (Figure 3). It also prefers the relatively cool waters of the upper and middle wadis and has been found in sites characterized by high conductivities [24].

Figure 3. Oued Kelaa a typical habitat of *Alainites* sp1.

***Alainites oukaimeden* (Thomas and Sartori, 1992)**

This Moroccan endemic has been mainly reported in springs and their emissaries on the northern and southern watersheds of the High Atlas [48,49,52,55,56].

This species inhabits a wide range of biotopes with a certain preference for those with abundant aquatic vegetation.

***Baetis berberus* Thomas, 1986**

The distribution area of this Moroccan endemic is limited to the High Atlas. It colonizes the highest streams and torrents and has an affinity for the crenel [54,56,91]. It is a strictly stenothermal cold water.

Baetis* gr *alpinus

Located in northern Morocco, this complex presents cryptic species [91,92] whose larvae are characterized by a reduced paracercus, a single row of denticles in the tarsal claws and a paraglossus with three rows of bristles. The genetic analysis showed high distances between the Moroccan populations and Spanish ones as well as with *Baetis maurus* [93]

***Baetis maurus* Kimmins,1938**

This Ibero-Maghrebian endemic seems to be absent only in Tunisia. In Morocco, it has been found in the Atlas and the Rif domains and is absent in the Central Plateau [39] and in the Oriental Morocco [32,33]. It has a wide altitudinal distribution [15,42,49,52,55–57,87,88,92] and preferred cold and fast waters of mountain streams.

***Baetis punicus* Thomas, Boumaiza and Soldán, 1983**

This Ibero-Maghrebian endemic [24,31,68,94] was reported only recently for the first time in Morocco [24,31] and is located in the western Rif where it was collected in fairly large altitudinal range. The first mention of this species in Europe was made by Ubero-Pascal et al. [94] who sampled it in South-East of Spain.

This species frequents permanent and temporary waters. Its ecological optimum and highest abundance are reached in the emissaries springs at high and medium altitude, but it develops further downstream, particularly during the winter period when the flow speed is quite high [24].

Baetis* gr *fuscatus

In North Africa, this group was first found in Algeria where it was identified as *B. bioculatus* [60]. In Morocco, it seems to be absent in the Central Plateau and Oriental Morocco [33,39]. This is probably due to water warming which limits its development.

This species has a clear preference for temperate waters with moderate to low current. It favors biotopes with a stony bottom rich in gravel and pebbles which constitute a good refuge for the larvae [24].

Baetis* gr *lutheri

In the Rif, specimens of *Baetis group lutheri* were preliminary identified as *B. meridionalis* [24,26,31]. They were sampled in 43 sites distributed over the lower wadis of Mediterranean and Atlantic hydrographic networks (Figure 4). Müller-Liebenau [95] mentioned the presence of *B. nigrescens* in Algeria, which also belongs to the *Baetis group lutheri*. Further studies are needed to know if the species is also present in Morocco.

Figure 4. Oued Laou a typical habitat of *Baetis* gr *lutheri*.

***Baetis pavidus* Grandi 1949**

This West Mediterranean species is known from south-western Europe and the Maghreb [24,29]. It has a wide distribution and frequents the main hydrographic networks of different Moroccan areas. This broad latitudinal distribution is associated with a large altitudinal amplitude since it was collected between 5 and 2600 m a.s.l.

This species mostly colonizes temperate waters from middle and lower streams but can be also found at higher altitudes during the summer season.

Baetis (Rhodobaetis)* gr *rhodani

This is an abundant and widespread species-group, made up of sibling species [96]. Three species belonging to the subgenus *Rhodobaetis* are reported by Soldán et al. from Algeria [97]. DNA barcodes are available for two of them and allow a secure identification [89]. In Morocco, *Baetis atlanticus* Soldán and Godunko 2006 has been generally reported under the name of *B. rhodani* in the Rif area where it has a wide distribution (El Yaagoubi et al., in preparation). *Baetis* gr *rhodani* is one of the most ubiquitous species complex and has a wide latitudinal distribution. It is also abundant in the cold waters with a fast to moderate velocity of current.

The revision of the identification of these species in different Moroccan areas will have to be carried out.

***Centroptilum* cf. *luteolum* (Müller, 1776)**

This genus is reported from the three Maghreb countries. A new species occurs in eastern Algeria [71]; the populations reported from Tunisia under *Centroptilum luteolum* [67,80] may correspond to this new species. In Morocco, the preliminary genetic and morphological studies revealed differences with populations of neighboring countries (Figure 5). Its distribution area is restricted in Morocco, as it is limited to the northern zone where it was collected in calm edge waters, loose substrate, low to moderate current, high temperatures, and sites rich in filamentous algae and mosses [24,26].

Figure 5. Habitus of *Centroptilum* cf. *luteolum* from the Rif.

***Cheleocloeon dimorphicum* (Soldán and Thomas, 1985)**

This Maghrebian endemic is distributed in the three Maghreb countries. In Morocco, it presents a rather discontinuous horizontal distribution since it was sampled in High Atlas [46,48,51,57], Oriental Morocco [33] and in the Rif [24,26,27,29]. It seems to be absent in the Middle Atlas and in the Central Plateau. This species deals with a wide range of habitats and occupies the permanent streams with stony bottom, moderate current and rich in plant debris.

***Cloeon* gr *dipterum* (Linné, 1761)**

Cloeon gr *dipterum* has a wide distribution [98]. In Morocco, it also has a large latitudinal and altitudinal distribution covering the Atlas and Rif areas, passing through the Central Plateau and Oriental Morocco. It mainly colonizes stagnant residual pools and pounds. This cosmopolitan taxon includes several cryptic species [98–101].

Recently, a morphological and genetic study made it possible to discover *Cloeon peregrinator* Gattolliat and Sartori, 2008 in Algeria [89]. In Morocco, this species has been identified under the name of *Cloeon dipterum* in the Rif area (El Yaagoubi et al., in preparation).

The revision of the identification of this species in different Moroccan areas will have to be carried out.

***Cloeon simile* Eaton, 1870**

This Palearctic species was reported by several authors from Algeria [61–63,78,84,85,102] and from north Tunisia [103]. In Morocco, this species seems more stenotopic than its congener *C. peregrinator* since it has been found in a restricted number of localities of the Rif [24] and of Oriental Morocco [33].

The larvae were found during the summer period when the temperatures are high enough. Its preferred biotopes are small streams with a moderate velocity, and with a dominant fine substrate (sand and gravel) and rich filamentous algae.

Labiobaetis neglectus **(Navás, 1913)**

Labiobaetis negletus is an Ibero-Maghrebian species, distributed in the three Maghreb countries. In Morocco, it has a fairly wide distribution area between the Rif [24,26,29], Oriental Morocco [33] and the Atlas [15,39,42,45,52]. This thermophilic species has a wide altitudinal and latitudinal distribution; it is abundant in permanent rivers at low and medium altitudes with moderate flow speeds. The larvae appear in spring, when the climate becomes temperate.

Nigrobaetis numidicus **(Soldán and Thomas, 1983)**

This North African endemic is known only from Algeria and Morocco. It was first reported from streams in the Rif under the name *N.* group *gracilis* [24]. The revision of this species revealed it should be named *N. numidicus.* Less rare in Morocco than its congener *N. rhithralis,* this species has been found in streams of the Middle Atlas [45].

Found only at low altitude, this species seems to have a thermophilic tendency and present a clear preference for wide streams.

Nigrobaetis rhithralis **(Soldán and Thomas, 1983)**

The distribution area of this Maghrebian endemic extends over the three Maghreb countries [24,89,103]. In Morocco, this species was first reported by El Alami et al. [31] under the name *Diphetor rhithralis.* It seems to present a limited ecological valence as it was only found in a restricted number of streams of the Rif.

The highest number of specimens was collected in a small stream with a quite cool water which crosses a fairly dense forest. The bed of the wadi is rich in sand; the riparian vegetation is abundant and prevents the heating of the water.

Procloeon **cf *bifidum*** **(Bengtsson, 1912)**

In Morocco, this species has been found sporadically in some sites in the Rif area [24,26]. This species presents a great similarity with the Algerian species *Procloeon stagnicola* Soldán and Thomas, 1983. The article I of the maxillary palp is as wide as the article II with a more rounded apex; it is also characterized by lateral spines on the segments V to the XI segments. Genetic and morphological study would be necessary to confirm the identification of this species.

This species presents a discontinuous distribution. It favors the lower courses of temperate waters with low velocity, and shallow depth. The bottom is generally stony dominated by gravel and pebbles with submerged vegetation [24].

Procloeon stagnicola **Soldán and Thomas, 1983**

In Morocco, the first mention of this North African endemic was made by Navás [11] who discovered it in the Tetouan region and who considered it as belonging to the Iberian species *Procloeon concinnum.* Similarly, studies that followed this discovery assigned the same name to this taxon [26,27,29,31]. Genetic analysis of Moroccan specimens revealed that it is indeed the species *P. stagnicola.* This species has a wide geographical distribution in Morocco since it has been found in the waterways of the different hydrographic networks.

Less altitudinal than its congener *P. bifidum,* it was collected in localities of the middle and lower reaches. It is a thermophilic species that has a clear preference for fine substrates rich in sand and gravel.

A genetic and morphologic analysis of *Procloeon* populations from Morocco; Algeria and Iberian Peninsula would be necessary to remove any ambiguity concerning the specific identification within this genus.

Procloeon pennulatum **(Eaton, 1870)**

Procloeon pennulatum has a Holarctic distribution. In Maghreb, this species is only reported from Morocco where it has a wider distribution compared to that of its congeners mentioned above. It has been reported in the waterways of the Middle Atlas [42], High Atlas [46,52], Central Plateau [38,39], Oriental Morocco [32,33,36] and the Rif [23,24], in fairly large altitudinal distribution.

This species favors lentic and temperate waters and streams characterized by a sandy bottom and abundant aquatic vegetation.

3.1.10. Family Prosopistomatidae Latreille, 1833

***Prosopistoma* sp.1**

This species was found for the first time in Morocco by Touabay et al. [88] who collected it in a Middle Atlas stream. So far, only one species has been mentioned in North Africa, *Prosopistoma alaini* Bojkova and Soldán 2015, described from Algeria. This Moroccan species is on the way to be described and has a fairly limited distribution.

3.2. *Biogeographical Affinities of Moroccan Ephemeroptera*

The analysis of Ephemeroptera species composition in Morocco, based on the chorological categories assigned to each taxon, shows that they can be divided essentially in Mediterranean species (68.7%), followed by Palaearctic elements (20.4%) and lastly, the elements with wide distribution (3.7%). Four taxa have an unknown distribution (7.4%) because their identification is doubtful and requires revision. This same ratio has been found by other authors and among neighboring countries [24,29,30,33,42,68,78,89].

Likewise, within the Mediterranean elements, four distribution categories have been observed and showed a clear dominance of the Moroccan endemics, followed by the Ibero-Maghrebian ones, the Maghrebian endemics and finally the west Mediterranean elements:

- Moroccan endemics: *A. oukaimeden*, *Alainites* sp1, *B. berberus*, *C.* gr *luteolum*, *B.* gr *lutheri*, *B.* gr *fuscatus*, *C.* (*Choroterpes*) *volubilis*, *H. assefae*, *H. vaillantorum*, *Habrophlebia* sp.1, *O. skoura*, *E. ifranensis*, *R. ayadi*, *R. giudicelliorum*, *R. mariae*, *R. ourika*, *R. ryszardi*; *Prosopistoma* sp1.
- Ibero-Maghrebian endemics: *B. maurus*, *B. punicus*, *P. concinnum*, *L. neglectus*, *A. almohades*, *O. skhounate*, *Habrophlebia* sp.2 and *S.* cf. *kabyliensis*.
- Maghrebian endemics: *C. dimorphicum*, *N. rhithralis*, *N. numidicus*, *P. stagnicola*, *C.* (*Choroterpes*) *atlas*, *C.* (*Euthraulus*) *lindrothi*, *Thraulus* sp1, *E. rothschildi* and *B.* gr *alpinus*.
- Atlanto-mediterranean species: *B. atlanticus*, *B. pavidus*, *C. peregrinator*.

This general pattern could change soon with the revision of some species whose identification is still doubtful.

4. Discussion

We have recorded 54 Ephemeroptera species belonging to 10 families, including some new taxa, still waiting to be described. The richness of this mayflies community is lower than that recorded in some regions of the northern Mediterranean countries neighboring Morocco [104,105]. The Mediterranean climate with strong fluctuations of temperature and rainfalls, the freshwater ecosystems with a temporary pattern [4,106], the considerable mineralization of lower courses and the wide annual thermal amplitudes [42,107–109] could explain this impoverishment.

Based on the present knowledge, the comparison of the mayflies diversity between different Moroccan biogeographical areas revealed that the Rif (37 species) has a more diversified fauna than Middle Atlas (31 species), High Atlas (30 species), Oriental Morocco (24 species) and finally Central Plateau (19 species). The coolness of the Rif climate would have favored the conservation and the colonization of some species of European origin (such as *Baetis* gr. *lutheri* and *Centroptilum* gr. *luteolum*) whose presence could not be detected in other Moroccan regions. Despite the high altitudes at which the High Atlas streams originate (2600–3000 m a.s.l), the diversity seems lower. This is probably due to climatic constraints combined with anthropogenic impacts on macroivertabrate populations that limit the development of orophilic and cold water stenothermal taxa [42,52,58]. In addition, the high number of temporary streams and springs in Central Plateau [39] and Oriental Morocco [32,33,35] does not represent suitable habitats for rheophilous species preferring the cool waters of upper streams. The High Atlas encompasses the highest number of endemics (10 spp) with five microendemics which are found only in this area (*A. oukaimeden*, *B. berberus*, *R. giudicellorum*, *R. ourika*, *H. vaillantorum*), followed by the Rif with nine endemics and three microendemics species (*Habrophlebia* sp.1, *Baetis* gr.

lutheri, Centroptilum cf. *luteolum*), Middle Atlas (8 spp) with three microendemics species (*R. ayadi, R. ryzardi, Prosopistoma* sp1), Oriental Morocco (5 spp) with, probably, only one microendemic species (*Thraulus* sp1) and finally Central Plateau (3 spp).

A comparison between Ephemeroptera fauna species richness from different regions of the Maghreb shows that Tunisia streams have a lower specific diversity [67,76,80,104] than Morocco and Algeria [24,42,65,71,77,89,109–111]. In addition, the analysis of the species similarity between neighboring countries shows that Morocco and Algeria have more species in common than Tunisia. They share 22 species,13 of them are endemics to Maghreb (*B. maurus, B. punicus, C. dimorphicum, L. neglectus, N.numidicus, N. rhithralis, E. ifranensis, E. rothschildi, C. (Choroterpes) atlas, C. (Euthraulus) lindrothi O. skhounate*) and nine are found beyond the Maghreb and the Iberian Peninsula (*B. atlanticus, B. pavidus, C. peregrinator, C. simile, P. pennulatum, P. cincta, P. luteus, E. virgo, E. glaucops, C. luctuosa, C. pusilla*). Morocco and Tunisia have 13 species in common which are also present in Algeria. The Rif area appears to be the Moroccan region with the most species in common with the Iberian Peninsula (*B. punicus, N. rhithralis, L. neglectus, A. almohades, P. concinnum*). This indicates that these neighboring regions have probably the same palaeogeographical evolution.

The Moroccan palaeogeographical history most probably explains the dominance of Palearctic components in the Moroccan mayfly fauna. Afrotropical components remain very limited (*Oligoneuriopsis* and *Chelecloeon*). The Mediterranean partition into subregions separated by strong reliefs has favored the speciation in other taxa groups such as mammals [112]. Thus, the rate of Moroccan endemism in the mayfly populations is higher than other macroinvertebrate groups where the Ibero-Maghrebian and/or Maghrebian elements prevail over the Moroccan endemics [24,29,30,33,35,113–115]. The present situation can be explained by the Mediterranean paleogeography. The formation of the Betic-Rifian massif [116–118] and the Messinian Crisis [119,120] permitted an important fauna interchanges between northern of Morocco and Western Europe [121]; the tropical and sub-tropical macroinvertebrates, that inhabited the Moroccan hydrographic networks at that time, possibly passed into Iberic Peninsula [42]. This passage would also, testify the presence of hydrographic networks [118] which favored the intercontinental exchanges of many aquatic macroinvertebrates, including Ephemeroptera larvae whose dispersal is limited. Mayflies imagines are fragile and have a short life, so their dispersion ability is rather limited and could be mostly passive by the winds [24]. Thus, the Rif received Iberian species, which explains the species with European origin, the important diversity and endemicity [122] in this part of the country. Also, the formation of barriers isolated the African continent from Europe and Asia and the quaternary climate changes has favorized the taxa speciation independently of their Eurasian congeners (*B. berberus, H. assefae*) ensuring the increase rate of mountainous endemism in the Mediterranean regions [24,42]. Similarly, the arid Saharan climate shift 12,000 years ago [123] has divided the Maghreb into two distinct zones, separated by the Atlas Mountains. The fauna had evolved there totally isolated, which explains, also, the high rate of Moroccan endemics in mountains [42].

5. Conclusions

Morocco can be considered as one of the hotspots of Ephemeroptera biodiversity in North Africa. The endemism rate could even increase with the intensification of prospections in areas not/or rarely studied such as the Anti-Atlas and the Moroccan Sahara. In addition, taxonomic revision and genetic analysis could validate the hypothesis of new species and elucidate the affiliation of some others.

In order to improve our knowledge of Moroccan fauna, the production of a distribution map, a red list of Moroccan Ephemeroptera would be interesting insofar as Moroccan endemics are located in vulnerable habitats, subject to strong anthropogenic pressures combined with increasing drought events. Thus, the protection measures and the conservation of these habitats are necessary to avoid their degradation. Furthermore, it would be critical to provide tools to preserve Moroccan biodiversity, particularly endemics, sensitive to global changes.

Author Contributions: Conceptualization, M.E.A.; methodology, M.E.A.; software, M.E.A.; validation, M.S., J.-L.G. and M.D.; formal analysis. M.E.A.; investigation, M.E.A., S.E.Y. and M.D.; resources, M.E.A. and M.D.; data curation, M.E.A.; writing—original draft preparation, M.E.A.; writing—review and editing, all co-authors.; visualization, M.E.A., J.-L.G. and M.S.; supervision, M.E.A., J.-L.G. and M.S.; project administration, M.E.A.; funding acquisition, M.E.A. All authors have read and agreed to the published version of the manuscript.

Funding: This research received no external funding.

Institutional Review Board Statement: Not applicable.

Informed Consent Statement: Not applicable.

Data Availability Statement: Data is available from the corresponding author upon request.

Acknowledgments: The first author want to thank Laurent Vuataz for his help in the genetic analysis of *Rhithrogena*. We are most grateful to Jallal Kassout for cartography support.

Conflicts of Interest: The authors declare no conflict of interest.

References

1. Ulbrich, U.; May, W.; Li, L.; Lionello, P.; Pinto, J.G.; Somot, S. Chapter 8: The Mediterranean climate change under global warming. In *Developments in Earth and Environmental Sciences*; Elsevier: Amsterdam, The Netherlands, 2006; pp. 399–415.
2. Carvalho, S.B.; Brito, J.C.; Crespo, E.J.; Possingham, H.P. From climate change predictions to actions–conserving vulnerable animal groups in hotspots at a regional scale. *Glob. Change Biol.* **2010**, *16*, 3257–3270. [CrossRef]
3. Bonada, N.; Resh, V.H. Mediterranean-climate streams and rivers: Geographically separated but ecologically comparable freshwater systems. *Hydrobiologia* **2013**, *719*, 1–29. [CrossRef]
4. Dakki, M.; Menioui, M. *Stratégie Nationale 2015–2024 pour les Zones Humides du Maroc: Diagnostic, Apport HCEFLCD/GIZ*; GIZ: Rabat, Morocco, 2016; p. 56.
5. Moisan, J.; Pelletier, L. *Protocole d'Échantillonnage des Macroinvertébrés Benthiques d'eau Douce du Québec Cours d'eau Peu profonds à Substrat Meuble, Ministère du Développement Durable, Plan Saint-Laurent pour un Développement Durable*; Développement Durable, Environnement et Parcs Québec: Québec, QC, Canada, 2011.
6. Tachet, H.; Richoux, P.; Bournaud, M.; Usseglio-Poltera, P. *Invertébrés d'Eau Douce*, 2nd ed.; CNRS Editions: Paris, France, 2002; 588p.
7. Saltveit, S.J.; Brittain, J.E.; Lillehammer, A. Stoneflies and River Regulation—A Review. *Regul. Streams* **1987**, *8*, 117–129.
8. Brittain, J.E.; Saltveit, S.J. A review of the effect of river regulation on mayflies (Ephemeroptera). *Regul. Rivers Res. Mgmt.* **1989**, *3*, 191–204. [CrossRef]
9. Studemann, D.; Landolt, P.; Sartori, M.; Hefti, D.; Tomka, I. *Insecta Helvetica Fauna 9: Ephemeroptera*; ConchBooks: Hackenheim, Germany, 1992.
10. Brittain, J.E.; Sartori, M. Ephemeroptera. In *Encyclopedia of Insects*; Resh, V.H., Cardé, R., Eds.; Academic Press: Cambridge, MA, USA, 2009; pp. 328–333.
11. Navás, S.L. *Insectos de la Excursion de D. Ascensio Cordina à Marruecos*; Museu de Ciències Naturals: Barcelona, Spain, 1922; Volume 4, pp. 119–127.
12. Lestage, J.A. Ephémères, Plécoptères et Trichoptères recueillis en Algérie et liste des espèces connues actuellement de l'Afrique du Nord. *Bull. Soc. Hist. Nat. Afr. Du Nord.* **1925**, *16*, 8–18.
13. Navas, S.L. Insectos de Berberia. *Bol. Soc. Ent. Esp. Madr.* **1935**, *18*, 77–100.
14. Kimmins, D.E. A new moroccan ephemeropteron. *Ann. Mag. Nat. Hist.* **1938**, *1*, 302–305. [CrossRef]
15. Dakki, M.; El Agbani, M.A. Ephéméroptères d'Afrique du Nord: 3. Eléments pour la connaissance de la faune marocaine. *Bull. Inst. Sci.* **1983**, *7*, 115–126.
16. Soldán, T.; Thomas, A.G.B. New and little-known species of mayflies (Ephemeroptera) from Algeria. *Acta Entomol. Bohemoslov.* **1983**, *80*, 356–376.
17. Thomas, A.G.B.; Mohati, A. Rhithrogena ourika n. sp., Éphéméroptère nouveau du Haut Atlas marocain (Heptageniidae). In *Annales de Limnologie-International Journal of Limnology*; EDP Sciences: Les Ulis, France, 1985; Volume 21, pp. 145–148.
18. Dakki, M.; Thomas, A.G.B. Rhithrogena ayadi n. sp., Éphéméroptère nouveau du Moyen Atlas marocain (Heptageniidae). In *Annales de Limnologie-International Journal of Limnology*; EDP Sciences: Les Ulis, France, 1986; Volume 22, pp. 27–29.
19. Sartori, M.; Thomas, A.B. Révision taxonomique du genre Habroleptoides Schönemund, 1929 (Ephemeroptera, Leptophlebiidae). I: Habroleptoides assefae, n. sp. du Haut-Atlas marocain. *Rev. Suisse Zool.* **1986**, *93*, 417–422. [CrossRef]
20. Thomas, A.G.B.; Vitte, B.; Soldán, T. Rhithrogena ryszardi n. sp., Éphéméroptère nouveau du Moyen Atlas (Maroc) et redescription de Rh. soteria Navás, 1917 (Heptageniidae). In *Annales de Limnologie-International Journal of Limnology*; EDP Sciences: Les Ulis, France, 1987; Volume 23, pp. 169–177.
21. Vitte, B. Rhithrogena mariae n. sp., Ephéméroptere nouveau du Rif marocain (Ephemeroptera, Heptageniidae). *Nouv. Rev. D'entomologie* **1991**, *8*, 89–96.

22. Thomas, A.G.B.; Vitte, B. Compléments et corrections à la faune des Ephéméroptères d'Afrique du Nord. 1. Le genre Choroterpes Eaton, sensu stricto (Ephemeroptera). In *Annales de Limnologie*; EDP Sciences: Les Ulis, France, 1988; Volume 24, pp. 61–65.
23. El Alami, M. Étude Hydrobiologique d'un Réseau Hydrographique Nord-Rifain, l'Oued Laou: Typologie, Ecologie et Biogéographie des Ephéméroptères. Ph.D. Thesis, Université Mohamed V, Rabat, Morocco, 1989; p. 187.
24. El Alami, M. Taxonomie, Ecologie et Biogéographie des Éphéméroptères du Rif (Nord du Maroc). Ph.D. Thesis, Université Abdelmalek Essaâdi, Tétouan, Morocco, 2002.
25. Dakki, M.; Giudicelli, J. Ephéméroptères d'Afrique du Nord: 2. Description d'Oligoneuriella skoura n. sp. et d'Oligoneuriopsis skhounate n. sp. avec notes sur leur écologie (Ephemeroptera, Oligoneuriidae). *Bull. Inst. Sci.* **1980**, *4*, 13–28.
26. Khadri, O.; Alami, M.; El Bazi, R.; Slimani, M. Ephemeroptera's diversity and ecology in streams of the ultramafic massif of Beni Bousera and in the adjacent non-ultramafic sites (NW, Morocco). *J. Mater. Environ. Sci.* **2017**, *8*, 3508–3523.
27. Guellaf, A.; El Alami, M.; Kassout, J.; Errochdi, S.; Khadri, O.; Kettani, K. Diversity and ecology of aquatic insects (Ephemeroptera, Plecoptera and Trichoptera) in the Martil basin (Northwestern Morocco). *Community Ecol.* **2021**, *22*, 331–350. [CrossRef]
28. El Alami, M.; Dakki, M. Peuplements d'Ephéméroptères et de Trichoptères de l'Oued Laou (Rif Occidentale, Maroc): Distribution longitudinale et biotypologie. *Bull. Inst. Sci.* **1998**, *21*, 51–70.
29. El Bazi, R.; El Alami, M.; Khadri, O.; Errochdi, S.; Slimani, M.; Bennas, N. Projet du parc naturel de Bouhachem (Nord-Ouest du Maroc) II: Ephemeroptera, Plecoptera, Trichoptera. *Boletín Soc. Entomol. Aragonesa* **2017**, *61*, 55–66.
30. Slimani, M. Etat Ecologique et Biodiversité des Macroinvertébrés Aquatiques du Haut Bassin Versant Loukkos (Nord-Ouest du Maroc) et Effet de l'Intermittence et des Altérations Anthropiques. Ph.D. Thesis, Université Abdelmalek Essaâdi, Tétouan, Morocco, 2018; p. 246.
31. El Alami, M.; Dakki, M.; Errami, M.; Alba-Tercedor, J. Nouvelles données sur les Baetidae du Maroc (Insecta: Ephemeroptera). *Zool. Baetica* **2000**, *11*, 105–113.
32. Berrahou, A.; Cellot, B.; Richoux, P. Distribution longitudinale des macroinvertébrés benthiques de la Moulouya et de ses principaux affluents (Maroc). *Ann. Limnol. Int. J. Lim.* **2001**, *37*, 223–235. [CrossRef]
33. Mabrouki, Y.; Taybi, A.F.; El Alami, M.; Berrahou, A. New and interesting data on distribution and ecology of Mayflies from Eastern Morocco (Ephemeroptera). *JMES* **2017**, *8*, 2839–2859.
34. Fagrouch, A.; Berrahou, A.; El Halouani, H. Impact d'un effluent urbain de la ville de Taourirt sur la structure des communautés de macroinvertébrés de l'oued Za (Maroc oriental). *J. Water Sci.* **2011**, *24*, 87–101.
35. Taybi, A.F.; Mabrouki, Y.; Dakki, M.; Berrahou, A.; Millán, A. Longitudinal distribution of macroinvertebrate in a very wet North African Basin: Oued Melloulou (Morocco). *Ann. Limnol. Int. J. Lim.* **2020**, *56*, 17. [CrossRef]
36. Lamri, D.; Hassouni, T.; Loukili, A.; Belghyti, D. Structure and macroinvertebrate diversity in the Moulouya river basin, Morocco. *JEZS* **2016**, *4*, 1116–1121.
37. Maamri, A.; Pattee, E.; Dolédec, S.; Chergui, H. The benthic macroinvertebrate assemblages in the Zegzel-Cherraa, a partly-temporary river system, Eastern Morocco. *Ann. Limnol. Int. J. Lim.* **2005**, *41*, 247–257. [CrossRef]
38. Qninba, A.J.; El Agbani, M.A.; Dakki, M.; Benhoussa, A. Evolution saisonnière de quelques peuplements d'invertébrés benthiques de l'Oued Bou Regreg (Maroc). *Bull. Inst. Sci.* **1988**, *12*, 149–156.
39. El Agbani, M.A.; Dakki, M.; Bournaud, M. Etude typologique du Bou Regreg (Maroc): Les milieux aquatiques et leurs peuplements en macroinvertébrés. *Bull. Ecol.* **1992**, *23*, 103–113.
40. Dakki, M. Recherches Hydrobiologiques sur un Cours d'eau du Moyen Atlas (Maroc). Ph.D. Thesis, Aix Marseille III, Marseille, France, 1979; p. 126.
41. Zerrouk, M.; Dakki, M.; El Agbani, M.A.; El Alami, M.; Bennas, N.; Qninba, A.; Himmi, O. Evolution of the benthic communities in a north-African river, the upper Sebou (middle atlas-Morocco) between 1981 and 2017: Effects of global changes. *Biologia* **2021**, *76*, 2973–2989. [CrossRef]
42. Dakki, M. Ecosystème d'eau courante du haut Sebou (Moyen Atlas); Etudes typologiques et analyses écologiques et biogéographie des principaux peuplements entomologiques. *Trav. Inst. Sci. Ser. Zool.* **1987**, *42*, 99.
43. Nechad, I.; Fadil, F.; Dakki, M. Changes in benthic communities in the Middle Atlas springs (Morocco) and their relationship with climate change. *J. Biodivers. Env. Sci.* **2018**, *12*, 96–108.
44. Giudicelli, J.; Dakki, M.; Dia, A. Caractéristiques abiotiques et hydrobiologiques des eaux courantes méditerranéennes. *Verh. Intern. Verein. Limnol.* **1985**, *22*, 2094–2101. [CrossRef]
45. Zerrouk, M.; Dakki, M.; Bennas, N.; El Agbani, M.A.; El Alami, M.; Ghamizi, M.; Ouassima, L.M.; Qninba, A.; Himmi, O. Nouvelles données sur les macroinvertebrés du bassin versant du Haut Sebou (Moyen Atlas, Maroc): Insectes, mollusques et crustacés. *Bol. SEA* **2021**, *69*, 29.
46. Pihan, J.C.; Mohati, A. Les peuplements benthiques du reseau permanent de l'Oued Ourika, Haut Atlas de Marrakech; qualité des eaux. *Int. Ver. Theor. Angew. Limnol. Verh.* **1985**, *22*, 2110–2113. [CrossRef]
47. Pihan, J.C.; Mohati, A. Etude hydrobiologique de deux torrents du Haut-Atlas de Marrakech l'assif Tiferguine et l'assif Oukaïrmeden. Impact des activités humaines. *Bull. Fac. Sci. Marrakech* **1983**, *2*, 23–61.
48. Ajakane, A. Etude Hydrobiologique du Bassin Versant de l'oued N'fis (Haut Atlas Marocain)—Biotypologie, Dynamique Saisonnière, Impact de 1'Assèchement sur les Communautés Benthiques. Ph.D. Thesis, University of Marrakech, Marrakech, Morocco, 1988; p. 189.

49. Mohati, A. Recherche Hydrobiologique sur un Cours d'eau du Haut Atlas de Marrakech (Maroc): L'oued Ourika, Écologie, Biotypologie et Impact des Activités Humaines sur la Qualité des Eaux. Ph.D. Thesis, Cadi Ayyad, Marrakech, Morocco, 1985; p. 108.
50. El Mezdi, Z. Les Khettaras de la région de Marrakech (Maroc): Un biotope hydrobiologique remarquable: Avec 2 figures dans le texte. *Int. Ver. Theor. Angew. Limnol. Verh.* **1985**, *22*, 2106–2109. [CrossRef]
51. El Mezdi, Z.; Giudicelli, J. Etude d'un écosystème limnique peu connu: Les khettaras de la région de Marrakech (Maroc) Habitats et peuplement. *Sci. Eau* **1987**, *6*, 281–297.
52. Bouzidi, A. Recherches Hydrobiologiques sur les Cours d'eau des Massifs du Haut Atlas (Maroc). Bio-Écologie des Macroin-Vertébrés et Distribution Spatiale des Peuplements. Ph.D. Thesis, Cadi Ayyad University, Marrakech, Morocco, 1989.
53. Bouzidi, A.; Giudicelli, J. Ecologie et distribution spatiale des invertébrés des eaux courantes du Haut Atlas marocain. *Rev. Fac. Sci. Marrakech* **1994**, *8*, 23–43.
54. Thomas, A.G.B.; Bouzidi, A. Trois Ephéméroptères nouveaux du Haut Atlas marocain (Heptageniidae, Baetidae, Leptophlebiidae). *Bull. Soc. Hist. Nat. Toulouse* **1986**, *122*, 7–10.
55. Ouahsine, H. Les Biocénoses d'Invertébrés Benthiques Dans un Torrent du Haut Atlas (Maroc): Le Tiferguine. Structure et Répartition du Peuplement—Régime Alimentaire, Dynamique des Populations et Production des Espèces Dominantes. Ph.D. Thesis, Cadi Ayyad University, Marrakech, Morocco, 1993.
56. Abdaoui, A.; El Moutaouakil, M.E.A.; Ghamizi, M. Diversité et distribution des Baetidae (Insecta, Ephemeroptera) du Parc National de Toubkal (Haut Atlas, Maroc). *Trav. Inst. Sci. Ser. Zool.* **2010**, *47*, 1–5.
57. Abessolo, J.Z.; Khebiza, M.Y.; Messouli, M. Contribution à la connaissance des éphéméroptères (Ephemeroptera) du bassin versant de l'Ourika (Haut-Atlas, Maroc). *Bol. Asoc. Esp. Entomol.* **2020**, *44*, 255–274.
58. Zuedzang Abessolo, J.-R.; Yacoubi Khebiza, M.; Messouli, M. Réponse des macroinvertébrés benthiques (éphéméroptères, plécoptères, trichoptères) aux pressions anthropiques dans un contexte de changement climatique sur le bassin versant de l'Ourika (Haut-Atlas du Maroc). *Hydroécol. Appl.* **2021**, *21*, 115–155. [CrossRef]
59. Mabrouki, Y.; Taybi, A.F.; El Alami, M.; Berrahou, A. Biotypology of stream macroinvertebrates from North African and semi arid catchment: Oued Za (Morocco). *Knowl. Manag. Aquat. Ecosyst.* **2019**, *420*, 17. [CrossRef]
60. Thomas, A.G.B. A provisional checklist of the mayflies of North Africa (Ephemeroptera). *Bull. Soc. Hist.* **1998**, *134*, 13–20.
61. Vitte, B.; Thomas, A.G.B. Compléments et corrections à la faune des Ephéméroptères d'Afrique du Nord. 2. Le genre Choroterpes Eaton, sous-genre Euthraulus Barnard (Ephemeroptera). *Ann. Limnol.* **1988**, *24*, 161–165. [CrossRef]
62. Benhadji, N.; Hassaine, K.A.; Sartori, M. Habrophlebia hassainae, a new mayfly species (Ephemeroptera: Leptophlebiidae) from North Africa. *Zootaxa* **2018**, *4403*, 557. [CrossRef] [PubMed]
63. Kechemir, L.H.; Sartori, M.; Lounaci, A. An unexpected new species of Habrophlebia from Algeria (Ephemeroptera, Leptophlebiidae). *ZooKeys* **2020**, *953*, 31–47. [CrossRef]
64. Thomas, A.; Gaino, E.; Marie, V. Complementary description of Habrophlebia vaillantorum Thomas, 1986 in comparison with H. fusca (Curtis, 1834) [Ephemeroptera, Leptophlebiiidae]. *Ephemera* **1999**, *1*, 9–21.
65. Gagneur, J.; Thomas, A.G.B. Contribution à la connaissance des Ephéméroptères d'Algérie. I. Répartition et écologie (1ère partie) (Insecta, Ephemeroptera). *Bull. Soc. Hist. Nat.* **1988**, *124*, 213–223.
66. Bae, Y.J.; McCafferty, W.P. Phylogenetic systematics of the Potamanthidae (Ephemeroptera). *Trans. Am. Entomol. Soc.* **1991**, *117*, 1–145.
67. Zrelli, S.; Boumaiza, M.; Bejaoui, M.; Gattolliat, J.-L.; Sartori, M. New reports of mayflies (Insecta: Ephemeroptera) from Tunisia. *Rev. Suisse Zool.* **2011**, *118*, 3–10.
68. Belfiore, C. *Guide per il Riconoscimento delle Specie Animali delle Acque Interne Italiane. Efemerotteri (Ephemeroptera)*; Consiglio Nazionale delle Ricerche: Verona, Italy, 1983.
69. Zamora-Muñoz, C.; Alba-Tercedor, J. *Caracterización y Calidad de las Aguas del río Monachil (Sierra Nevada, Granada), Facores Fisico-Qímicos y Comunidades de Macroinvertabrados Acuáticos*; Agencia del Medio Ambiente, Ed.; Anel: Granada, Spain, 1992; p. 171.
70. Bebba, N.; El-Alami, M.; Arigue, S.F.; Arab, A. Etude mésologique et biotypologique du peuplement des éphéméroptères de l'Oued Abdi (Algérie). *JMES* **2015**, *6*, 1164–1177.
71. Samraoui, B.; Bouhala, Z.; Chakri, K.; Márquez-Rodríguez, J.; Ferreras-Romero, M.; El-Serehy, H.A.; Samraoui, F.; Sartori, M.; Gattolliat, J.-L. Environmental determinants of mayfly assemblages in the Seybouse River, north-eastern Algeria (Insecta: Ephemeroptera). *Biologia* **2021**, *76*, 2277–2289. [CrossRef]
72. Soldán, T. A revision of the Caenidae with ocellar tubercles in the nymphal stage (Ephemeroptera). *Acta Univ. Carol. Biol.* **1986**, *1982*, 289–362.
73. Arifi, K.; Tahri, L.; Hafiane, F.Z.; Elblidi, S.; Fekhaoui, M. Diversité des macroinvertébrés aquatiques de la retenue du barrage Sidi Mohammed Ben Abdellah à la confluence avec les eaux de l'oued Grou et bio- évaluation de la qualité de ses eaux (Région de Rabat, Maroc). *Entomol. Faun. Faun. Entomol.* **2019**, *72*, 13–20.
74. Gallardo-Mayenco, A. Distribucion espacial de los Efemeropteros (Insecta: Ephemeroptera) en dos cuencas mediterraneas a diferentes altitudes. Spatial distribution of mayfly (Insecta: Ephemeroptera) in two Mediterranean river basins at different altitudes. *Zool. Baetica* **2003**, *13*, 93–110.

75. Badri, A. Etude Hydrobiologique d'un Cours d'eau de Plaine en Zone Semi-Aride: Le Tensift. Impact des Crues sur la Biocénose. Ph.D. Thesis, Faculté des Sciences et Techniques Marrakech, Marrakech, Morocco, 1985.
76. Boumaïza, M.; Thomas, A.G.B. Répartition et écologie des Ephéméroptères de Tunisie, 1ère partie (Insecta, Ephemeroptera). *Archs. Inst. Pasteur Tunis.* **1986**, *63*, 567–599.
77. Lounaci, A.; Brosse, S.; Ait Mouloud, S.; Lounaci-Daoudi, D.; Mebarki, N.; Thomas, A. Current knowledge of benthic invertebrate diversity in an Algerian stream: A species checklist of the Sebaou River basin (Tizi-Ouzou). *Bull. Soc. Hist. Nat. Toulouse* **2000**, *136*, 43–55.
78. Thomas, A.G.B.; Gagneur, J. Compléments et corrections à la faune des Ephéméroptères d'Afrique du Nord. 6. Alainites sadati n. sp. d'Algérie (Ephemeroptera, Baetidae). *Bull. Soc. Hist. Nat.* **1994**, *130*, 43–45.
79. Bouhala, Z.; Márquez-Rodríguez, J.; Chakri, K.; Samraoui, F.; El-Serehy, H.A.; Ferreras-Romero, M.; Samraoui, B. The life history of the Ibero-Maghrebian endemic Oligoneuriopsis skhounate Dakki and Guidicelli (Ephemeroptera: Oligoneuriidae). *Limnologica* **2020**, *81*, 125761. [CrossRef]
80. Zrelli, S.; Gattolliat, J.-L.; Boumaïza, M.; Thomas, A. First record of *Alainites sadati* Thomas, 1994 (Ephemeroptera: Baetidae) in Tunisia, description of the larval stage and ecology. *Zootaxa* **2012**, *3497*, 60. [CrossRef]
81. Barber-James, H.M.; Zrelli, S.; Yanai, Z.; Sartori, M. A reassessment of the genus Oligoneuriopsis Crass, 1947 (Ephemeroptera, Oligoneuriidae, Oligoneuriellini). *ZooKeys* **2020**, *985*, 15–47. [CrossRef]
82. Giudicelli, J.; Dakki, M. Les sources du Moyen Atlas et de Rif (Maroc): Faunistique (description de deux espèces nouvelles de Trichoptères), écologie, intérêt biogéographique. *Bijdr. Tot Dierkd.* **1984**, *54*, 83–100. [CrossRef]
83. Ouahsine, H.; Lavandier, P.; Céréghino, R. Influence of temperature and macrophytes development on the larval population dynamics of Epeorus sylvicolaPict (Ephemeroptera) in a torrential river of the «Haut-Atlas de Marrakech» (Morocco). *Ann. Limnol. Int. J. Limnol.* **1996**, *32*, 27–31. [CrossRef]
84. Soldán, T.; Gagneur, J. Ecdyonurus rothschildi Navás, 1929: Description de la larve (Ephemeroptera, Heptageniidae). In *Annales de Limnologie-International Journal of Limnology*; EDP Sciences: Les Ulis, France, 1985; Volume 21, pp. 141–144.
85. Boumaïza, M.; Thomas, A.G.B. Distribution and ecological limits of Baetidae vs the other mayfly families in Tunisia: A first evaluation (Insecta, Ephemeroptera). *Bull. Soc. Hist. Nat.* **1995**, *131*, 27–33.
86. Eaton, A.E. List of Ephemeridae hitherto observed in Algeria, with localities. *Ent. Month. Mag.* **1899**, *35*, 4–5.
87. Badri, A. Influence des Crues sur les Ecosystèmes Lotiques du Haut Atlas: Étude des Perturbations et des Mécanismes de Recolonisation à Travers les Peuplements d'Algues et d'Invertébrés. Ph.D. Thesis, Cadi Ayyad University, Marrakech, Morocco, 1993.
88. Touabay, M.; Aouad, N.; Mathieu, J. Etude hydrobiologique d'un cours d'eau du Moyen-Atlas: L'oued Tizguit (Maroc). *Ann. Limnol.* **2002**, *38*, 65–80. [CrossRef]
89. Benhadji, N. Caractérisation taxonomique, phylogénétique des Ephéméroptères Schistonotes du bassin versant de la Tafna (Ouest algérien). Ph.D. Thesis, Faculté des sciences de la Nature et de la Vie et des Sciences de la Terre et de l'Université Abou Bekr Belkaid, Tlemcen, Algeria, 2020; p. 159.
90. Alba-Tercedor, J.; EL Alami, M. Description of the Nymphs and Eggs of Acentrella almohades sp. n. from Morocco and Southern Spain (Ephemeroptera: Baetidae). *Aquat. Insects* **1999**, *21*, 241–247. [CrossRef]
91. Peru, N.; Thomas, A. Complements et corrections a la faune des ephemeropteres d'Afrique du Nord. 7. Description complementaire de Baetis berberus Thomas, 1986 (Ephemeroptera, Baetidae). *Ephemera* **2001**, *3*, 75–82.
92. Godunko, R.J.; Soldán, T.; Staniczek, A.H. Baetis (Baetis) cypronyx sp. n., a new species of the Baetis alpinus species-group (Insecta, Ephemeroptera, Baetidae) from Cyprus, with annotated checklist of Baetidae in the Mediterranean islands. *ZooKeys* **2017**, *644*, 1–32.
93. Múrria, C.; Bonada, N.; Vellend, M.; Zamora-Muñoz, C.; Alba-Tercedor, J.; Sainz-Cantero, C.E.; Garrido, J.; Acosta, R.; Alami, M.E.; Barquín, J.; et al. Local environment rather than past climate determines community composition of mountain stream macroinvertebrates across Europe. *Mol. Ecol.* **2017**, *26*, 6085–6099. [CrossRef]
94. Ubero-Pascual, N.A.; Soler, A.G.; Puig, M.A. Primera cita de Baetis punicus Thomas, Boumaiza & Soldán, 1983 (Ephemeroptera: Baetidae) para el continente europeo. *Bol. Asoc. Esp. Entomol.* **1996**, *20*, 128.
95. Müller-Liebenau, I. Ephemeroptera (Insecta) von den Kanarischen Inseln. *Gewässer Abwässer* **1971**, *50*, 7–40.
96. Williams, H.C.; Ormerod, S.J.; Bruford, M.W. Molecular systematics and phylogeography of the cryptic species complex Baetis rhodani (Ephemeroptera, Baetidae). *Mol. Phylogenetics Evol.* **2006**, *40*, 370–382. [CrossRef] [PubMed]
97. Soldán, T.; Godunko, R.; Thomas, A. Baetis chelif n. sp., a new mayfly from Algeria with notes on B. sinespinosus Soldán & Thomas, 1983, n. stat. (Ephemeroptera: Baetidae). *Genus* **2005**, *16*, 155–165.
98. Bauernfeind, E.; Soldán, T. *The mayflies of Europe (Ephemeroptera)*; Apollo Books: Ollerup, Denmark, 2012; p. 783.
99. Rutschmann, S.; Gattolliat, J.-L.; Hughes, S.J.; Báez, M.; Sartori, M.; Monaghan, M.T. Evolution and island endemism of morphologically cryptic *Baetis* and *Cloeon* species (Ephemeroptera, Baetidae) on the Canary Islands and Madeira. *Freshw. Biol.* **2014**, *59*, 2516–2527. [CrossRef]
100. Rutschmann, S.; Detering, H.; Simon, S.; Funk, D.H.; Gattolliat, J.-L.; Hughes, S.J.; Raposeiro, P.M.; DeSalle, R.; Sartori, M.; Monaghan, M.T. Colonization and diversification of aquatic insects on three Macaronesian archipelagos using 59 nuclear loci derived from a draft genome. *Mol. Phylogenetics Evol.* **2017**, *107*, 27–38. [CrossRef]

101. Yano, K.; Takenaka, M.; Tojo, K. Genealogical Position of Japanese Populations of the Globally Distributed Mayfly Cloeon dipterum and Related Species (Ephemeroptera, Baetidae): A Molecular Phylogeographic Analysis. *Zool. Sci.* **2019**, *36*, 479–489. [CrossRef]
102. Verrier, M.L. Ephéméroptères récoltés par M. Paul Rémy au Hoggar et au Tidikelt. *Bull. Soc. Zool. Fr.* **1952**, *77*, 292–304.
103. Zrelli, S.; Boumaiza, M.; Bejaoui, M.; Gattolliat, J.L.; Sartori, M. New data and revision of the Ephemeroptera of Tunisia. *Inland Water Biol.* **2016**, *3*, 99–106.
104. Thomas, A.G.B. Liste faunistique des Éphémères de France. 2021. Available online: https://www.opie-benthos.fr/opie/pages_dyna.php?idpage=751.pdf (accessed on 30 May 2022).
105. Alba-Tercedor, J.; Jáimez-Cuéllar, P. Checklist and historical evolution of the knowledge of Ephemeroptera in the Iberian Peninsula, Balearic and Canary Islands. In Proceedings of the 10th International Conference on Ephemeroptera & Plecoptera, Perugia, Italy, 8–11 August 2001; Research Update on Ephemeroptera & Plecoptera 2003. Gaino, E., Ed.; University of Perugia: Perugia, Italy, 2003; pp. 91–97.
106. Gutiérrez-Cánovas, C.; Arribas, P.; Naselli-Flores, L.; Bennas, N.; Finocchiaro, M.; Millán, A.; Velasco, J. Evaluating anthropogenic impacts on naturally stressed ecosystems: Revisiting river classifications and biomonitoring metrics along salinity gradients. *Sci. Total Environ.* **2018**, *658*, 912–921. [CrossRef]
107. Zrelli, S.; Boulaaba, S.; Bejaoui, M.; Boumaïza, M.; Sartori, M. Description et répartition de Potamanthus luteus Linnaeus 1767 (Ephemeroptera, Potamanthidae) en Tunisie. *Entomol. Faun. Faun. Entomol.* **2015**, *68*, 223–228.
108. Sartori, M.; Brittain, J. Order Ephemeroptera. In *Thorp and Covich's Freshwater Invertebrates: Ecology and General Biology*, 4th ed.; Academic Press: Cambridge, MA, USA, 2015; pp. 873–891.
109. Lounaci, A.; Brosse, S.; Thomas, A.; Lek, S. Abundance, diversity and community structure of macroinvertebrates in an Algerian stream: The Sébaou wadi. *Ann. Limnol. Int. J. Lim.* **2000**, *36*, 123–133. [CrossRef]
110. Mebarki, M.; Taleb, A.; Arab, A. Environnemental factors influencing the composition and distribution of Mayfly larvae in Northern Algerian Wadis (Regional Scale). *Nouv. Rev. Entomol.* **2017**, *8*, 89–96.
111. Yalles-Satha, A.; Alami, M.E.; Kechemir, L.H.; Desvilettes, C.; Chenchouni, H. Diversity, phenology and distribution of mayfly larvae (Ephemeroptera) along an altitudinal gradient in two permanent Wadis of Algeria. *Orient. Insects* **2022**, *56*, 14–46. [CrossRef]
112. Cheylan, G. Endémisme et spéciation chez les Mammifères méditerranéens. *Vie Milieu Life Environ.* **1990**, *40*, 137–143.
113. Bennas, N.; L'mohdi, O.; El-Haissoufi, M.; Charfi, F.; Ghlala, A.; El-Alami, M. New data on the aquatic insect fauna of Tunisia. *Trans. Am. Entomol. Soc.* **2018**, *144*, 575–592. [CrossRef]
114. Benamar, L.; Bennas, N.; Sánchez, A.M. Les Coléoptères aquatiques du Parc National de Talassemtane (Nord-Ouest du Maroc): Biodiversité, degré de vulnérabilité et état de conservation. *Bol. SEA* **2011**, *49*, 231–242.
115. Errochdi, S. Biodiversité, Typologie et Qualité des Eaux de Deux Réseaux Hydrographiques Nord Marocains Laou et Tahaddart et Taxonomie, Biogéographie et Atlas des Pléocoptères du Maroc. Ph.D. Thesis, Université Abdelmalek Essaadi, Tetouan, Morocco, 2015; p. 495.
116. Benson, R.H.; Bied, K.R.-E.; Bonaduce, G. An important current reversal (influx) in the Rifian Corridor (Morocco) at the Tortonian-Messinian boundary: The end of Tethys Ocean. *Paleoceanography* **1991**, *6*, 165–192. [CrossRef]
117. Dercourt, J.; Zonenshain, L.; Ricou, L.-E.; Kazmin, V.; Le Pichon, X.; Knipper, A.; Grandjacquet, C.; Sbortshikov, I.; Geyssant, J.; Lepvrier, C.; et al. Geological evolution of the tethys belt from the atlantic to the pamirs since the LIAS. *Tectonophysics* **1986**, *123*, 241–315. [CrossRef]
118. Hsü, K.J.; Ryan, W.B.F.; Cita, M.B. Late Miocene Desiccation of the Mediterranean. *Nature* **1973**, *242*, 240–244. [CrossRef]
119. Krijgsman, W.; Hilgen, F.J.; Raffi, I.; Sierro, F.; Wilson, D.S. Chronology, causes and progression of the Messinian salinity crisis. *Nature* **1999**, *400*, 652–655. [CrossRef]
120. Trájer, A.J.; Sebestyén, V.; Padisák, J. The impacts of the Messinian Salinity Crisis on the biogeography of three Mediterranean sandfly (Diptera: Psychodidae) species. *Geobios* **2021**, *65*, 51–66. [CrossRef]
121. Lavergne, S.; Hampe, A.; Arroyo, J. In and out of Africa: How did the Strait of Gibraltar affect plant species migration and local diversification? *J. Biogeogr.* **2013**, *40*, 24–36. [CrossRef]
122. Jaskuła, R. The Maghreb—One more important biodiversity hot spot for tiger beetle fauna (Coleoptera, Carabidae, Cicindelinae) in the Mediterranean region. *ZooKeys* **2015**, *482*, 35–53. [CrossRef] [PubMed]
123. Quade, J.; Dente, E.; Armon, M.; Ben Dor, Y.; Morin, E.; Adam, O.; Enzel, Y. Megalakes in the Sahara? A Review. *Quat. Res.* **2018**, *90*, 253–275. [CrossRef]

MDPI

Article

Taxonomy, Distribution and Life Cycle of the Maghrebian Endemic *Rhithrogena sartorii* (Ephemeroptera: Heptageniidae) in Algeria

Boudjéma Samraoui [1,2,*], Laurent Vuataz [3,4], Michel Sartori [3,4], Jean-Luc Gattolliat [3,4], Fahad A. Al-Misned [5], Hamed A. El-Serehy [5] and Farrah Samraoui [1,6]

1 Laboratoire de Conservation des Zones Humides, Université 8 Mai 1945 Guelma, Guelma 24000, Algeria; fsamraoui@gmail.com
2 Department of Biology, University Badji Mokhtar Annaba, Annaba 23000, Algeria
3 Museum of Zoology, Palais de Rumine, Place Riponne 6, CH-1014 Lausanne, Switzerland; laurent.vuataz@vd.ch (L.V.); michel.sartori@vd.ch (M.S.); jean-luc.gattolliat@vd.ch (J.-L.G.)
4 Department of Ecology and Evolution, Biophore University of Lausanne, CH-1015 Lausanne, Switzerland
5 Department of Zoology, College of Science, King Saud University, P.O. Box 2455, Riyadh 11451, Saudi Arabia; almisned@ksu.edu.sa (F.A.A.-M.); helserehy@ksu.edu.sa (H.A.E.-S.)
6 Department of Ecology, Université 8 Mai 1945 Guelma, Guelma 24000, Algeria
* Correspondence: bsamraoui@gmail.com

Abstract: Despite being recorded in Algeria since the nineteenth century, the genus *Rhithrogena* has never been the object of a taxonomical study and no identified species is known from this country. Investigations of the relict mountain streams of El Kala, north-eastern Algeria, have led to the discovery of a *Rhithrogena* population. Morphological and molecular analyses identified the species as the Maghrebian endemic *Rhithrogena sartorii*, so far known only from neighboring Tunisia. We report on the species' distribution, status, and life cycle and discuss its potential role as a bioindicator in environmental monitoring.

Keywords: aquatic insects; conservation; life cycle; limnology; mayfly; North Africa; rivers; streams

Citation: Samraoui, B.; Vuataz, L.; Sartori, M.; Gattolliat, J.-L.; Al-Misned, F.A.; El-Serehy, H.A.; Samraoui, F. Taxonomy, Distribution and Life Cycle of the Maghrebian Endemic *Rhithrogena sartorii* (Ephemeroptera: Heptageniidae) in Algeria. *Diversity* **2021**, *13*, 547. https://doi.org/10.3390/d13110547

Academic Editor: Michael Wink

Received: 29 September 2021
Accepted: 25 October 2021
Published: 29 October 2021

Publisher's Note: MDPI stays neutral with regard to jurisdictional claims in published maps and institutional affiliations.

1. Introduction

Rhithrogena, a member of the subfamily Rhithrogeniinae (Heptageniidae), is a Holarctic genus with numerous species in the Palearctic region, occupying mainly cold, fast-flowing, and well-oxygenated headwaters [1,2]. Furthermore, isolated populations of *Rhithrogena* in mountainous rivers and streams display high levels of endemism and are often on the IUCN Red List [3].

Despite its ecological importance, the taxonomic status of many *Rhithrogena* species remains a challenge, even in Europe, where taxonomic studies of mayflies are relatively well advanced [4,5]. Based on various nymphal and adult characters, species are lumped into "species groups" [6,7]. However, this grouping remains controversial, marred by cryptic diversity and taxonomic oversplitting [4,5].

The first record of *Rhithrogena* in North Africa occurred at the edge of the Sahara, when, on 19 March 1895, Eaton [8] collected an immature male at Biskra, Algeria. Eaton went on to speculate that the species might have flown south from the deep canyons descending from the Aures Mountains. This specimen and others encountered in various localities across the Maghreb remained unidentified for several decades until the almost synchronous descriptions from Morocco of a series of new species: *Rh. ourika* (High Atlas: 1500 and 2600 m) [9], *Rh. ayadi* (Middle Atlas: 2150 m) [10], *Rh. giudicelliorum* (High Atlas: 2800 m) [11], and *Rh. ryszardi* (Middle Atlas: 1260 m) [12].

Subsequently, Vitte [13] described an additional species, *Rh. mariae*, from the Moroccan Rif. In particular, *Rh. mariae* differed from other North African *Rhithrogena* species by occurring at a much lower altitude (160 m). Finally, two decades later, Zrelli et al. [14] described a new *Rhithrogena* species, *Rh. sartorii* from Tunisia, which gives a total of six known *Rhithrogena* species in North Africa. Most of the described Maghrebian *Rhithrogena* species are only known as imagos, whereas *Rh. mariae* is known at the nymphal and adult stages, and *Rh. sartorii* at the nymphal and subimaginal stages.

As part of a long-term limnological survey of Algeria, we collected mayfly nymphs from various regions of the country [15], and, in this study, we report the discovery of *Rh. sartorii* in the relict mountain streams of El Kala, the first record for Algeria, and provide information on its distribution. Because knowledge of immature stages, voltinism, and larval growth patterns provide insights into basic life-history traits and is essential to developing and implementing appropriate conservation strategies [16], we also identified the last three nymphal stages and inferred the species' life cycle.

2. Materials and Methods

2.1. Study Area

The Algero-Tunisian border is flanked on its northern part by a mountain range known as Kroumiria, where the Kebir-East River emanates. The watershed of the Kebir-East River is second in size only to the Seybouse River in north-eastern Algeria. Further north, the Oued el Eurg basin drains the hills sandwiched between Kroumiria and the Mediterranean Sea (Figure 1). The climate is typically Mediterranean, with an alternating hot, dry period (May–October) and a rainy season (November–April).

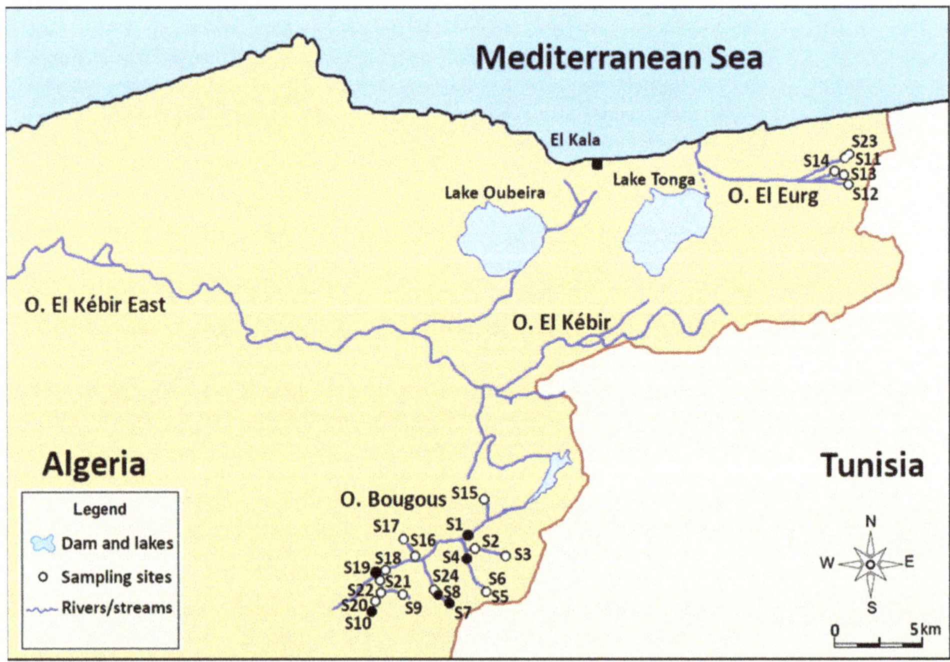

Figure 1. Study area with sampling sites. Dark circles indicate localities where *Rhithrogena sartorii* has been recorded.

2.2. Sampling

A set of 24 localities (S1-S24), distributed across both O. Bougous, the main tributary of O. El Kebir, and the O. El Eurg watershed, were sampled monthly from November 2018 to June 2021 [15,17]. Mayfly nymphs were collected using a dipnet (500 µm mesh size, 35 cm diameter) and by walking slowly and repeatedly across all micro-habitats (aquatic vegetation, rocks, leaf litters, riffles, runs, pools, flats, etc.) for ten minutes at each locality, as described in [18–20].

2.3. Molecular Analyses

To complement morphological examinations, we compared mitochondrial DNA sequences of specimens from this study to Tunisian topotype specimens. Specifically, we generated a 658-bp fragment of the cytochrome c oxidase subunit I (COI) gene from five newly-sequenced specimens (two from Algeria and three from the type locality in Tunisia) using LCO1490 and HCO2198 primers [21]. For all specimens, we followed the non-destructive DNA extraction procedure described in [4]. The DNA was extracted using the BioSprint 96 extraction robot (Qiagen Inc., Hilden, Germany). Polymerase Chain Reaction (PCR) was conducted in a volume of 30 μL, consisting of 9 μL (unknown concentration) of template DNA, 1.5 μL (10 μM) of each primer, 0.24 μL (25 mM) of dNTP solution (Promega, Madison, WI, USA), 6 μL of 10X buffer (Promega, Madison, WI, USA) containing 7.5 mM of $MgCl_2$, 3 μL (25 mM) of $MgCl_2$, 1.5 U of Taq polymerase (Promega, Madison, WI, USA), and 8.46 μL of sterile ddH_2O. Optimized PCR conditions included initial denaturation at 95 °C for 5 min, 38 cycles of denaturation at 95 °C for 40 s, annealing at 50 °C for 40 s, and extension at 72 °C for 40 s, with a final extension at 72 °C for 7 min. The purification and automated sequencing were carried out in Microsynth (Balgach, Switzerland). We further included one published COI sequence from [5], also corresponding to a topotype specimen (Table 1). The sequences were aligned using MAFFT [22] as implemented in Jalview 2.11.1.4 [23]. MEGAX [24,25] was used to visualize the alignment, calculate the number of variable sites, define two groups (one group with the two sequences from Algeria, one group with the four topotype sequences from Tunisia), and calculate K2P [26] mean distances within and between groups.

Table 1. Codes and origin of specimens examined in the COI analysis. For each specimen, the GBIF code, the sampling information (country, locality, coordinates, and date of sampling), the GenBank accession number of the COI sequence, and the corresponding publication source are provided. All specimens from Tunisia are from the type locality (topotypes).

GBIF Code	Country	Locality	Latitude	Longitude	Date	GenBank ID	Source
GBIFCH00671210	Tunisia	Ennour	36.8018	8.6568	28.IV.2010	LN868554	Vuataz et al. (2016)
GBIFCH00671211	Tunisia	Ennour	36.8018	8.6568	28.IV.2010	MZ433256	This study
GBIFCH00671212	Tunisia	Ennour	36.8018	8.6568	28.IV.2010	MZ433257	This study
GBIFCH00671213	Tunisia	Ennour	36.8018	8.6568	28.IV.2010	MZ433258	This study
GBIFCH00673108	Algeria	Guitna inf	36.6379	8.3652	06.VI.2019	MZ433260	This study
GBIFCH00673114	Algeria	Guitna inf	36.6181	8.3462	06.VI.2019	MZ433259	This study

2.4. Morphometry

Rhithrogena nymphs from two localities, Guitna sup (S7) (Figure 2) and Guitna inf (S8), were selected for measurements. With one exception (see results), measurements were lumped together after inspection of density plots, and Mann–Whitney U tests did not reveal any differences between the two localities. Body length (BL), head width (HW), and length of the mesonotum + wing pad (mn + wsl) were measured using a Precision Steel Rule to the nearest 0.1 mm. The criteria for instar assignment were BL, HW, mn + wsl, and the ratio (mn + wsl)/HW, hereafter referred to as the "Ratio" [27]. Only the last three instars (F-0, F-1, and F-2) were identified; all other stadia were designated as "smaller nymphs". Instars were determined through graphical plots and statistical analyses. The sex of each nymph in the last two instars was determined according to the presence (male) or absence of genital forceps (gonostyli) on the ventral surface of the ninth abdominal segment. Presence in F-0 nymphs of dark wing pads was assumed as evidence of imminent emergence.

2.5. Statistical Analysis

A fast, density-based clustering analysis of BL, HW, mn + wsl, and Ratio was performed using DBSCAN (density-based spatial clustering of applications with noise) [28] to identify the last three instars, F-0, F-1, and F-2. The algorithm attempts to identify the structure in the spatial data set by aggregating objects into similar subgroups [29]. All statistical tests were conducted using R software [30].

Figure 2. View of Guitna sup., a typical habitat of *Rhithrogena sartorii* during winter (**a**) and summer (**b**).

3. Results

3.1. Distribution and Phenology

During the study period, *Rhithrogena sartorii* nymphs were recorded between January and June at six localities: Pont Bougous (S1), Zitoun Meftah (S4), Guitna sup (S7), Guitna inf (S8), Nouazi (S10), and Kherrata (S19). Nymphs were found in streams that had a substrate made up of cobbles, stones, and boulders and in microhabitats with a relatively cold, fast-flowing current. The nymphal growth and development occurred in winter and spring with marked differences between years: Both in 2019 and 2021, nymphs were first recorded in March, whereas, in 2020, nymphs were first collected in January. In all years, no nymphs were recorded beyond June (Figure 3).

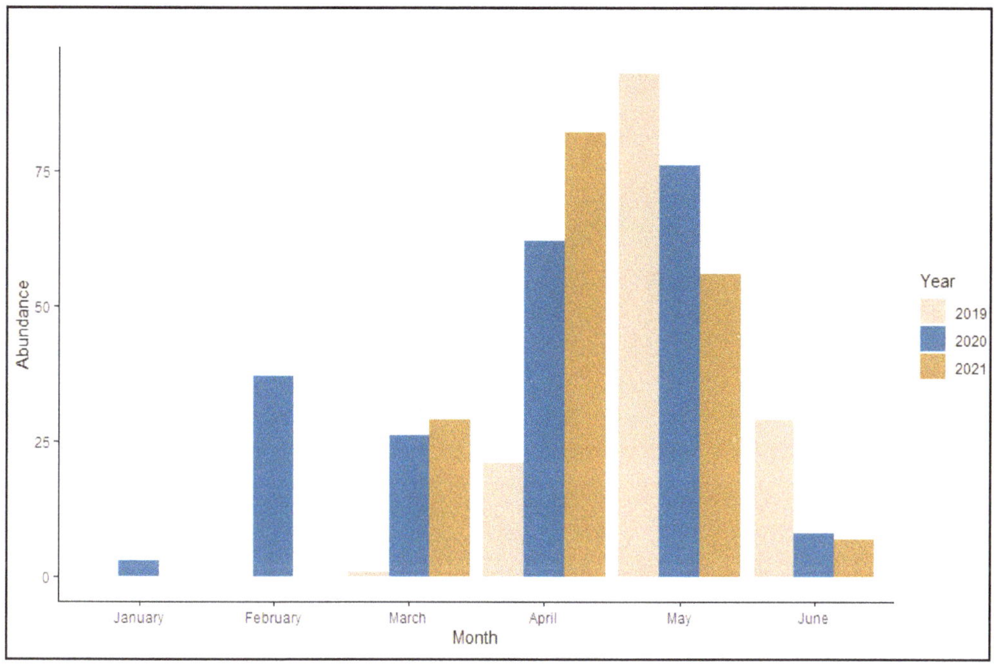

Figure 3. Histogram of the number of *Rhithrogena sartorii* specimens sampled each year in the study area (January 2019–June 2021).

3.2. Taxonomy

Rhithrogena sartorii nymphs are characterized by the following combination of characters: (1) All gills are crenulated (Figure 4A–C); (2) compared to *Rh. insularis* (Figure 5A), the plica of the dorsal face of the first gill is well expressed, clearly triangular, the leading edge somewhat concave (Figure 5B); (3) the lateral sclerites of the first sternite are slightly turned backward, sometimes perpendicular to the body axis (Figure 4C); (4) the upper face of femora of all legs has a well-expressed rounded blackish hypodermal macula (Figure 4B); (5) the crown of the galea-lacinia has 9–11 comb-shaped bristles, each with 6–7 teeth (Figure 5C).

3.3. Molecular Analyses

There were no missing data, gaps, or ambiguous sites in the COI alignment, and a total of four variable sites were recorded. The K2P mean distances within groups were 0.25% and 0.15% for Tunisian (topotype) and Algerian sequences, respectively. The K2P mean distance between groups was 0.23% (maximum distance: 0.61%). Two Tunisians and one Algerian specimen shared the same COI haplotype.

Figure 4. *Rhithrogena sartorii*, habitus in dorsal view (**A**), lateral view (**B**), and ventral view (**C**). The arrow points to the lateral sclerites of the first sternite. Scale bar: 1 mm.

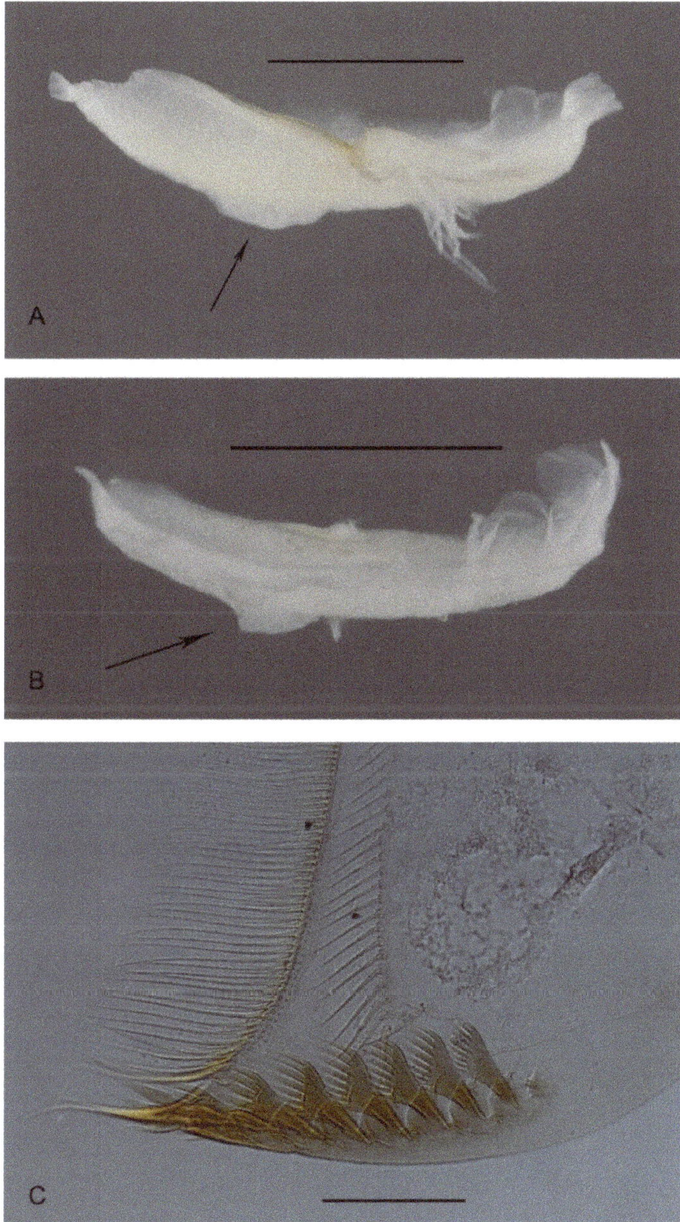

Figure 5. *Rhithrogena* spp. Latero-dorsal view of the first gill to emphasize the shape of the plica (arrow) in *Rh. insularis* (**A**) and *Rh. sartorii* (**B**), Scale bar: 1 mm. Crown of the galea-lacinia of *Rh. sartorii* (**C**). Scale bar 0.1 mm.

3.4. Morphometry

A total of 524 nymphs were measured. Overall, nymphal body length (BL) ranged from 2.0 to 10.8 mm, while the ranges of head width (HW) and mesonotum length + wing pad length (mn + wsl) were 0.8–3.4 and 0.2–5.2 mm, respectively. Females' BL were marginally longer than males' (one-way ANOVA: $F_{1,164} = 3.6$, $p = 0.06$). All other measured morphometric characters did not differ between the sexes. Assignment of the last three final instars suggested the presence of three clusters corresponding to F-0: Ratio $\geq$ 1.1, F-1: 1.1 > Ratio $\geq$ 0.75, F-2: 0.75 > Ratio $\geq$ 0.5. The rest may be grouped into the category "smaller instars" (Figure 6a). The allometric growth of wing pads (mn + wsl) at the F-0 instar contrasting sharply with the isometric growth of BL and HW (Figure 6a,b) is noteworthy.

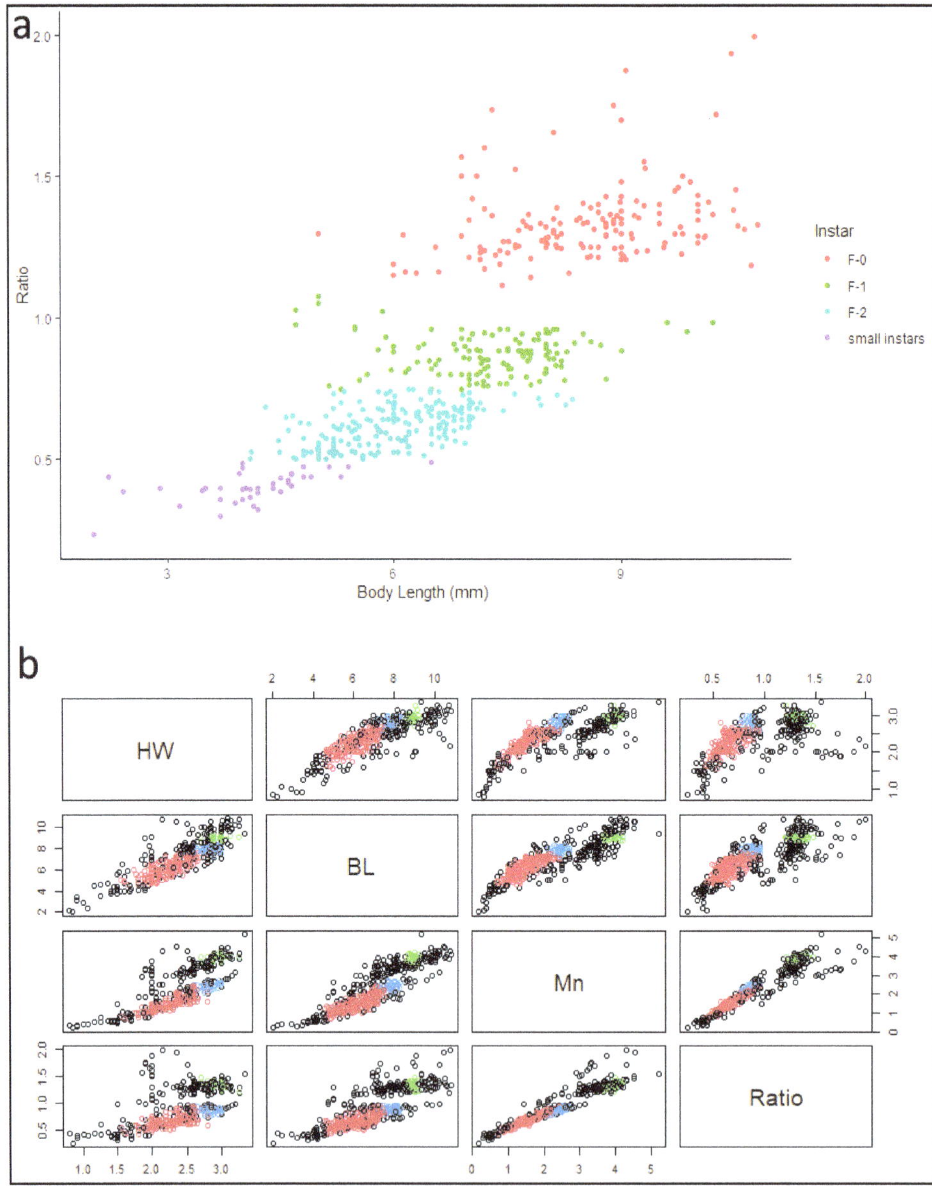

Figure 6. (**a**) Plot of Ratio (mesonotum length + wing pad length (mn + wsl)/head width (HW)) versus body length (BL) showing the assignment of F-0 (green), F-1 (blue), and F-2 (red); (**b**) Multiplot of the DBSCAN clustering indicating three classes (colored dots) corresponding to the last three nymphal instars. BL, HW, and Mn (mn + wsl) units are in mm.

The density-based clustering algorithm, DBSCAN, using the following parameters: eps (maximum radius between two neighbors belonging to the same cluster) = 0.35, and MinPts (minimum number of neighbors required to form a cluster) = 15, and the variables: BL, HW, mn + wsl, and Ratio, confirmed the preliminary visual inspection by assigning 318 nymphs into three clusters corresponding to F-0, F-1, and F-2 (Figure 6b). The rest (206) corresponded to "smaller nymphs" and noise.

3.5. Life Cycle

Both in 2019 and 2021, the nymphs first appeared in March, and development proceeded quickly with F-0 nymphs with pigmented wing pads occurring from April to June (Figure 7a–d). In 2020, nymphal development was more protracted, with nymphs collected from January to June, but in all three years, the emergence spanned April to June, coinciding with the drying up of streams in early summer. The size difference between the two sampling sites in May 2020 (BL: $\log_{e\,(\text{Wmann-Whitney})} = 5.44$, N = 76, $p = 2.09 \times 10^{-6}$; HW:$\log_{e\,(\text{Wmann-Whitney})} = 5.20$, N = 76, $p = 1.19 \times 10^{-7}$) and the persistence of "small instars" nymphs in June 2020 at Guitna inf (S8) while Guitna sup (S7) dried up is noteworthy.

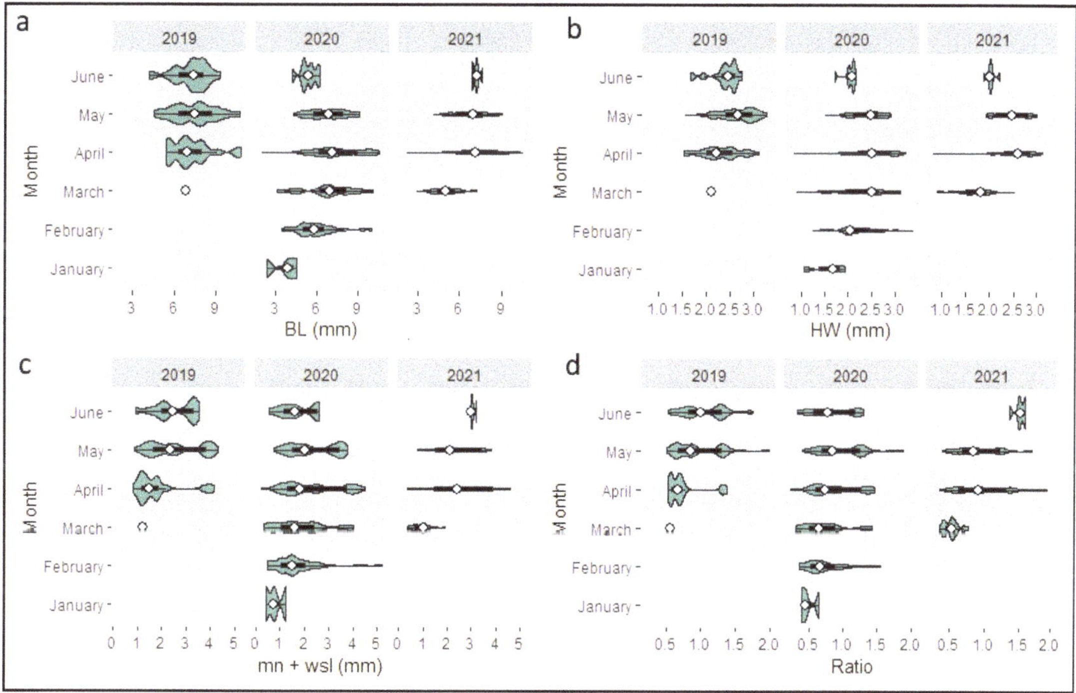

Figure 7. Seasonal changes in size–frequency distribution of *Rhithrogena sartorii* (2019–2021) for BL (**a**), HW (**b**), mn + wsl (**c**), and Ratio (**d**).

4. Discussion

4.1. Distribution

Both in Tunisia [14] and in Algeria, the distribution of *Rhithrogena sartorii* was restricted to the Kroumiria mountain range. Overall, the species seems confined to the metarhithral and parapotamal river reaches. The lower end of the altitudinal range of the species' habitats (200–650 m), almost matching *Rh. mariae,* which is able to colonize lower stretches (160 m) in Morocco, is noteworthy [13].

4.2. Taxonomy

Recently described from Tunisia, *Rh. sartorii* was thought to belong to the *insularis*-species group [14]. However, after preliminary investigations reported in [5], a careful re-examination of these nymphs conducted here confirmed that this species is more related to species of the so-called *sowai*-group [6]. Specifically, the shape of the plica, which is clearly concave (always less prominent and convex in species of the *hybrida*-group—Figure 5A), and the lateral sclerites of the first sternite, which can be slightly turned backward (always perpendicular to body axis in *hybrida*-group). Species of the *sowai*-group are poorly known; seven species have been described with only one in the nymphal stage [31], although an unnamed species has been described at the nymphal stage from Portugal [32]. All of these species are restricted to the Mediterranean basin. Finally, the lack of true affinities with species of the *hybrida*-group is demonstrated by the quite isolated position *Rh. sartorii* occupies in the phylogeny of European species of the genus based on mitochondrial and nuclear markers [5].

With a maximum of 0.61%, the COI K2P distances between sequences from Algeria and topotypes from Tunisia are very low, typically corresponding to intraspecific divergence in previous mayfly barcoding studies (e.g., [33–36]). Moreover, given that one of our specimens has the same COI haplotype as the two topotype specimens and that the combination of morphological characters fully fits the *Rh. sartorii* description, we can be confident in our identification. This is not surprising, as both populations are located in the same mountain range (Kroumiria), only c. 30 km distant from each other.

4.3. Eaton's Rhithrogena

In contrast to the relatively straightforward identification of *Rh. sartorii*, Eaton's *Rhithrogena* still remains shrouded in mystery. Unless the specimen is retrieved, we will probably never know which species of *Rhithrogena* Eaton [8] collected in Algeria, but we can safely rule out many of the known North African species on the basis of their limited distribution and ecology. Indeed, four species (*Rh. ourika*, *Rh. ayadi*, *Rh. giudicelliorum*, and *Rh. ryszardi*) occupy an altitudinal range between 1260 and 2800 m. Likewise, based on the location, Biskra, where Eaton has recorded the specimen, we can safely assume that the species was able to stand high temperatures in one phase of its life cycle. North African rheophilic *Rhithrogena* species, such as *Rh. mariae*, are present at low altitudes, but due to their localized distribution in the Rif, this latter is unlikely to represent a good candidate (but see [13]) for Eaton's *Rhithrogena*, which once inhabited the edge of the desert.

In addition, based on the flight period (late winter) of Eaton's specimen, we can also exclude that it was *Rh. sartorii* that emerges in late spring. The delayed nymphal growth and development of *Rh. sartorii* are suggestive of a univoltine winter/spring life cycle, whereas nymphal development in Eaton's species probably occurred in late autumn and winter, similar to the life cycle of *Rh. germanica* [37,38]. Based on all these elements and the extensive surveys of mayflies of the Aures Mountains (unpublished), we hypothesize that Eaton's *Rhithrogena* has probably gone extinct.

4.4. Life Cycle

Nymphal development of *Rh. sartorii* occurred during winter and spring, but there was considerable annual variation (January–March) in the first records of nymphs, probably linked to the close relationship between egg development and water temperature [39,40]. For instance, eggs of the cold stenothermal *Rh. loyolaea* and *Rh. nivata* rarely hatch at temperatures above 10 °C, thus restricting the species to cold streams [41,42]. If this was the case for *Rh. sartorii*, this threshold would limit hatching to winter months. In addition, once hatching is underway, the differential growth rates in small nymphs may be responsible for the extended period of their presence [41].

Rhithrogena sartorii managed a single generation per year, with the nymphal stage spread over winter and spring. According to Clifford's classification [43], the species exhibited a seasonal univoltine cycle (Us-Uw) where the egg stage and part of the nymphal stage overwinters. Nevertheless, the univoltine life cycle of *Rh. sartorii* is quite unusual, with a winter and vernal growth and a long embryonic diapause during the warm months. It is somewhat distinct from the life cycle of *Rh. germanica*, a univoltine winter species, which emerges in Central Europe between February and April, undergoes a summer embryonic diapause with eggs hatching once the temperature drops in October [37].

Furthermore, the presence of small nymphs of *Rh. sartorii* in June may either suggest a protracted egg hatching period or a proclivity for the species to undertake a second generation if environmental conditions are adequate. In all three years, the habitats dried up, and thus, this question deserves further investigation. Although *Rhithrogena* species are known to be mainly univoltine [43], plasticity in voltinism has been demonstrated, ranging from semivoltinism for *Rh. loyalae* [44] to partial bivoltinism [45], and even bivoltinism [46] for *Rh. semicolorata* and *Rh. diaphana*, respectively.

4.5. Conservation

Although species of the genus *Rhithrogena* may be perceived as less threatened, as their rhithral habitats may be contending with lesser anthropogenic pressures than downstream habitats, they are highly sensitive to various environmental factors [47,48]. In addition, due to historical factors (transboundary region and previous war zone), the El Kala district has been relatively maintained as a hotspot of freshwater biodiversity. However, despite its status as a Man and the Biosphere Reserve, the area is now under severe anthropogenic pressures fueled by a burgeoning population [49,50]. With its restricted distributional range encompassing the Tunisian and Algerian Kroumiria, *Rh. sartorii* is clearly an endemic species of conservational concern. Moreover, in most sites and during the three-year study, the species was never abundant. Thus, the limited range, low abundance, and narrow ecological niche (rheobiont associated with riffles) make this threatened species and its habitat vulnerable to various natural and anthropogenic stressors (climate change, pollution, land conversion, etc.). *Rhithrogena sartorii* may act as a useful bioindicator of such scarce habitats and an umbrella species for the conservation of the unique freshwater biodiversity hosted by the Kroumiria mountain range that spans north-eastern Algeria and north-western Tunisia [15,51,52]. Unless urgent steps are taken to lessen human encroachment on its habitats, this imperiled Maghrebian microendemic may rapidly go extinct.

5. Conclusions

A survey of the highlands of the El Kala region, north-eastern Algeria, has led to the discovery of a species of *Rhithrogena* that occupied the hyporhithral and parapotamal river reaches. Molecular and morphological analyses identified the species as *Rh. sartorii*, a Maghrebian microendemic confined to the Kroumiria mountain range and environs on the Algero-Tunisian border. The species exhibited a univoltine life cycle (Us-Uw) with emergence spread between April and June. *Rhithrogena sartorii* is threatened due to the species' limited range and the mounting anthropogenic pressures (water abstraction, fire, pollution, etc.) in the region.

Author Contributions: Conceptualization, B.S. and F.S.; methodology, B.S. and L.V.; software, B.S. and L.V. validation, M.S.; formal analysis, B.S. and L.V.; investigation, B.S., M.S. and L.V.; resources, M.S., F.A.A.-M. and H.A.E.-S.; data curation, J.-L.G.; writing—B.S. and L.V.; writing—review and editing, all co-authors; visualization, B.S. and L.V.; supervision, B.S.; project administration, F.S.; funding acquisition, F.A.A.-M. and H.A.E.-S. All authors have read and agreed to the published version of the manuscript.

Funding: This research was funded by King Saud University, Riyadh, Saudi Arabia, through the Researchers Supporting Project Number (RSP-2021/19) and the APC was funded by Biophore University of Lausanne.

Institutional Review Board Statement: The study was conducted according to the guidelines of the Declaration of Helsinki, and approved by the Algerian Ministère de l'Enseignement Supérieur et de la Recherche Scientifique (M.E.S.R.S.).

Data Availability Statement: Data is available from the corresponding author upon request.

Acknowledgments: We are indebted to two anonymous reviewers for helpful comments. We are also most grateful to Manuel Ferreras-Romero for logistic support. Sonia Zrelli (Bizerte) graciously deposited the Tunisian specimens in the collections of the Museum of zoology, Lausanne.

Conflicts of Interest: The authors declare no conflict of interest.

References

1. Barber-James, H.M.; Gattolliat, J.-L.; Sartori, M.; Hubbard, M.D. Global diversity of mayflies (Ephemeroptera, Insecta) in freshwater. *Hydrobiologia* **2007**, *595*, 339–350. [CrossRef]
2. Webb, J.M.; McCafferty, W.P. Heptageniidae of the world. Part II. Key to the genera. *Can. J. Arthropod Identif.* **2008**, *7*, 1–55.
3. Jacobus, L.M.; Macadam, C.R.; Sartori, M. Mayflies (Ephemeroptera) and their contributions to ecosystem services. *Insects* **2019**, *10*, 170. [CrossRef] [PubMed]
4. Vuataz, L.; Sartori, M.; Wagner, A.; Monaghan, M.T. Toward a DNA taxonomy of Alpine *Rhithrogena* (Ephemeroptera: Heptageniidae) using a mixed yule-coalescent analysis of mitochondrial and nuclear DNA. *PLoS ONE* **2011**, *6*, e19728. [CrossRef]
5. Vuataz, L.; Rutschmann, S.; Monaghan, M.T.; Sartori, M. Molecular phylogeny and timing of diversification in Alpine *Rhithrogena* (Ephemeroptera: Heptageniidae). *BMC Evol. Biol.* **2016**, *16*, 194. [CrossRef]
6. Sowa, R. Contribution à la connaissance des espèces européennes de *Rhithrogena* Eaton (Ephemeroptera, Heptageniidae) avec le rapport particulier aux espèces des Alpes et des Carpates. In Proceedings of the Fourth International Conference on Ephemeroptera, Bechyne, Czechia, 4–10 September 1983; Landa, V., Soldán, T., Tonner, M., Eds.; CSAV: Bechyne, Czechia, 1984; pp. 37–52.
7. Zurwerra, A.; Metzler, M.; Tomka, I. Biochemical systematic and evolution of the European Heptageniidae (Ephemeroptera). *Arch. Hydrobiol.* **1987**, *109*, 481–510.
8. Eaton, A.E. List of Ephemeridae hitherto observed in Algeria with localities. *Ent. Mon. Mag.* **1899**, *35*, 4–5.
9. Thomas, A.G.; Mohati, A. Rhithrogena ourika n. sp., Ephéméroptère nouveau du Haut Atlas marocain (Heptageniidae). *Ann. Limnol.-Int. J. Limnol.* **1985**, *21*, 145–148. [CrossRef]
10. Dakki, M.; Thomas, A.G. *Rhithrogena ayadi* n. sp., Ephéméroptère nouveau du Moyen Atlas marocain (Heptageniidae). *Ann. Limnol.-Int. J. Limnol.* **1986**, *22*, 27–29. [CrossRef]
11. Thomas, A.G.B.; Bouzidi, A. Trois Ephéméroptères nouveaux du Haut Atlas marocain (Heptageniidae, Baetidae, Leptophlebidae). *Bull. Soc. Hist. Nat. Toulouse* **1986**, *122*, 7–11.
12. Thomas, A.G.; Vitte, B.; Soldán, T. *Rhithrogena ryszardi* n. sp., Ephéméroptère nouveau du Moyen Atlas (Maroc) et redescription de *Rh. soteria* Navás, 1917 (Heptageniidae). *Ann. Limnol.-Int. J. Limnol.* **1987**, *23*, 169–177. [CrossRef]
13. Vitte, B. *Rhithrogena mariae* n. sp. Ephéméroptère nouveau du Rif marocain (Ephemeroptera, Heptageniidae). *Nouv. Rev. D'Entomologie* **1991**, *8*, 89–96.
14. Zrelli, S.; Sartori, M.; Bejaoui, M.; Boumaiza, M. *Rhithrogena sartorii*, a new mayfly species (Ephemeroptera: Heptageniidae) from North Africa. *Zootaxa* **2011**, *3139*, 63–68. [CrossRef]
15. Samraoui, B.; Márquez-Rodríguez, J.; Ferreras-Romero, M.; El-Serehy, H.A.; Samraoui, F.; Sartori, M.; Gattolliat, J. Biogeography, ecology, and conservation of mayfly communities of relict mountain streams, north-eastern Algeria. *Aquat. Conserv. Mar. Freshw. Ecosyst.* **2021**. [CrossRef]
16. Samways, M.J. *Insect Conservation Biology*; Chapman & Hall: London, UK, 1984.
17. Samraoui, B.; Márquez-Rodríguez, J.; Ferreras-Romero, M.; Sartori, M.; Gattolliat, J.-L.; Samraoui, F. Life history and ecology of the Maghrebian endemic *Choroterpes atlas* Soldán & Thomas, 1983 (Ephemeroptera: Leptophlebiidae). *Limnologica* **2021**, *89*, 125887. [CrossRef]
18. Bouhala, Z.; Márquez-Rodríguez, J.; Chakri, K.; Samraoui, F.; El-Serehy, H.A.; Ferreras-Romero, M.; Samraoui, B. The life history of the Ibero-Maghrebian endemic *Oligoneuriopsis skhounate* Dakki and Guidicelli (Ephemeroptera: Oligoneuriidae). *Limnologica* **2020**, *81*, 125761. [CrossRef]
19. Bouhala, Z.; Márquez-Rodríguez, J.; Chakri, K.; Samraoui, F.; El-Serehy, H.A.; Ferreras-Romero, M.; Samraoui, B. The life cycle of the Maghrebian endemic *Ecdyonurus rothschildi* Navás, 1929 (Ephemeroptera: Heptageniidae) and its potential importance for environmental monitoring. *Limnology* **2021**, *22*, 17–26. [CrossRef]
20. Samraoui, B.; Bouhala, Z.; Chakri, K.; Márquez-Rodríguez, J.; Ferreras-Romero, M.; El-Serehy, H.A.; Samraoui, F.; Sartori, M.; Gattolliat, J.-L. Environmental determinants of mayfly assemblages in the Seybouse River, north-eastern Algeria (Insecta: Ephemeroptera). *Biologia* **2021**, *76*, 2277–2289. [CrossRef]
21. Folmer, O.; Black, M.; Hoeh, W.; Lutz, R.; Vrijenhoek, R. DNA primers for amplification of mitochondrial cytochrome c oxidase subunit I from diverse metazoan invertebrates. *Mol. Mar. Biol. Biotechnol.* **1994**, *3*, 294–299. [PubMed]
22. Katoh, K.; Standley, D.M. MAFFT Multiple sequence alignment software version 7: Improvements in performance and usability. *Mol. Biol. Evol.* **2013**, *30*, 772–780. [CrossRef] [PubMed]
23. Waterhouse, A.M.; Procter, J.B.; Martin, D.M.A.; Clamp, M.; Barton, G.J. Jalview Version 2—A multiple sequence alignment editor and analysis workbench. *Bioinformatics* **2009**, *25*, 1189–1191. [CrossRef]

24. Kumar, S.; Stecher, G.; Li, M.; Knyaz, C.; Tamura, K. MEGA X: Molecular Evolutionary Genetics Analysis across computing platforms. *Mol. Biol. Evol.* **2018**, *35*, 1547–1549. [CrossRef] [PubMed]
25. Stecher, G.; Tamura, K.; Kumar, S. Molecular Evolutionary Genetics Analysis (MEGA) for macOS. *Mol. Biol. Evol.* **2020**, *37*, 1237–1239. [CrossRef]
26. Kimura, M. A simple method for estimating evolutionary rates of base substitutions through comparative studies of nucleotide sequences. *J. Mol. Evol.* **1980**, *16*, 111–120. [CrossRef]
27. Tennessen, K. A method for determining stadium number of late stage Dragonfly Nymphs (Odonata: Anisoptera). *Èntomol. News* **2017**, *126*, 299–306. [CrossRef]
28. Ester, M.; Kriegel, H.P.; Sander, J.; Xu, X. A density-based algorithm for discovering clusters in large spatial databases with noise. In Proceedings of the 2nd International Conference on Knowledge Discovery and Data Mining KDD-96, Portland, OR, USA, 2–4 August 1996; pp. 226–231.
29. Hahsler, M.; Piekenbrock, M.; Doran, D. dbscan: Fast Density-Based Clustering with R. *J. Stat. Softw.* **2019**, *91*, 1–30. [CrossRef]
30. Team, R.C. *A Language and Environment for Statistical Computing*; R Foundation for Statistical Computing: Vienna, Austria, 2021.
31. Bauernfeind, E.; Soldan, T. *The Mayflies of Europe (Ephemeroptera)*; Apollo Books: Ollerup, Denmark, 2012.
32. Sartori, M.; Hughes, S.J. Description of a peculiar *Rhithrogena* nymph from the Iberian Peninsula (Ephemeroptera, Heptageniidae). *Limnetica* **2007**, *26*, 415–434.
33. Ball, S.L.; Hebert, P.D.N.; Burian, S.K.; Webb, J.M. Biological identifications of mayflies (Ephemeroptera) using DNA barcodes. *J. North Am. Benthol. Soc.* **2005**, *24*, 508–524. [CrossRef]
34. Webb, J.M.; Jacobus, L.M.; Funk, D.H.; Zhou, X.; Kondratieff, B.; Geraci, C.J.; DeWalt, R.E.; Baird, D.J.; Richard, B.; Phillips, I.; et al. A DNA barcode library for North American Ephemeroptera: Progress and prospects. *PLoS ONE* **2012**, *7*, e38063. [CrossRef] [PubMed]
35. Cardoni, S.; Tenchini, R.; Ficulle, I.; Piredda, R.; Simeone, M.C.; Belfiore, C. DNA barcode assessment of Mediterranean mayflies (Ephemeroptera), benchmark data for a regional reference library for rapid biomonitoring of freshwaters. *Biochem. Syst. Ecol.* **2015**, *62*, 36–50. [CrossRef]
36. Morinière, J.; Hendrich, L.; Balke, M.; Beermann, A.J.; König, T.; Hess, M.; Koch, S.; Müller, R.; Leese, F.; Hebert, P.D.N.; et al. A DNA barcode library for Germany's mayflies, stoneflies and caddisflies (Ephemeroptera, Plecoptera and Trichoptera). *Mol. Ecol. Resour.* **2017**, *17*, 1293–1307. [CrossRef]
37. Lubini, V.; Sartori, M. Current status, distribution, life cycle and ecology of *Rhithrogena germanica* Eaton, 1885 in Switzerland: Preliminary results (Ephemeroptera, Heptageniidae). *Aquat. Sci.* **1994**, *56*, 388–397. [CrossRef]
38. Demarteau, B. *Rhithrogena germanica* (Eaton, 1885) nouvelle espèce pour la faune belge (Ephemeroptera: Heptageniidae). *Bull. Soc. R. Belg. d'Entomol.* **2015**, *151*, 243–249.
39. Sweeney, B.W. Factors influencing life-history patterns of aquatic insects. In *The Ecology of Aquatic Insects*; Resh, V.H., Rosenberg, D.M., Eds.; Praeger: New York, NY, USA, 1984; pp. 56–100.
40. Brittain, J.E.; Campbell, I.C. The effect of temperature on egg development in the Australian mayfly genus *Coloburiscoides* (Ephemeroptera: Coloburiscidae) and its relationship to distribution and life history. *J. Biogeogr.* **1991**, *18*, 231. [CrossRef]
41. Humpesch, U.H.; Elliott, J.M. Effect of temperature on the hatching time of eggs of three *Rhithrogena* spp. (Ephemeroptera) from Austrian Streams and an English Stream and River. *J. Anim. Ecol.* **1980**, *49*, 643. [CrossRef]
42. Knispel, S.; Sartori, M.; Brittain, J.E. Egg development in the mayflies of a Swiss glacial floodplain. *J. N. Am. Benthol. Soc.* **2006**, *25*, 430–443. [CrossRef]
43. Clifford, H.F. Life cycles of mayflies (Ephemeroptera), with special reference to voltinism. *Quaest. Entomol.* **1982**, *18*, 15–90.
44. Sowa, R. Ecology and biogeography of mayflies (Ephemeroptera) of running waters in the Polish part of the Carpathians, 1. Distribution and quantitative analysis. *Acta Hydrobiol.* **1975**, *17*, 223–297.
45. Thibault, M.; Valdes, H.; Bosviel, A.; Vignes, J.-C. Le développement des éphéméroptères d'un ruisseau à truites des Pyrénées-Atlantiques, le lissuraga. *Ann. Limnol.-Int. J. Limnol.* **1971**, *7*, 53–120. [CrossRef]
46. Neveu, A.; Lapchin, L.; Vignes, J.C. Le macrobenthos de la basse Nivelle, petit fleuve côtier des Pyrennées-Atlantiques. *Ann. Zool. Ecol. Anim.* **1979**, *11*, 1293–1370.
47. Bauernfeind, E.; Moog, O. Mayflies (Insecta: Ephemeroptera) and the assessment of ecological integrity: A methodological approach. *Assess. Ecol. Integr. Run. Waters* **2000**, *422*, 71–83. [CrossRef]
48. Brittain, J.E.; Sartori, M. Ephemeroptera. In *Encyclopedia of Insects*; Elsevier: Amsterdam, The Netherlands, 2009; pp. 328–334.
49. Benslimane, N.; Chakri, K.; Haiahem, D.; Guelmami, A.; Samraoui, F.; Samraoui, B. Anthropogenic stressors are driving a steep decline of hemipteran diversity in dune ponds in north-eastern Algeria. *J. Insect Conserv.* **2019**, *23*, 475–488. [CrossRef]
50. Morghad, F.; Samraoui, F.; Touati, L.; Samraoui, B. The times they are a changin': Impact of land-use shift and climate warming on the odonate community of a Mediterranean stream over a 25-year period. *Vie Milieu* **2019**, *69*, 25–33.
51. Zrelli, S.; Boumaïza, M.; Béjaoui, M.; Gattolliat, J.-L.; Sartori, M. New data and revision of the Ephemeroptera of Tunisia. *Inland Water Biol. Suppl.* **2016**, *3*, 99–106.
52. Korbaa, M.; Ferreras-Romero, M.; Bejaoui, M.; Boumaiza, M. Two species of Odonata newly recorded from Tunisia. *Afr. Èntomol.* **2014**, *22*, 291–296. [CrossRef]

 diversity

Article

Is Coloburiscidae (Ephemeroptera) Monophyletic? A Comparison of Datasets

Jarod Meecham and T. Heath Ogden *

Department of Biology, Utah Valley University, Orem, UT 84058, USA; jp.meecham@gmail.com
* Correspondence: heath.ogden@uvu.edu

Abstract: Coloburiscidae consists of three living genera with a Gondwanan distribution—*Coloburiscoides* from Australia, *Coloburiscus* from New Zealand, and *Murphyella* from Chile. Molecular-based phylogenetic analyses of Ephemeroptera (mayflies) have been somewhat successful in resolving many higher-level relationships in the order. Most of these analyses, however, have been ambiguous with respect to the family Coloburiscidae. This study presents the first phylogenetic analysis specific to Coloburiscidae using data generated from 448 phylogenomic sequences and data generated from the Sanger sequencing of five genes: 12S, 16S, 18S, 28S, and H3. Bayesian and likelihood analyses were conducted on each dataset and, ultimately, on a combined dataset of the two. Coloburiscidae was shown to be supported as monophyletic in each instance where the phylogenomic data were included. *Coloburiscoides* was shown as sister to *Murphyella* + *Coloburiscus*.

Keywords: mayflies; phylogenomics; phylogenetics; systematics

Citation: Meecham, J.; Ogden, T.H. Is Coloburiscidae (Ephemeroptera) Monophyletic? A Comparison of Datasets. *Diversity* **2022**, *14*, 505. https://doi.org/10.3390/d14070505

Academic Editors: Marina Vilenica, Zohar Yanai, Laurent Vuataz and Luc Legal

Received: 17 May 2022
Accepted: 19 June 2022
Published: 22 June 2022

1. Introduction

Mayflies represent an ancient order of winged insects that date back 300 million years [1–3]. The extant lineages of the order are found in freshwater ecosystems worldwide, except for the continent of Antarctica and some remote islands [4,5]. Currently, mayfly diversity constitutes around 40 families, with approximately 460 genera, and almost 3700 described species [6]. While some families are believed to be of Gondwanan origin, today there are only four families that exhibit a strict Gondwanan distribution: Amelotopsidae, Coloburiscidae, Nesameletidae, and Oniscigastridae [4]. Of these, the monophyly of Coloburiscidae and Ameletopsidae remains elusive.

1.1. Review of Taxonomy

The family Coloburiscidae, sometimes described as the spinose-gilled mayfly family, is relatively small in comparison to other mayfly families and has only three genera: *Coloburiscoides* Lestage (1935), *Coloburiscus* Eaton (1888), and *Murphyella* Lestage (1930). *Coloburiscoides* and *Coloburiscus* are only found in Australia and New Zealand, respectively, while *Murphyella* is endemic to Chile, thus displaying a Gondwanan distribution [4,7–9]. The family is not currently believed to have a presence in the remaining Gondwanan land masses, but this could be due to a comparatively decreased effort to explore mayfly taxonomy in underdeveloped countries. In total, the family has seven described species (see Table 1).

Higher-level classifications within Ephemeroptera have been problematic for decades. Edmunds [10] considered Coloburiscidae as a subfamily of Siphlonuridae. Riek [11] instead proposed a subfamily status under Oligoneuridae, while Landa [12] proposed Coloburiscidae as its own family. Later, McCafferty [13,14] developed a classification system of Ephemeroptera, where Coloburiscidae was proposed as a sister to the families Isonychiidae, Oligoneuridae, and Heptageniidae (including *Arthroplea* and *Pseudiron*). Together, these four families constituted the suborder Setisura due to several putative apomorphies within the group [13]. Kluge [15] developed a separate nomenclature system than McCafferty, in which Coloburiscidae had a family status; however, Kluge does not refer to any formal analysis of characteristics in his review [15,16].

Table 1. The location of each taxon of Coloburiscidae.

Genus	Specific Name	Location
Coloburiscoides	*giganteus* Tillyard (1933)	Australia
Coloburiscoides	*haleuticus* Eaton (1871)	Australia
Coloburiscoides	*munionga* Tillyard (1933)	Australia
Coloburiscus	*humeralis* Walker (1853)	New Zealand
Coloburiscus	*tonnoiri* Lestage (1935)	New Zealand
Coloburiscus	*remota* Walker (1853)	New Zealand
Murphyella	*needhami* Lestage (1930)	Chile

1.2. Review of Relationships

Molecular data brought new insights into the relationships of mayflies. For example, the position of Ephemeroptera to all other extant pterygotes [17] was examined. The relationships of all the families of Ephemeroptera were investigated in a combined analysis [5]. The morphological data were composed of 101 characters and the molecular data came from the Sanger sequencing of nuclear and mitochondrial genes: 12S, 16S, 18S, 28S, and H3. Significant findings from this analysis include strong support for the monophyly of Ephemeroptera, as well as for several other major lineages proposed by McCafferty [14] and Kluge [15], while others were not corroborated as monophyletic. Thus, it was recognized that future robust phylogenetic analyses were needed to resolve previously ambiguous relationships. With respect to Coloburiscidae, the findings failed to support both the suborder Setisura and the family Coloburiscidae as monophyletic groups (Figure 1). The monophyly of Coloburiscidae was never supported in any of the molecular data trees in the 2009 analysis [5]. However, it was supported as monophyletic in the morphology tree, with *Coloburiscus* + *Murphyella* being sisters to *Coloburiscoides*.

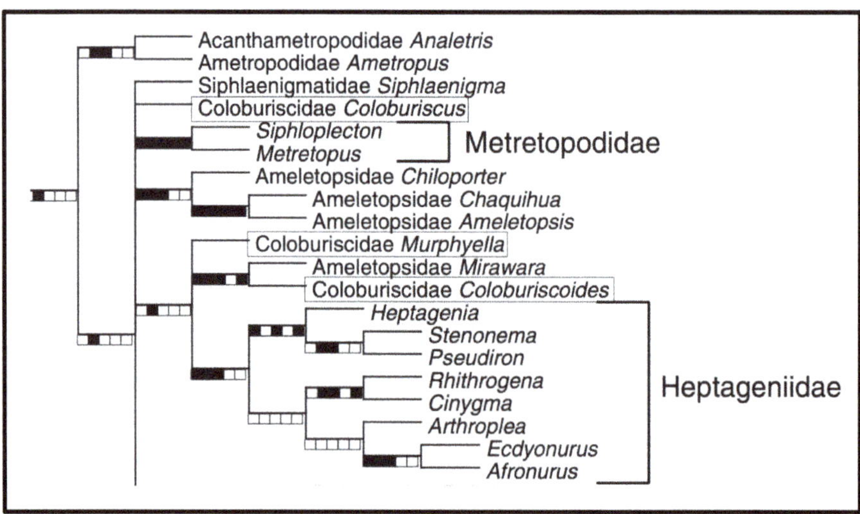

Figure 1. Maximum likelihood results from Figure 2 of Ogden et al., (2009) [5], showing non-monophyly of Coloburiscidae. The boxes are filled in (from left to right) if bootstrap value > 90 (box1), Bremer support value > 2 (box 2), Bayes posterior probability > 90 (box 3), node present when gaps + 5th state character (box 4), and node present in POY analysis (box 5).

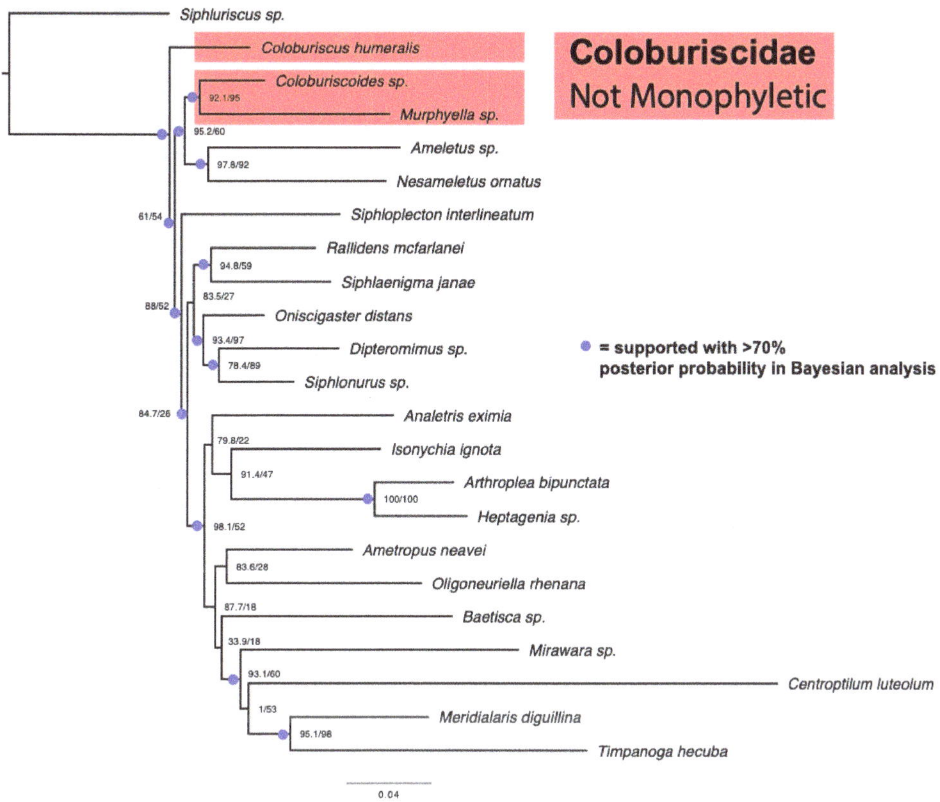

Figure 2. Maximum likelihood analysis of 5-gene Sanger dataset. Results show the non-monophyly of Coloburiscidae, highlighted in red. The blue circles indicate the nodes that were supported with >70% posterior probability in the Bayesian analysis. The numbers on the nodes are SH-aLRT support %/ultrafast bootstrap support %.

The most recent "initial evaluation" phylogenetic analysis on Ephemeroptera was conducted using targeted capture and next-generation sequencing of 448 protein-coding regions [18]. Thirty-five families of mayflies were represented in the 105 taxa dataset. This analysis represented the largest ever phylogenetic analysis of the order and brought new insights into many higher-level relationships with strong support values. However, this work was a preliminary proposal of relationships, presented at the international meeting for Ephemeroptera, and emphasized the importance of large datasets for resolving relationships. This analysis only used amino acids and did not examine nucleotide datasets. Of note, Coloburiscidae was found to be monophyletic, with *Coloburiscoides* + *Coloburiscus* being sisters to *Murphyella*. Hence, this study contradicted the molecular data analyses of Ogden et al. (2009) [5]. Furthermore, the 2019 [18] analysis did not recover the same arrangement for the three genera of Coloburiscidae. The 2019 analysis only used an amino acid dataset in a Bayesian framework. Hence, additional scrutiny of these data is necessary to elucidate the monophyly and relationships of the genera of this family.

Considering the contradictory results between the Ogden et al., 2009 analysis [5] and the Ogden et al., 2019 analysis [18], and that the 2019 analysis was a preliminary approach, this study aims to further test the monophyly of Coloburiscidae and the relationships between the three genera. To this end, this research will: (1) conduct Bayesian and maximum likelihood (ML) analyses of Coloburiscidae using the same five genes (five-gene Sanger dataset) from the 2009 [5] analysis, with some additional sequenced data; (2) conduct Bayesian and ML analyses using the phylogenomic dataset generated as part of the analysis in Ogden et al., 2019 [18]; and (3) conduct Bayesian and ML analyses of both datasets combined as a single dataset.

2. Materials and Methods

2.1. *Taxonomic Sampling*

The total dataset consists of 23 taxa. Ingroup taxa include 1 representative from each of the three genera of Coloburiscidae and 19 other closely related taxa used as representatives for key lineages and families. *Siphluriscus* has been supported as the most basal taxa within the order Ephemeroptera and will be used as the outgroup [5,18].

2.2. *5-Gene Sanger Dataset and Analysis*

The Ogden et al. 2009 dataset had some missing data in the five Sanger genes. In order to complete the dataset further, DNA extraction was performed on the thorax or legs of each specimen using the Qiagen DNeasy Blood & Tissue kit (Valencia, CA, USA) and following the animal tissue protocol. Sequences were targeted for amplification via the standard polymerase chain reaction using the BioRad® T100 Thermo Cycler (Hercules, CA, USA). The primers for each gene are the same as used in the Ogden et al. [5] molecular analysis. Gel electrophoresis was used to confirm the successful amplification of genes. The purification of DNA was conducted using the QIAquick PCR Purification kit (Valencia, CA, USA) and following the standard protocol. The purified DNA was sequenced at Psomagen Inc. (Rockville, MD, USA). The sequences were manually curated with Sequencher® 5.2.4 [19]. The newly generated data for genes 12S, 16S, 18S, 28S, and H3 (Supplementary Materials Table S1) were combined with the data from the 2009 dataset, and MUSCLE software was used to align the DNA sequences [20,21] using the default settings. Aligned gene regions were concatenated using Sequence Matrix 1.8 [22]. A Bayesian analysis was conducted using MrBayes [23,24] with the nst = 6 invgamma model for 10,000,000 generations. From the cold chain, the first 25% of the sample was discarded as the burn-in. The analysis resulted in a final split frequency of 0.0055. IQTREE software [25] was used to run an ML analysis with 1000 ultrafast bootstraps using the best-fit model selected by IQTREE: GTR + F + I + G4.

2.3. *Phylogenomic Dataset and Analysis*

Molecular data for each taxon (see Table 2) were generated in the Ogden et al., 2019 analysis [18]. Detailed information on protocols and workflow is specified elsewhere [3,18]. In summary, probe kits were designed for orthologous protein-coding loci across the genomes of all taxa. Library preparation, hybrid enrichment, and sequencing were conducted at RAPiD Genomics (Gainesville, FL, USA) using Illumina HiSeq 3000. Assembly and data cleanup were conducted using the anchored phylogenomics pipeline of Breinholt et al. [26]. The alignment of each locus was performed using MAFFT with default parameters. The phylogenomic data were constructed in two ways and analyzed in Bayesian and ML frameworks. First, only the first and second positions of each codon were included due to their conserved nature (DNA12 dataset) [3]; there was evidence of third position saturation. Second, the nucleotides were translated into amino acid sequences (AA dataset). The Bayesian analyses were conducted using MrBayes [23,24] for 10,000,000 generations with four chains. The first 25% of the sample was discarded as the burn-in for all runs. The model for the DNA12 dataset was nst = 6 and invgamma, and for the AA dataset, the Aamodel was used to provide a mixture of models with fixed rate matrices. The MrBayes analyses resulted in final split frequencies <0.003. IQTREE software [25] was used to run an ML analysis with 1000 ultrafast bootstraps. The best-fit model selected by IQTREE for the DNA12 dataset was GTR + F + I + G4, and for the AA dataset, the mtZOA + F + I + G4 was the best-fit model. To further test branch support, an SH-like aLRT with 1000 replicates was also carried out in IQTREE.

2.4. Combined Dataset and Analysis

The aligned sequences from the 5-gene Sanger dataset and the DNA12 dataset were concatenated using Sequence Matrix 1.8 [22]. Bayesian and ML analyses were conducted as described above. The nst = 6 invgamma model was used in MrBayes and the GTR + F + I + G4 model was used in IQTREE for the ML analysis.

3. Results

The alignment for the five-gene Sanger dataset was 5321 bp in length with 1045 parsimony informative sites. The ML IQTREE phylogenetic reconstruction results are shown in Figure 2. Coloburiscidae was not recovered as monophyletic, but *Coloburiscoides* was shown to be a sister to *Murphyella*. The alignment for the phylogenomic dataset was 61,116 bp in length with 4859 parsimony informative sites. The ML IQTREE phylogenetic reconstruction of the DNA12 results is shown in Figure 3. Coloburiscidae was strongly supported (100% SH-aLRT and 100% bootstrap) as monophyletic, and *Coloburiscoides* was somewhat supported (88% SH-aLRT and 90% bootstrap) as being a sister to *Coloburiscus* + *Murphyella*. The combined dataset tree shown in Figure 4 also strongly supported (100% SH-aLRT and 100% bootstrap) the monophyly of Coloburiscidae, and somewhat supported (88% SH-aLRT and 93% Bootstrap) *Coloburiscoides* as being a sister to *Coloburiscus* + *Murphyella*.

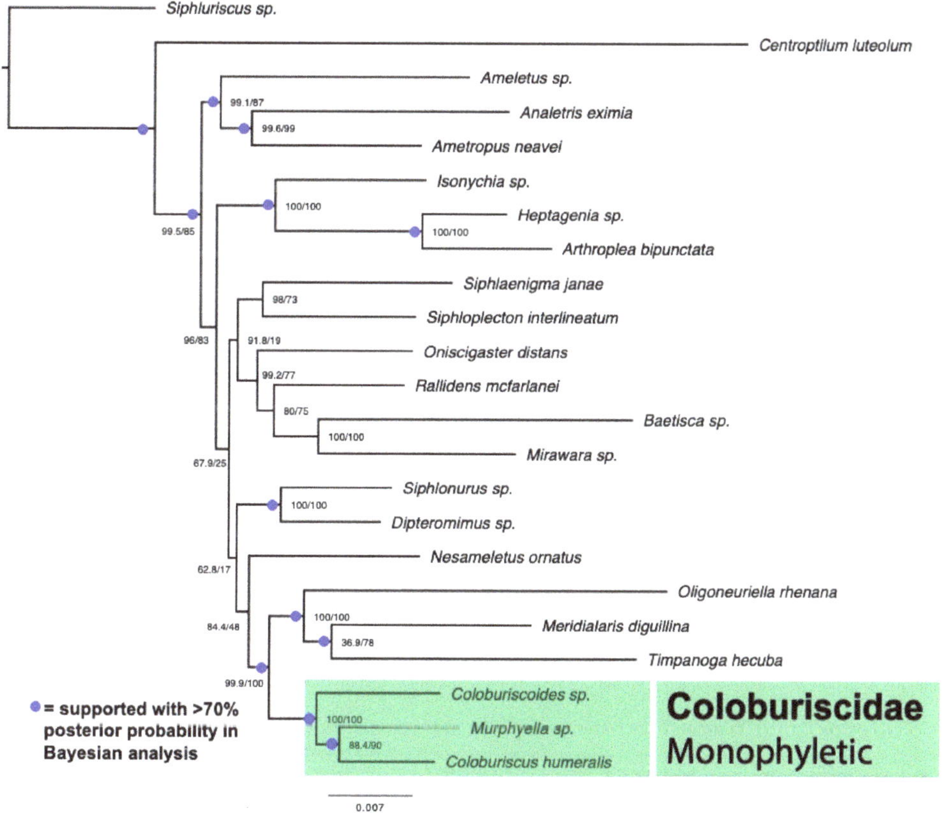

Figure 3. Maximum likelihood analysis of the phylogenomic DNA12 dataset. Results strongly support the monophyly of Coloburiscidae, highlighted in green. The blue circles indicate the nodes that were supported with >70% posterior probability in the Bayesian analysis. The numbers on the nodes are SH-aLRT support %/ultrafast bootstrap support %.

The AA dataset analysis (not in the figures as it was similar to the 2019 analysis) in ML and Bayesian analyses resulted in a strongly supported monophyletic Coloburiscidae; however, there was weaker support for the *Murphyella–Coloburiscoides* + *Coloburiscus* relationship, contradicting the relationship between the three genera found in the DNA12 dataset results.

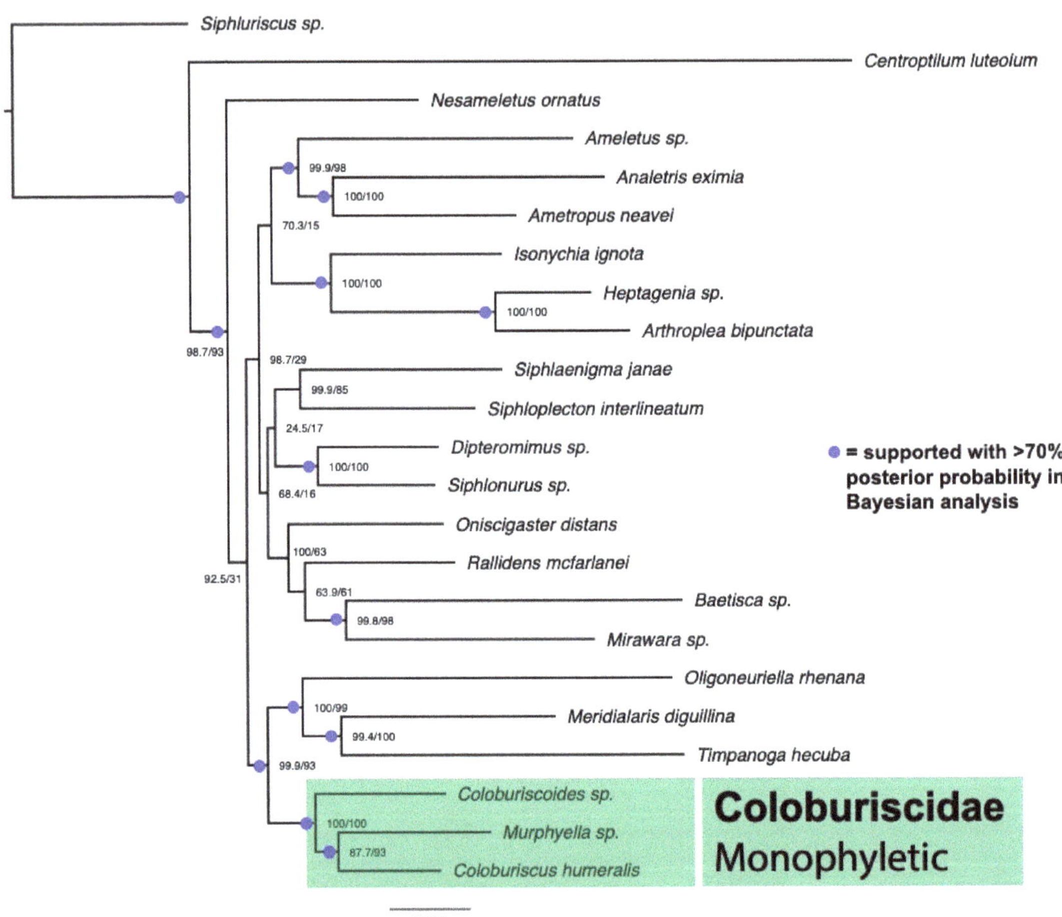

Figure 4. **Maximum** likelihood analysis of the combined 5-gene Sanger dataset + phylogenomic DNA12 dataset. Results strongly support the monophyly of Coloburiscidae, highlighted in green. The blue circles indicate the nodes that were supported with >70% posterior probability in the Bayesian analysis. The numbers on the nodes are SH-aLRT support %/ultrafast bootstrap support %.

Table 2. Taxon sampling for 5-gene Sanger and phylogenomic datasets.

Family	Genus	Specific Name	Number of Exons Captured
Baetiscidae	*Baetisca*	sp.	421
Acanthametropodidae	*Analetris*	*exima*	422
Ameletidae	*Ameletus*	sp.	435
Ameletopsidae	*Mirawara*	sp.	429

Table 2. *Cont.*

Family	Genus	Specific Name	Number of Exons Captured
Ametropodidae	*Ametropus*	*naevi*	431
Arthropleidae	*Arthroplea*	*bipunctata*	399
Baetidae	*Centroptilum*	*luteolum*	414
Coloburiscidae	*Murphyella*	sp.	438
Coloburiscidae	*Coloburiscoides*	sp.	442
Coloburiscidae	*Coloburiscus*	*humeralis*	435
Dipteromimidae	*Dipteromimus*	sp.	432
Ephemerellidae	*Timpanoga*	sp.	414
Heptageniidae	*Heptagenia*	sp.	431
Isonychiidae	*Isonychia*	sp.	444
Leptophlebiidae	*Meridialaris*	*diguillina*	413
Metropodidae	*Siphloplecton*	*interlineatum*	432
Nesameletidae	*Nesameletus*	*ornatus*	420
Oligoneuridae	*Oligoneuriella*	*rhenana*	415
Oniscigastridae	*Oniscigaster*	*distans*	441
Rallidentidae	*Rallidens*	*mcfarlanei*	358
Siphlaenigmatidae	*Siphlaenigma*	*janae*	390
Siphloneuridae	*Siphlonurus*	sp.	445
Siphluriscidae	*Siphluriscus*	sp.	340

4. Discussion

The purpose of this research was to further investigate the relationships of Coloburiscidae through molecular-based phylogenetic analysis. Coloburiscidae was shown to be monophyletic each time the phylogenomic data were included in any methodological framework, dataset (morphological, DNA12, or AA), or analysis method (ML or Bayesian). Hence, it can be strongly concluded that Coloburiscidae is a monophyletic lineage. Of the other taxa selected for this analysis, it is also strongly supported that the Coloburiscidae is sister to the lineages Leptophlebiidae, Oligoneuridae, and Ephemerellidae, which aligns with the 2019 tree.

However, the relationships between the three genera are not as concordant across all the analyses. The DNA12 dataset supports *Coloburiscoides* as a sister to *Murphyella* + *Coloburiscus*, with fairly high support values (100% posterior probability in the Bayesian analysis, >90% bootstrap, and >87% SH-aLRT in the ML analysis) and agrees with the morphological tree (Figure 4 of Ogden et al., 2009) [5]. The AA dataset results somewhat weakly support (92% posterior probability in the Bayesian analysis, 65% bootstrap, and 12% SH-aLRT) *Murphyella* + *Coloburiscus* + *Coloburiscoides*. Not surprisingly, this was the same result from the Ogden et al., 2019 [18] analysis that also used amino acid sequences as the base dataset.

The question remains, which relationship of the three genera is correct? The morphology tree from 2009 and the DNA12 dataset of this study (with its relatively higher support values than the AA dataset results) support the *Coloburiscoides* as a sister to *Murphyella* + *Coloburiscus* as the most likely proposal. However, the AA dataset supports the sister relationship of *Coloburiscus* + *Coloburiscoides*, which aligns better with the Gondwana breakup consensus that Australia and New Zealand would have had land bridges in the more recent past. The fragmentation of Gondwana, which began approximately 150 Mya [27,28], continues to be examined within the field of biogeography as a growing number of studies point to organismal distribution patterns that can be explained by such a process [29–32]. Perhaps the most famous example of Gondwanan distribution is the southern beeches (*Nothofagus)* found throughout Australasia and South America [33], with a fossil record dating back 80 million years [34]. Gondwanan vicariance is widely accepted to have played a major role in distribution and speciation; however, several studies caution against the tendency to invoke these geologic events as the only possible explanation for them [33,34]. An alternate hypothesis for observable patterns of distribution among

Coloburiscidae includes dispersal events. While it has been generally hypothesized that mayflies are poor dispersers, some oceanic and volcanic islands have been colonized with subsequent in situ radiation [6].

Thus, the intrafamilial relationships remain mostly inconclusive; however, these results and the 2019 analysis firmly support the monophyly of the family Coloburiscidae and its placement relative to other mayfly families. The five-gene Sanger dataset failed to support Coloburiscidae as monophyletic (individual analyses of each gene likewise did not support monophyly) and showed low support values across many nodes. Therefore, while these genes have been touted as somewhat successful in estimating relationships in previous analyses, one can only infer that they are not informative for all depths in an evolutionary tree, especially for relationships as old as the ones being examined in these lineages of mayflies. This point further illustrates the importance and effectiveness of robust datasets (i.e., more loci and more taxa) and analyses in resolving higher-level relationships.

Supplementary Materials: The following are available online at https://www.mdpi.com/article/10.3390/d14070505/s1. Table S1. Table of the GenBank accession numbers for the Sanger sequencing data for the taxa.

Author Contributions: Conceptualization, J.M. and T.H.O.; formal analysis, J.M. and T.H.O.; funding acquisition, J.M. and T.H.O.; investigation, J.M.; methodology, J.M. and T.H.O.; resources, T.H.O.; software, T.H.O.; writing—original draft, J.M.; writing—review and editing, T.H.O. All authors have read and agreed to the published version of the manuscript.

Funding: This research was funded primarily by the NSF EAGER grant-1545281. Utah Valley University also funded this research through the Undergraduate Research Scholarly and Creative Activities (URSCA) Grant.

Data Availability Statement: Sequence data is available as indicated from Ogden et al., 2019 (18) and Genbank accession numbers (Supplementary Materials Table S1).

Acknowledgments: The authors would like to thank Danielle Taylor and Paul Dunn for their recommendations in the development of this manuscript. Thanks to Utah Valley University (UVU) Department of Biology, UVU College of Science and Health, and UVU Office of Engaged Learning.

Conflicts of Interest: The funders had no role in the design of the study; in the collection, analyses, or interpretation of data; in the writing of the manuscript, or in the decision to publish the results.

References

1. Kukalová-Peck, J. Ephemeroid wing venation based upon new gigantic Carboniferous mayflies and basic morphology, phylogeny, and metamorphosis of pterygote insects (Insecta, Ephemerida). *Can. J. Zool.* **1985**, *63*, 933–955. [CrossRef]
2. Bauernfeind, E.; Soldan, T. *The Mayflies of Europe (Ephemeroptera)*; Brill NV: Ollerup, Denmark, 2012; pp. 513–515. [CrossRef]
3. Miller, D.B.; Bartlett, S.; Sartori, M.; Breinholt, J.W.; Ogden, T.H. Anchored phylogenomics of burrowing mayflies (Ephemeroptera) and the evolution of tusks. *Syst. Entomol.* **2018**, *43*, 692–701. [CrossRef]
4. Barber-James, H.M.; Gattolliat, J.L.; Sartori, M.; Hubbard, M.D. Global Diversity of Mayflies (Ephemeroptera, In-secta) in Freshwater. In *Freshwater Animal Diversity Assessment*; Springer: Berlin/Heidelberg, Germany, 2017; pp. 339–350.
5. Ogden, T.; Gattolliat, J.L.; Sartori, M.; Staniczek, A.; Soldán, T.; Whiting, M. Towards a new paradigm in mayfly phylogeny (Ephemeroptera): Combined analysis of morphological and molecular data. *Syst. Entomol.* **2009**, *34*, 616–634. [CrossRef]
6. Jacobus, L.M.; Macadam, C.R.; Sartori, M. Mayflies (Ephemeroptera) and Their Contributions to Ecosystem Services. *Insects* **2019**, *10*, 170. [CrossRef] [PubMed]
7. Blakey, R.C.; Fielding, C.R.; Frank, T.D.; Isbell, J.L. Gondwana paleogeography from assembly to breakup—A 500 my odyssey. *Geol. Soc. Am. Spec. Pap.* **2008**, *441*, 1–28. [CrossRef]
8. Derkarabetian, S.; Baker, C.M.; Giribet, G. Complex patterns of Gondwanan biogeography revealed in a dispersal-limited arachnid. *J. Biogeogr.* **2021**, *48*, 1336–1352. [CrossRef]
9. Harley, S.; Fitzsimons, I.; Zhao, Y. Antarctica and supercontinent evolution: Historical perspectives, recent advances and unresolved issues. *Geol. Soc. Lond. Spec. Publ.* **2013**, *383*, 1–34. [CrossRef]
10. Edmunds, G.F.; Allen, R.K.; Peters, W.L. *An Annotated Key to the Nymphs of the Families and Subfamilies of May-Flies (Ephemeroptera)*; University of Utah Biology: Salt Lake City, UT, USA, 1963; Volume 13, pp. 1–49.
11. Riek, E.F. The classification of the Ephemeroptera. In Proceedings of the First International Conference on Ephemeroptera, Ephemer, Tallahassee, 17–20 August 1970; pp. 160–178.

12. Landa, V. A contribution to the evolution of the order Ephemeroptera based on comparative anatomy. In Proceedings of the First International Conference on Ephemeroptera, Ephemer, Tallahassee, 17–20 August 1970; pp. 155–159.
13. McCafferty, W.P. The Cladistics, Classification, and Evolution of the Heptagenioidea (Ephemeroptera). *Ephemer. Plecoptera* **1991**, 87–102.
14. McCafferty, W.P. Toward a Phylogenetic Classification of the Ephemeroptera (Insecta): A Commentary on Systematics. *Ann. Entomol. Soc. Am.* **1991**, *84*, 343–360. [CrossRef]
15. Kluge, N. *The Phylogenetic System of Ephemeroptera*; Kluwer Academic: Dordrecht, The Netherlands, 2004. [CrossRef]
16. Ogden, T.H.; Whiting, M.F. Phylogeny of Ephemeroptera (mayflies) based on molecular evidence. *Mol. Phylogenetics Evol.* **2005**, *37*, 625–643. [CrossRef]
17. Ogden, T.; Whiting, M.F. The problem with "the Paleoptera Problem:" sense and sensitivity. *Cladistics* **2003**, *19*, 432–442. [CrossRef]
18. Ogden, T.H.; Breinholt, J.W.; Bybee, S.M.; Miller, D.B.; Sartori, M.; Shiozawa, D.; Whiting, M.F. Mayfly phylo-genomics: Initial evaluation of anchored hybrid enrichment data for the order Ephemeroptera. *Zoosymposia* **2019**, *16*, 167–181. [CrossRef]
19. Genecodes. *Sequencher*; Version 5.2.4; Genecodes Co.: Ann Arbor, MI, USA, 1999.
20. Edgar, R.C. MUSCLE: A multiple sequence alignment method with reduced time and space complexity. *BMC Bioinform.* **2004**, *5*, 113. [CrossRef]
21. Edgar, R.C. MUSCLE: Multiple sequence alignment with high accuracy and high throughput. *Nucleic Acids Res.* **2004**, *32*, 1792–1797. [CrossRef] [PubMed]
22. Vaidya, G.; Lohman, D.J.; Meier, R. SequenceMatrix: Concatenation software for the fast assembly of multi-gene datasets with character set and codon information. *Cladistics* **2011**, *27*, 171–180. [CrossRef] [PubMed]
23. Huelsenbeck, J.P.; Ronquist, F. MRBAYES: Bayesian inference of phylogenetic trees. *Bioinformatics* **2001**, *17*, 754–755. [CrossRef]
24. Ronquist, F.; Huelsenbeck, J.P. MrBayes 3: Bayesian phylogenetic inference under mixed models. *Bioinformatics* **2003**, *19*, 1572–1574. [CrossRef]
25. Nguyen, L.-T.; Schmidt, H.A.; Von Haeseler, A.; Minh, B.Q. IQ-TREE: A Fast and Effective Stochastic Algorithm for Estimating Maximum-Likelihood Phylogenies. *Mol. Biol. Evol.* **2015**, *32*, 268–274. [CrossRef]
26. Breinholt, J.W.; Earl, C.; Lemmon, A.R.; Lemmon, E.M.; Xiao, L.; Kawahara, A.Y. Resolving Relationships among the Megadiverse Butterflies and Moths with a Novel Pipeline for Anchored Phylogenomics. *Syst. Biol.* **2017**, *67*, 78–93. [CrossRef]
27. Veevers, J. Gondwanaland from 650–500 Ma assembly through 320 Ma merger in Pangea to 185–100 Ma breakup: Supercontinental tectonics via stratigraphy and radiometric dating. *Earth-Science Rev.* **2004**, *68*, 1–132. [CrossRef]
28. Yoshida, M.; Santosh, M. Voyage of the Indian subcontinent since Pangea breakup and driving force of supercontinent cycles: Insights on dynamics from numerical modeling. *Geosci. Front.* **2018**, *9*, 1279–1292. [CrossRef]
29. Barker, N.P.; Weston, P.; Rutschmann, F.; Sauquet, H. Molecular dating of the 'Gondwanan' plant family Proteaceae is only partially congruent with the timing of the break-up of Gondwana. *J. Biogeogr.* **2007**, *34*, 2012–2027. [CrossRef]
30. Toussaint, E.F.A.; Bloom, D.; Short, A.E.Z. Cretaceous West Gondwana vicariance shaped giant water scavenger beetle biogeography. *J. Biogeogr.* **2017**, *44*, 1952–1965. [CrossRef]
31. Gamble, T.; Bauer, A.M.; Greenbaum, E.; Jackman, T. Evidence for Gondwanan vicariance in an ancient clade of gecko lizards. *J. Biogeogr.* **2007**, *35*, 88–104. [CrossRef]
32. Noben, S.; Kessler, M.; Quandt, D.; Weigand, A.; Wicke, S.; Krug, M.; Lehnert, M. Biogeography of the Gondwanan tree fern family Dicksoniaceae-A tale of vicariance, dispersal and extinction. *J. Biogeogr.* **2017**, *44*, 2648–2659. [CrossRef]
33. Waters, J.M.; Craw, D. Goodbye Gondwana? New Zealand Biogeography, Geology, and the Problem of Circularity. *Syst. Biol.* **2006**, *55*, 351–356. [CrossRef]
34. Knapp, M.; Stöckler, K.; Havell, D.; Delsuc, F.; Sebastiani, F.; Lockhart, P.J. Relaxed Molecular Clock Provides Evidence for Long-Distance Dispersal of Nothofagus (Southern Beech). *PLOS Biol.* **2005**, *3*, e14. [CrossRef]

Article

Taxonomy and Distribution of the Gomphid Dragonfly *Orientogomphus minor* (Laidlaw, 1931) (Odonata: Gomphidae) in Thailand

Damrong Chainthong and Boonsatien Boonsoong *

Animal Systematics and Ecology Speciality Research Unit (ASESRU), Department of Zoology, Faculty of Science, Kasetsart University, Bangkok 10900, Thailand; damrong.cha@ku.th
* Correspondence: fscibtb@ku.ac.th

Abstract: The taxonomy and distribution of *Orientogomphus minor* (Laidlaw, 1931) were investigated in Thailand. Gomphid nymphs were collected from 28 sampling sites in streams in eastern, western, and southern Thailand. The nymph of *O. minor* is described for the first time and the male is re-described and illustrated based on a reared specimen. The taxonomic characteristics of the nymphs of the genus *Orientogomphus* are discussed. The nymph of *O. minor* differs from that of *O. armatus* Chao & Xu, 1987, the only other *Orientogomphus* species with a described nymphal stage, by the presence of lateral spines on abdominal segments six to nine and by a slender, stick-shaped third antennal segment. Multivariate analyses revealed a strong correlation between the distribution of *O. minor* and other three gomphid species with restricted distribution in Thailand (*Nychogomphus duaricus* (Fraser, 1924), *Onychogomphus louissiriusi* Fleck, 2020 and *Stylogomphus thongphaphumensis* Chainthong, Sartori & Boonsoong, 2020). Those species were recorded solely in streams in the western part of the country. Nymphs of *O. minor* were predominantly associated with stony substrates.

Keywords: gomphid nymphs; *Orientogomphus*; Thailand

Citation: Chainthong, D.; Boonsoong, B. Taxonomy and Distribution of the Gomphid Dragonfly *Orientogomphus minor* (Laidlaw, 1931) (Odonata: Gomphidae) in Thailand. *Diversity* **2022**, *14*, 291. https://doi.org/10.3390/d14040291

Academic Editors: Michael Wink, Marina Vilenica, Zohar Yanai and Laurent Vuataz

Received: 15 March 2022
Accepted: 10 April 2022
Published: 12 April 2022

Publisher's Note: MDPI stays neutral with regard to jurisdictional claims in published maps and institutional affiliations.

1. Introduction

The Gomphidae (clubtail dragonflies), a well-known family in the Odonata, comprise about 87 genera and 1000 species worldwide [1,2]. With the exception of the Libellulidae, the species diversity is likely higher for the Gomphidae than for any other family of Anisoptera [3,4]. The gomphid nymphs have several notable morphological features: (1) the antennae have four segments with the third larger than the others, and the fourth very small, (2) the prementum and palpal lobes of the labium are flattened (not scoop-shaped), and (3) the body is diverse in form, cylindrical, broad and slender to extremely flattened like a leaf. Most gomphid nymphs are lotic species that are commonly components of benthic communities and contribute to ecosystem services (e.g., they are highly predaceous and serve as food for humans and as indicators of environmental changes) [5,6]. In the past decade, taxonomic studies of the Gomphidae in Thailand have continued to increase, and many new taxa have been described. The discovery of gomphid dragonflies adds 54 species and 27 genera (e.g., *Anisogomphus* [7,8], *Burmargomphus* [9], *Stylogomphus* [10], *Onychogomphus* [11] and *Microgomphus* [12]) from Thailand to the known species.

To date, most studies of odonates in Thailand have focused on the taxonomy of the adult stage. So far, eight families (Aeshnidae, Chorogomphidae, Cordulegastridae, Corduliidae, Gomphidae, Libellulidae, Macromiidae and Synthemistidae), 97 genera and 207 species of dragonflies have been recorded in Thailand. Studies on their biology and ecology are scarce, and only limited data are available on the nymphal stages and their distributions in lotic ecosystems. However, gomphid nymphs have been continuously described from Thailand [12–18], and the number of nymphal descriptions of Gomphidae species will steadily increase in the future because the diversity and taxonomy of the adult stage are well known.

The genus *Orientogomphus* was established by Chao & Xu [19], with species characterised as small to medium-sized, with divergent inferior appendages that are much shorter than the superior appendages (usually about half the length) and with apical margins shallowly concave. The superior appendages are long, bracket-like in dorsal view, and abruptly curved apically in lateral view, with a minute peg-like process at the tip. The prepuce is absent. The genus, distributed in Southeast Asia and China, currently comprises seven known species [20]. Of these, only *O. minor* (Laidlaw, 1931) has been recorded from Thailand. This small species is distributed throughout Thailand and extends to Peninsular Malaysia. A distribution map of adult specimens has been published [21]; however, knowledge of the distribution of the nymphal stage in Thai streams is sparse [21]. To date, *O. armatus* Chao & Xu, 1987 is the only species with a described nymph within the genera [22].

This paper provides the first description and illustration of the final stadium nymphs of *O. minor*, based on reared specimens, and compares and discusses the morphological characteristics of the nymphs of related species and genera. We also investigated the distribution and microhabitat of this species within lotic ecosystems.

2. Materials and Methods

2.1. Study Area and Sampling

The gomphid nymphs were collected from first order to third-order streams (28 sampling sites) in eastern, western, and southern Thailand (Figure 1). The nymphs were collected using a D-frame net in a variety of microhabitats, including sweepings of pool-litter, weeds, roots of riverside trees, mud and margin litter, or kick samples from riffles, sand, gravel, and pebbles. Gomphid nymphs were recorded at all microhabitats. Nymphs were identified to the species level using the published literature [10–18,23]. A distribution map was generated with SimpleMappr software (https://simplemappr.net) (accessed on 4 March 2022) [24].

2.2. Rearing and Identification

Nymphs of *O. minor* were found in three sampling sites. Of these, nymphs were collected from a sandy substrate in one locality at Huai Khayeng, Thong Pha Phum district, Kanchanaburi Province, in western Thailand (Figure 2). Full-grown nymphs were transferred to the laboratory for rearing. The nymphs were reared in potable water in an earthenware pot (a rearing device for a single nymph with a netted cover) with a mixture of sand and gravel as substrate. Each rearing chamber was connected to an air supply via aquarium tubing. Chironomid larvae were offered as prey, which *Orientogomphus* nymphs fed on readily. The nymphs were reared in the laboratory until they emerged as adults. The exuviae were preserved in 80% ethanol, and the adults were pinned and dried 3 days after their emergence. The species identification was confirmed based on Asahina [25] and Wilson [21]. All drawings were illustrated with the aid of a camera lucida. Measurements (mm) and photographs were taken with a NIKON SMZ800 stereoscopic microscope (NIKON Corporation, Tokyo, Japan). All dragonfly specimens are deposited in the Zoological Museum, Kasetsart University (ZMKU), Bangkok, Thailand (Aquatic Insects Collection section). The terminology for the nymphal mandibular formula followed that of Watson [26].

Figure 1. Map showing the sampling sites in eastern, western, and southern Thailand.

Figure 2. Nymphal habitat of *Orientogomphus minor* (Huai Khayeng, Thong Pha Phum district, Kanchanaburi Province).

2.3. Data Analysis

Gomphid nymph assemblages (presence/absence data) in response to spatial change were visualised by performing a principal component analysis (PCA), which identifies independent axes of variability and relates species samples to each axis. The relationship between gomphid nymphal species composition and microhabitat was investigated using two-way cluster analysis (Jaccard distance measure and the Group average linkage method). Multivariate analyses were performed using PC-ORD software version 7.01 [27].

3. Results

3.1. Taxonomy

3.1.1. Description of the Last Stadium Nymph

Material examined. THAILAND: 1 (exuvia) and 5 nymphs, Huai Khayeng, Thong Pha Phum district, Kanchanaburi Province, 14°36′20″ N 98°34′38″ E, 206 m a.s.l., 14.XII.2014, D. Chainthong leg; 3 nymphs, Huai Sat Lek; Kaeng Krachan district, Petchaburi Province, 12°38′14′′ N 99°30′59′′ E; 162 m a.s.l., 25.II.2018, D. Chainthong leg.

The general appearance and detailed structures are shown in Figures 3–5.

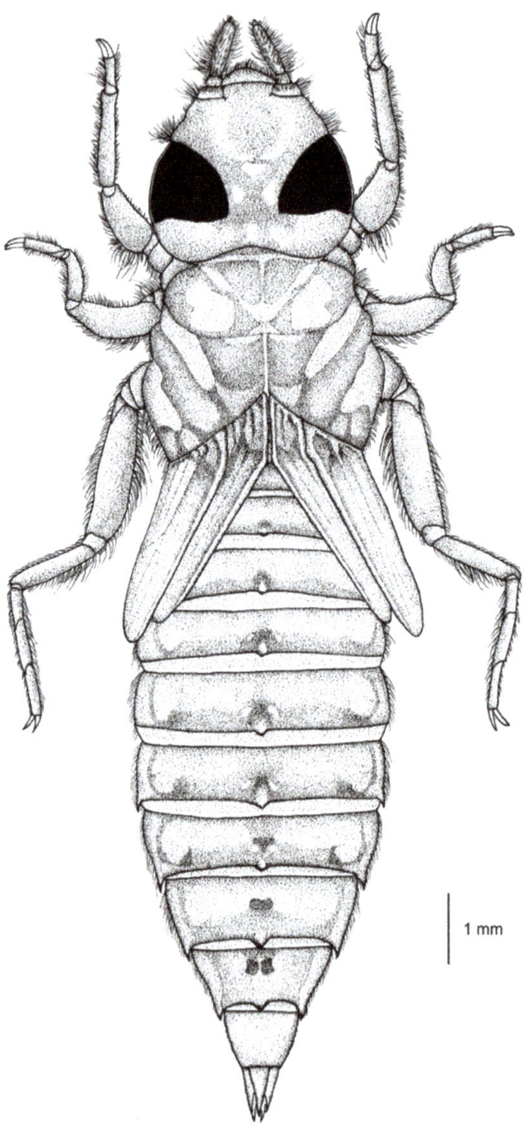

Figure 3. Dorsal view of nymph of *Orientogomphus minor*.

Figure 4. Habitus of *Orientogomphus minor* nymph.

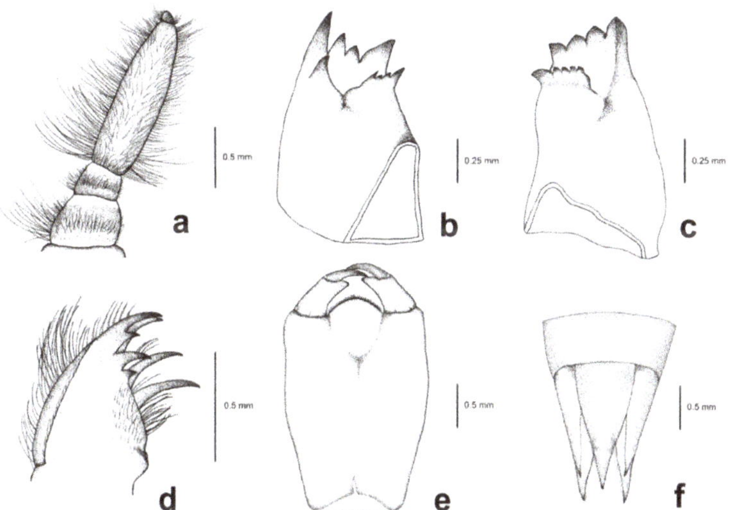

Figure 5. Morphological features of *Orientogomphus minor* nymph: (**a**) dorsal view of antenna; (**b**) ventral view and internal view of right mandible; (**c**) ventral view and internal view of left mandible; (**d**) ventral view of right maxilla; (**e**) ventral view of prementum; (**f**) dorsal view of anal appendages.

Colouration. Nymphs uniformly bright yellow. Body broadly lanceolate and covered with hair-like setae, dorsal surface strongly convex, ventral surface flat (Figure 4).

Head. Head broad and flat, frontal part with triangular appearance in dorsal view, posterior lobe of the head shorter than the eye length; eyes large and broadest across, with three large ocelli. Antennae four-segmented, first two segments small and rather circular; third segment slender, stick-shaped, slightly dorso-ventrally flattened and slightly upcurved; fourth segment vestigial, knob-like. All four segments bear long and dense hairs (Figure 5a). Mandibles as in Figure 5b,c, with mandibular formula: L 1234 0 a(m1–3)b/R 1234 y a(m1–2)b with a > b in both mandibles (Figure 5b,c). Maxillae: galeolacinia with seven moderately incurved teeth, three dorsal teeth nearly equal in length, four ventral teeth of different sizes, apical one largest; stipes and palp setose (Figure 5d).

Labium. Flat and not protruding when at rest (Figure 5e). Prementum-postmentum articulation reaching the posterior margin of the procoxa. Prementum subrectangular, longer than wide, in a ratio of 3:2, sides convex, convergent basally, with small teeth and minute setae at lateral margins; apical margin convex, with ventral row of 45–50 short, subquadrate reddish brown teeth, and dorsal rows of whitish piliform setae on apical border; labial palp with uniformly inflexed inner edge, yellowish brown, apical lobe reddish, rounded, internal margin arched inward, feebly serrulate. Movable hook reddish brown, sharp and moderately incurved.

Thorax. Small; prothorax narrower than head, dorsal portion raised at sides forming two mushroom-shaped ridges. Wing sheaths strongly divergent, reaching S4 (inner wing pad length 4.25 mm, outer wing pad length 3.75 mm). Legs short and stout, fore and middle legs strongly curved; protibia (length 2.0 mm) decidedly longer than profemora (length 1.25 mm); mesotibia (length 2.25 mm) slightly longer than mesofemora (length 1.75 mm); metafemur slightly longer than metatibia. Tarsal formula 2-2-3, tarsi yellowish. Rows of minute setae scattered along the femur, tibia, and tarsus of all six legs.

Abdomen. Broadly lanceolate, uniformly bright yellow, mid-dorsal black markings on S7–9. Mid-dorsal spines, absent on S1 and most prominent on S2–9, largest mid-dorsal spine on S8 (Figure 3). Lacking lateral spines on S2–5. Lateral edges of abdomen serrated, with spine projections that become more protrusive on S6–9 (Figure 3). Anal appendages elongated. Epiproct and cerci subequal in length; paraprocts longer than the other appendages (Figure 5f).

Measurements (in mm, $n = 9$): Length of total body 19.86–20.38; abdomen length 11.06–11.54; abdominal maximum width 4.90–5.08; head maximum width 4.08–4.23; length of hind femur 2.26–2.53; length of antennae third segment 1.50–1.69; length of antennae fourth segment 0.10–0.14; length of epiproct 1.26–1.44; length of cerci 1.18–1.23; length of paraprocts 1.48–1.56.

3.1.2. Taxonomy of the Adult

Material examined. THAILAND: 1 male adult (reared), Huai Khayeng stream, Thong Pha Phum district, Kanchanaburi province, 14°36′20″ N 98°34′38″ E, 206 m a.s.l., nymph collected on 14 July 2014, adult emerged on 18 February 2015, D. Chainthong leg.

In this study, we reared the *Orientogomphus* nymph until emergence of the male adult. The identification as a male adult of *O. minor* was confirmed based on Wilson [21], using the following diagnostic characters: head, pterothoracic, and caudal appendages (Figure 6a–d). A brief description of the male adult is presented based on our reared specimen following Wilson [21].

Diagnosis. Wilson [21] revised the known *Orientogomphus* specimens from northern Myanmar, China, Vietnam and Thailand as four species (*O. armatus*, *O. circularis* (Selys, 1894), *O. minor* and *O. naninus* (Förster, 1905), respectively). A distributional map has been provided by Wilson [21]. In Thailand, the adults of *O. minor* were recorded in Sakon Nakhon, Chiang Mai, Tak, Phatthalung, Krabi and Songkhla provinces [21,28,29].

Head (Figure 6a). Black with yellow markings; labrum with a pair of transverse ellipsoid yellow spots; genae black; anteclypeus yellow; postclypeus black, with a large yellow spot laterally; postfrons with a broad yellow band, antefrons black.

Thorax. Pattern of colouration as shown in Figure 6b. Prothorax black with yellow laterally; pterothorax dorsal suture with a yellow streak; mesothoracic collar yellow, except on middle; black stripe along first lateral suture disconnected to humeral, mesepimeron ventral margin yellow; legs black.

Wing. Hyaline, venation dark brown, pterostigmata very dark brown, anal triangle 4-celled with the smallest cell a well-defined rectangle; anal field 2-celled, with A2 arising from the subtriangle rather than directly from anal vein between cu-a and the subtriangle.

Abdomen. Abdomen predominantly black, with bright yellow markings; S1 mostly yellow laterally, S2 yellow around auricle, dorsal yellow spots on S2–S6, S4–S7 with dorsolateral yellow markings at base, S8–S9 with yellow lateral markings, epiproct black, cerci yellow outside and brownish to the proximal 2/3 and yellowish to the distal 1/3 (Figure 6c–d).

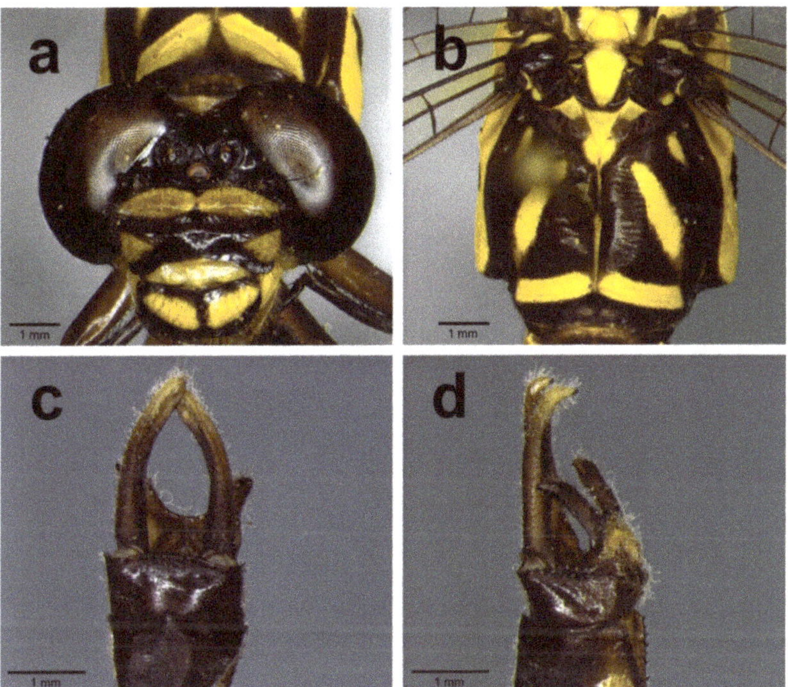

Figure 6. Morphological features of *Orientogomphus minor* adult male: (**a**) frontal view of head; (**b**) dorsal view of pterothoracic pattern; (**c**) dorsal view of caudal appendages; (**d**) lateral view of caudal appendages.

Accessory genitalia. The anterior hamulus is only slightly hooked; the posterior one does not bend caudad, but the anterior one is as high as the posterior one.

3.2. Distribution

3.2.1. Spatial Distribution of Gomphidae Species

Sixteen genera and 18 species of gomphid nymphs were found among the 28 sampling sites in eastern, western, and southern Thailand. PCA analysis revealed that most gomphid nymphs were strongly correlated with axis 1 (20% of the total variance explained). Among the gomphid species, the distribution of *O. minor* was strongly associated with the distribution of *Nychogomphus duaricus* (Fraser, 1924), *Onychogomphus louissiriusi* Fleck, 2020 and *Stylogomphus thongphaphumensis* Chainthong, Sartori & Boonsoong, 2020 (Figure 7). Those species were recorded solely in streams in the western part of Thailand.

3.2.2. Substrate Preference of *O. minor* Nymphs

A two-way cluster analysis showed two groups (I and II) of substrate types (microhabitat) and two groups (A and B) of gomphid species (Figure 8). Gomphid species (group A) *Megalogomphus sumatranus* (Krüger, 1899), *Phaenandrogomphus asthenes* Lieftinck, 1964, *Lamelligomphus castor* (Lieftinck, 1941), *Paragomphus capricornis* (Förster, 1914), *O. minor*, *Nepogomphus walli* (Fraser, 1924), *N. duaricus*, *O. louissiriusi*, *S. thongphaphumensis* and *S. malayanus* Sasamoto, 2001 were found predominantly associated with stony (pebble, gravel, and sand) substrates. In contrast, gomphid species (group B) *Heliogomphus selysi*

Fraser, 1925, *Microgomphus svihleri* (Asahina, 1969), *Gomphidia abbotti* Williamson, 1907, *Gomphidictinus perakensis* (Laidlaw, 1902), *Burmagomphus williamsoni* Förster, 1914, *B. divaricatus* Lieftinck, 1964, *Merogomphus paviei* Martin, 1904 and *Macrogomphus kerri* Fraser, 1932 were associated with litter (pool/marginal) and mud substrate types. The nymphs of *O. minor* were found in a substrate with mixture of sand and gravel, together with *P. asthenes*, *L. castor*, *P. capricornis*, *N. walli*, *N. duaricus* and *O. louissiriusi*.

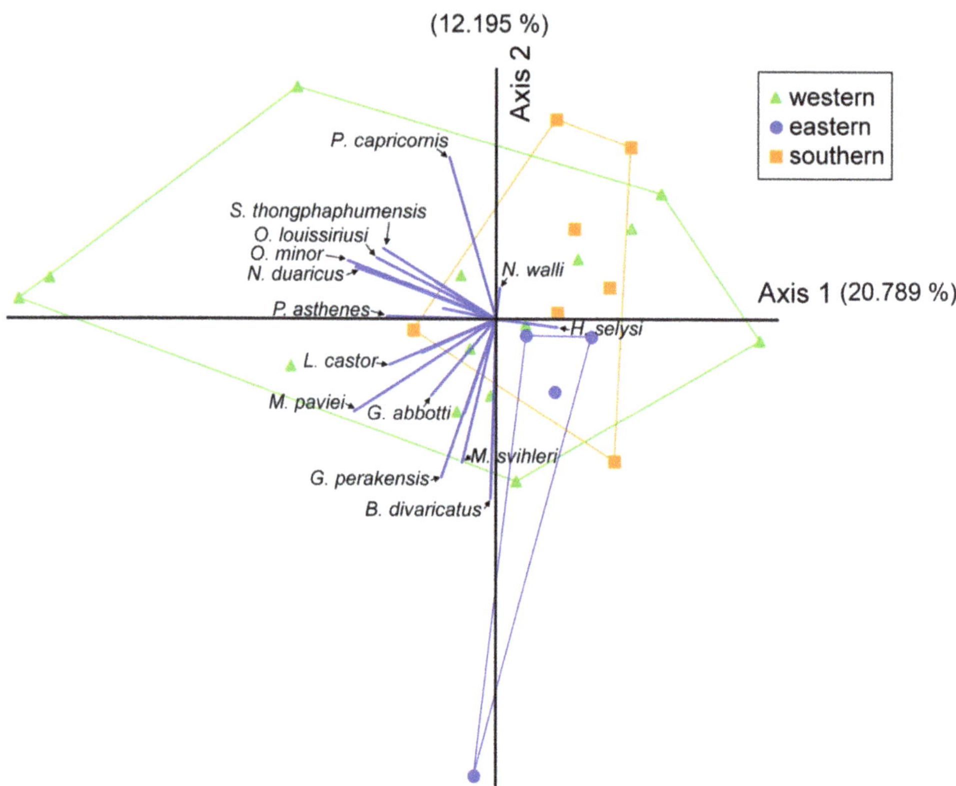

Figure 7. Principal component analysis (PCA) ordination biplots with sample and species scores of Gomphidae species (vectors of *S. thongphphumensis*, *O. louissiriusi* and *N. duaricus* are related to *O. minor*). Percentages of variance explained on the first two axes are indicated.

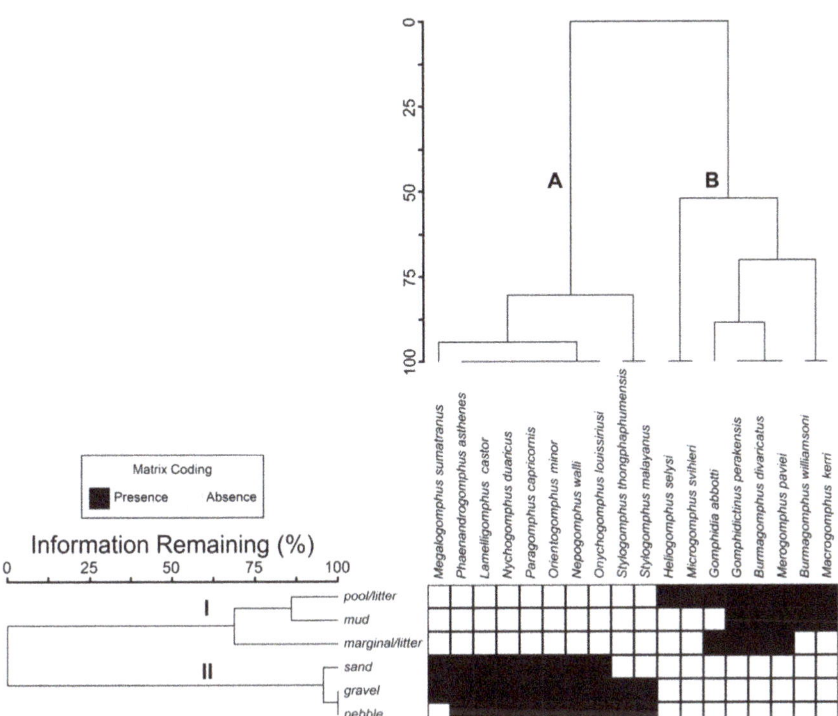

Figure 8. Two-way cluster analysis dendrogram of six microhabitats and 18 Gomphidae species. The two groups (I and II) of microhabitats and the two groups (A and B) of Gomphidae species are indicated.

4. Discussion

Based on the description of the nymph of *Amphigomphus hansoni* Chao, 1954 by Xu [30] and comparison with the nymphs of *Nihonogomphus lieftincki* Chao, 1954 and *O. armatus*, Xu [30] concluded that the nymphal morphological characters of the genus *Amphigomphus* are closer to those of *Orientogomphus* than of *Nihonogomphus*. Therefore, we selected two species of the genus *Amphigomphus* for comparison in this study. The evidence afforded by the characters of four gomphid nymphs species listed in Table 1 shows that the nymphs of *O. minor* can be distinguished from those of the other three species by a front margin of the median lobe furnished with about 50 finger-shaped serrations and the presence of lateral spines on S6–9. We also found that the *O. minor* nymph is similar to the *O. armatus* nymph only in the prementum length-to-width ratio, wing length, and mid-dorsal spines on the abdomen. The nymphs of *O. minor* share similarities with *Amphigomphus* nymphs in terms of wing pad length, mid-dorsal position on the abdomen and the shape of the third antennal segment [18,30].

We showed that *O. minor* nymphs were distributed in the western streams of Thailand and associated with other gomphid species (*N. duaricus*, *O. louissiriusi* and *S. thongphaphumensis*), which are restricted in their geographic distributions [10,11,16]. The nymphs of *O. minor* were usually found together with those of *Lamelligomphus*, *Nepogomphus*, *Onychogomphus*, *Paragomphus* and *Phaenandrogomphus*. These nymphs burrow deeply into the pebble, gravel, and sand substrates in streams. The nymphal microhabitat preference varies for Gomphidae [6], resulting in related morphological adaptations (e.g., burrowers in sand and mud (*Anisogomphus*, *Burmagomphus* and *Onychogomphus*) and in detritus accumulations (*Heliogomphus* and *Microgomphus*)).

Table 1. Comparison of morphological characters of four gomphid species (two *Orientogomphus* and two *Amphigomphus*, modified from [18,30]).

Characters/Species	***O. minor***	***O. armatus***	***A. somnuki* Hämäläinen, 1996**	***A. hansoni***
Third antennal segment of nymph	slender, stick-shaped, longer than antennal S1 + 2	spindle-shaped, shorter than antennal S1 + 2	cylindrical, parallel-sided, longer than antennal S1 + 2	stick-shaped, longer than antennal S1 + 2
Prementum length-to-width ratio	about 3:2	about 3:2	about 1:0.83	about 3:2
Front margin of median lobe	furnished with about 50 finger-shaped serrations	furnished with about 60 finger-shaped serrations	furnished with about 27 or 28 small teeth	furnished with about 40 finger-shaped serrations
Wing pads length	strongly divergent, reaching middle of S4	strongly divergent, reaching middle of S4	reaching basal half and posterior margin of S4,	strongly divergent, reaching middle of S4
Mid-dorsal spines on abdomen	present on S2–9	present on S2–9	present on S2–9	present on S2–9
Lateral spines on abdomen	present on S6–9	present on S7–9	present on S7–9	present on S7–9

Anthropogenic threats, such as deforestation, erosion, riparian vegetation removal, channelisation, and flow regulation, have effects on macroinvertebrate communities, including the odonate species composition [6]. Disturbance of the forest status is causing a decline in dragonfly species diversity, community composition, and structure [31]. The spatial distribution is influenced mainly by the presence of coarse detritus and by sediment particle size [32]. Removal of riparian vegetation also has a strong effect on odonate species composition and is associated with the loss of some species (*Dicterias atrosanguinea* Selys, 1853 and *Chalcopteryx scintillans* McLachlan, 1870) in Amazonia [33]. In Thailand, the need of protecting rivers and streams is increasing due to increasing human activities. For example, alterations in water flow by damming have affected the characteristics of stream ecosystems, resulting in altered microhabitats, water flows, and even changes from running water to standing water. The changes in microhabitat composition due to check dams also affect the community of dragonflies by changing the types and numbers of prey species, thereby affecting the food chain [34]. Therefore, knowledge of the microhabitat preferences of gomphid genera, which contain one or a few species (e.g., *Amphigomphus*, *Anisogomphus*, *Asahinagomphus*, *Asiagomphus*, *Davidius*, *Ethygomphus*, *Heliogomphus*, *Mattigomphus*, *Nihonogomphus*, *Siebodius*, *Stylogomphus*, *Sinictiogomphus*) can provide insight into the conservation issues of gomphid dragonfly nymphs in Thailand.

5. Conclusions

The taxonomic characteristics of *O. minor* are presented, and the nymph is described and illustrated for the first time from a reared specimen collected in streams of western Thailand. Morphological characteristics and distribution of *Orientogomphus* nymphs were discussed and compared to the related species and genera. The geographic distribution of *O. minor* is restricted to the western streams of Thailand and is associated with several other gomphid species with restricted distribution in the country (i.e., *N. duaricus*, *O. louissiriusi* and *S. thongphaphumensis*). Nymphs of the studied species burrow deeply into the pebble, gravel, and sandy substrates in streams.

Author Contributions: Conceptualization, B.B. and D.C.; methodology, D.C. and B.B.; software, B.B.; validation, D.C. and B.B.; investigation, D.C.; resources, B.B. and D.C.; specimen curation, D.C. and B.B.; data curation, D.C. and B.B.; writing—original draft preparation, D.C.; writing—review and editing, B.B.; funding acquisition, B.B. All authors have read and agreed to the published version of the manuscript.

Funding: This research was funded by the Faculty of Science, Kasetsart University Postgraduate Studentship (ScKUPGS), supported by the Center of Excellence on Biodiversity (BDC) Office of Higher Education Commission (BDC-PG4-160020).

Institutional Review Board Statement: This research was approved by the Institutional Animal Care and Use Committee, Faculty of Science, Kasetsart University, Thailand under Project number ACKU61-SCI-030.

Data Availability Statement: Data is available from the corresponding author upon request.

Acknowledgments: We are most grateful to our colleagues for assistance during field trips. We would like to thank the Department of Zoology and International SciKU Branding (ISB), the Faculty of Science at Kasetsart University in Bangkok for their assistance and use of their facilities. Finally, we are thankful to the guest editors and reviewers of our manuscript for their valuable comments.

Conflicts of Interest: The authors declare no conflict of interest.

References

1. Suhling, F.; Sahlén, G.; Gorb, S.; Kalkman, V.J.; Dijkstra, K.D.B.; van Tol, J. Oder Odonata. In *Ecology and General Biology: Thorp and Covich' Freshwater Invertebrates*, 4th ed.; Thorp, J.H., Rogers, D.C., Eds.; Academic Press: New York, NY, USA, 2015; pp. 894–932.
2. Novelo-Gutiérrez, R.; Ramírez, A.; González-Soriano, E. Superfamily Gomphidae. In *Keys to Neotropical Hexapoda: Thorp and Covich' Freshwater Invertebrates*, 4th ed.; Hamada, N., Thorp, J.H., Rogers, D.C., Eds.; Academic Press: New York, NY, USA, 2018; pp. 367–375.
3. Carle, F.L.; Kjer, K.M.; May, M.L. A molecular phylogeny and classification of Anisoptera (Odonata). *Arthropod Syst. Phylog.* **2015**, *73*, 281–301.
4. Kalkman, V.J.; Clausnitzer, V.; Dijkstra, K.D.B.; Orr, A.G.; Paulson, D.R.; van Tol, J. Global diversity of dragonflies (Odonata) in freshwater. *Hydrobiologia* **2008**, *595*, 351–363. [CrossRef]
5. May, M.L. Odonata: Who They Are and What They Have Done for Us Lately: Classification and Ecosystem Services of Dragonflies. *Insects* **2019**, *10*, 62. [CrossRef]
6. Dudgeon, D.D. Tropical Asian Streams Zoobenthos, Ecology and Conservation. Hong Kong University Press: Hong Kong, China, 1999.
7. Sasamoto, A. *Anisogomphus yanagisawai* sp. nov., a new gomphid dragonfly from northern Thailand (Odonata: Anisoptera: Gomphidae). *Zootaxa* **2015**, *3904*, 421–426. [CrossRef] [PubMed]
8. Makbun, N. *Anisogomphus yingsaki* (Odonata: Gomphidae) sp. nov., a new gomphid species from Thailand. *Zootaxa* **2017**, *4306*, 437–443. [CrossRef]
9. Makbun, N. A new species of the genus *Burmagomphus* Williamson (Odonata: Gomphidae) from Northern Thailand. *Zootaxa* **2017**, *4269*, 133–136. [CrossRef]
10. Chainthong, D.; Sartori, M.; Boonsoong, B. *Stylogomphus thongphaphumensis* (Odonata: Anisoptera: Gomphidae), a new gomphid dragonfly and the first record of *S. malayanus* Sasamoto, 2001 from Thailand. *Zootaxa* **2020**, *4763*, 231–245. [CrossRef] [PubMed]
11. Fleck, G. *Onychogomphus (Siriusonychogomphus) louissiriusi*, a new species and new subgenus from Thailand (Odonata: Anisoptera: Gomphidae). *Faunitaxys* **2021**, *8*, 1–9.
12. Makbun, N.; Fleck, G. Description of *Microgomphus farrelli* sp. nov. (Odonata: Anisoptera: Gomphidae) based on adults of both sexes and larvae from Northern Thailand. *Zootaxa* **2018**, *4422*, 442–450. [CrossRef]
13. Ferro, M.L.; Sites, R.W. Description of the *Gomphidictinus perakensis* (Laidlaw) (Odonata: Gomphidae) with distributional notes. *Proc. Entomol. Soc. Wash.* **2006**, *108*, 76–81.
14. Boonsoong, B.; Chainthong, D. Description of the final-instar larva of *Heliogomphus selysi* Fraser (Odonata: Gomphidae). *Zootaxa* **2014**, *3764*, 482–488. [CrossRef] [PubMed]
15. Boonsoong, B.; Chainthong, D. Description of the last stadium larva and female of *Microgomphus thailandica* Asahina, 1981 (Odonata: Gomphidae). *Zootaxa* **2014**, *3811*, 271–279. [CrossRef] [PubMed]
16. Chainthong, D.; Boonsoong, B. Description of two final stadium *Onychogomphus* larvae from Thailand (Odonata: Gomphidae). *Zootaxa* **2016**, *4066*, 561–570. [CrossRef] [PubMed]
17. Novelo-Gutiérrez, R.; Sites, R.W. The larvae of *Phaenandrogomphus* Lieftinck, 1964 in Thailand, including the description of *P. tonkinicus* (Fraser, 1926) with a larval diagnosis and new province records of *P. asthenes* Lieftinck, 1964 (Odonata: Gomphidae). *Zootaxa* **2019**, *4700*, 377–384. [CrossRef] [PubMed]
18. Novelo-Gutiérrez, R.; Sites, R.W. The larva of *Amphigomphus somnuki* Hämäläinen, 1996 and the first records of the genus *Stylogomphus* Fraser, 1922 for Thailand (Odonata: Gomphidae). *Zootaxa* **2019**, *4555*, 121–126. [CrossRef] [PubMed]
19. Chao, H.F.; Xu, J.F. Descriptions of a new genus and species of gomphid dragonfly reared from nymph in Fujian Province, with notes on allied species (Gomphidae: Onychogomphinae). *J. Fujian Agric. Coll.* **1987**, *16*, 259–266.
20. World Odonata List. Available online: https://www2.pugetsound.edu/academics/academic-resources/slater-museum/biodiversity-resources/dragonflies/world-odonata-list2/ (accessed on 9 March 2022).
21. Wilson, K.D.P. Crepuscular activity in *Orientogomphus minor* (Laidlaw) comb. nov. from Thailand and clarification of the taxonomic status of closely related species. *Agrion* **2008**, *12*, 15–21.

22. Chao, H.F. *The Gomphid Dragonflies of China (Odonata: Gomphidae)*; The Science and Technology Publishing House: Fuzhou, China, 1990.
23. Lieftinck, M.A. Synonymic notes on East Asiatic Gomphidae with descriptions of two new species (Odonata). *Zool. Meded.* **1964**, *39*, 89–110.
24. Shorthouse, D.P. SimpleMappr, an Online Tool to Produce Publication-Quality Point Maps. Available online: https://www.simplemappr.net (accessed on 4 March 2022).
25. Asahina, S.A. list of the Odonata from Thailand. Thailand. Part XIV. Gomphidae-2. *Tombo* **1986**, *29*, 7–53.
26. Watson, M.C. The utilization of mandibular armature in taxonomic studies of anisopterous nymphs. *Trans. Am. Entomol. Soc.* **1956**, *81*, 155–202.
27. McCune, B.; Mefford, M.J. *PC-ORD. Multivariate Analysis of Ecological Data*; MjM Software Design: Gleneden Beach, OR, USA, 2011.
28. Hämäläinen, M.; Pinratana, A. Distribution maps by provinces. In *Atlas of the Dragonflies of Thailand*; Gabriel in Thailand: Bangkok, Thailand, 1999.
29. Kosterin, O.E. Notes on intraspecific variation of some Gomphidae (Odonata) species in Cambodia. *Int. Dragonfly Fund* **2014**, *68*, 1–16.
30. Xu, Q.H. Description of the last instar larva of *Amphigomphus hansoni* Chao, with notes on the systematic status of the genus *Amphigomphus chao* (Anisoptera: Gomphidae). *Odonatologica* **2012**, *41*, 55–59.
31. Rith-Najarian, J. The influence of forest vegetation variables on the distribution and diversity of dragonflies in a northern Minnesota forest landscape: A preliminary study (Anisoptera). *Odonatologica* **1998**, *27*, 335–351.
32. Suhling, F. Spatial distribution of the larvae of *Gomphus pulchellus* Selys (Anisoptera: Gomphidae). *Advan. Odonatol.* **1994**, *6*, 101–111.
33. da Silva Monteiro Júnior, C.; Couceiro, S.R.M.; Hamada, N.; Juen, L. Effect of vegetation removal for road building on richness and composition of Odonata communities in Amazonia, Brazil. *Int. J. Odonatol.* **2013**, *16*, 135–144. [CrossRef]
34. Kositpon, T.; Phalaraksh, C. Effects of check dams on water quality and macroinvertebrate diversity of Hom Jom Stream, Lamphun Province, Thailand. *Suranaree J. Sci. Technol.* **2012**, *19*, 113–123.

 diversity

MDPI

Article

Diversity, Status and Phenology of the Dragonflies and Damselflies of Cyprus (Insecta: Odonata)

David J. Sparrow [1], Geert De Knijf [2,*] and Rosalyn L. Sparrow [1]

1 Cyprus Dragonfly Study Group, P.O. Box 62624, Paphos 8047, Cyprus; davidrospfo@hotmail.com (D.J.S.); dragonfliescyprus@gmail.com (R.L.S.)
2 Research Institute for Nature and Forest (INBO), Havenlaan 88 bus 73, 1000 Brussels, Belgium
* Correspondence: geert.deknijf@inbo.be; Tel.: +32-476-403-454

Abstract: Based on literature data, unpublished material and the results of the year-round monitoring at selected sites island-wide by the Cyprus Dragonfly Study Group since 2013, we acquired an excellent knowledge of the diversity and status of the Odonata of Cyprus. Altogether, 37 species are on the island's checklist. *Ischnura pumilio, Aeshna affinis and Brachythemis impartita* were only very rarely recorded in the past but are considered to be no longer present. The single record of *Calopteryx virgo* from 1930 is in our opinion a misidentification and has been removed from the checklist. The island has a rather impoverished odonate fauna compared to neighbouring countries. There are no endemic species, but the island is home to some range of restricted species of which *Ischnura intermedia* is the most important. Flight seasons determined for the 31 species with sufficient data were generally found to be longer than reported for other countries in the Eastern Mediterranean. This may be due to intensive year-round monitoring but could also result from Cyprus' warmer climate. Very wide annual variations were found in the abundance of all species over the seven years and show an almost immediate response to the wide fluctuations in Cyprus' annual rainfall levels.

Keywords: odonate; flight period; checklist; Eastern Mediterranean; citizen science; climate

Citation: Sparrow, D.J.; De Knijf, G.; Sparrow, R.L. Diversity, Status and Phenology of the Dragonflies and Damselflies of Cyprus (Insecta: Odonata). *Diversity* **2021**, *13*, 532. https://doi.org/10.3390/d13110532

Academic Editors: Luc Legal, Marina Vilenica, Zohar Yanai and Laurent Vuataz

Received: 29 September 2021
Accepted: 22 October 2021
Published: 25 October 2021

1. Introduction

Biodiversity monitoring is an essential first step in being able to track changes in ecosystems and species distributions and abundances locally and contributes to an important global understanding of trends. Records collected through structured monitoring schemes, as well as opportunistic observations, can assist in developing conservation measures aimed at halting or reversing declines in species and habitat quality. Data collected through citizen science projects are increasingly important for assessing biodiversity at global and regional scales [1]. For dragonflies and damselflies in Europe, such detailed information is only available for northern and central localities, but not in any depth for any country or the area in general in the Eastern Mediterranean [2,3]. Kalkman and van Pelt [4] published data on the distribution and flight season of species in Turkey but acknowledged that their records were not evenly distributed over the year. In the European Atlas of Odonata [5], flight season data for the Maghreb, Turkey and Greece, were included but it was stated that the data were limited since very little recording had been conducted outside the main summer holiday season. The recently published Atlas of the dragonflies and damselflies of West and Central Asia [6] also contains phenology data, which includes all species present on Cyprus. The data, however, need to be interpreted with care since flight seasons for species for which there are few records may be longer than indicated and for widespread species may be longer than that found at any given location.

As a first step towards filling the gap in our knowledge in the Eastern Mediterranean, the Cyprus Dragonfly Study Group (CDSG) was established to carry out year-round monitoring of the island's Odonata, starting at the beginning of 2013. In this paper, we present the results from the first seven years of this monitoring programme. This has allowed us to determine the status of each species and the flight season of all but the island's rarest species with a high degree of confidence. Although not as thorough as the CDSG programme, an expedition in June 1994 [7] and an intensive survey from June 2003 to September 2004 on the northern side of the island [8] gave us insights into how the status of species has changed and the impact of climate change over this 26-year timeline.

2. Study Area

Cyprus, the third-largest Mediterranean island, lying at the eastern extremity of the Mediterranean basin, sits at the intersection of the Middle East (Asia), Europe and Africa. The island (Figure 1) is defined by two mountain ranges, the Troodos massif and the Kyrenia (Pentadaktylos) range, separated by a flat, broad east–west plain, the Mesaoria. The Troodos massif dominates the south, west and central part of the island, rising to 1951 m a.s.l. The Kyrenia range, which rises to a maximum of 1024 m a.s.l., is a narrow, largely unbroken ridge that runs for approximately 160 km along the north coast. A coastal plain up to five km in width separates it from the sea.

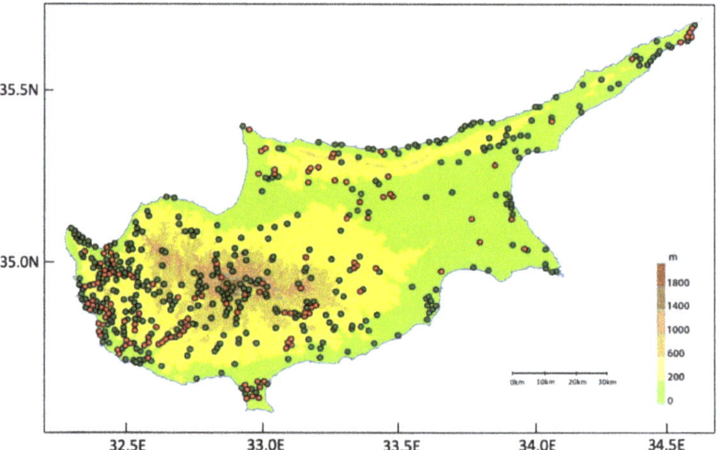

Figure 1. Map of localities from which dragonflies and damselflies have been recorded by the CDSG on Cyprus (n = 703). The red dots relate to sites (n = 136) that have been monitored for at least two years, and the green dots to other non-regularly surveyed sites.

Cyprus has an intense Mediterranean climate with hot rainless summers from mid-May to mid-September and generally mild and rainy winters from November to mid-March. Consequently, annual rainfall is measured for hydrometeorological years (1-x to 30-ix). The average annual rainfall is considered to be 503 mm (based on the average for the thirty-year period from 1961/1962 to 1990/1991) [9], but the island has a high regional variation with the wettest area (1100 mm per annum) at the top of the Troodos massif, dropping to a low of 300 to 350 mm per annum on the Mesaoria plain and in the east. Hence, the main rivers arise on the Troodos massif and the lotic Odonata species are restricted to this region. In response to the water stress levels, numerous reservoirs and water storage tanks have been created with 108 listed [10]. All the major rivers are dammed and the resulting reservoirs, along with other water storage tanks, have created new habitats for lentic species, but have had a major impact on the downstream habitats. Annual rainfall levels vary significantly and during the study period varied from 309 mm in the hydrometeorological year 2015/2016 to 795 mm in 2018/2019 [9]. This results in many habitats being highly unstable. Reservoir levels can change very rapidly. Rivers can go from being in flood to having no flow and also becoming overgrown with vegetation.

The most stable habitats are those fed by springs, which may form small pools, streams or feed village tanks and swimming pools that attract several species.

3. Methods

A monitoring programme of the dragonflies and damselflies of Cyprus by the CDSG was set up at the start of 2013. Just over 50 sites were initially selected for either monthly or twice monthly dragonfly recording. Given the limited resources (ca. 10 regular and active volunteers), the selection was made to ensure that all the then-known main species and all habitat types were included and that there was good geographic coverage. A transect was defined for each site, varying in length from a point count to a line transect of 400 m. When counting dragonflies, the recorder walks in one direction, counting individuals observed in front of them, over a section of water and the bank. Any individual that flies in from behind the observer is not counted. Counting dragonflies from a single point was performed where it was not possible to count them on a line transect. In this case, the water and bank vegetation were visually scanned for the presence and abundance of each species of Odonata. Details of all species present along these transects were noted, including evidence of breeding behaviour. With relatively few odonates, all of which can be easily identified in the field, netting is rarely needed. It was, therefore, possible to train inexperienced volunteers in dragonfly recording quite rapidly. Some initial problems were encountered with the pruinose-blue *Orthetrum* species, but when in doubt, the recorders were encouraged to photograph the specimens for later confirmation. We were fortunate in having a team of dedicated volunteers since year-round recording required a high degree of commitment.

The results presented here relate to the years 2013 to 2019 (the study period). During this period, several sites became no longer worthwhile for continued monitoring (see Section 2). However, the CDSG has constantly sought out new localities for monitoring, and once a location became no longer viable, it could be replaced by a new site. This resulted in records from 703 sites island-wide, 136 of which were monitored for a significant time during the study period (Figure 1).

From the data collected, we were able to determine the status of each of the odonate species on Cyprus, based on the number of localities from where it was recorded. Species occurring at more than 150 of the 703 sites are considered to be 'very common'; those above 100 to be 'common'; above 50 are 'rather scarce'; more than 10 sites are 'scarce'; and less than 10 are 'extremely rare'. Species on the checklist for which we have no records since 2013 or the recent study period were considered to be no longer present on the island, given the intensity of monitoring. We then compared these data to the previous studies in June 1994 [7] and June 2003 to September 2004 [8].

Flight season data were derived from the CDGS database and were limited to the period 2013–2019. The data for each species were grouped into 10-day periods (decades) for each month, logging the number of records (one species observed at one site on one day) and abundance (count of adults) into each monthly decade. Data of records of just larvae or exuviae were excluded in this analysis. These results were then graphed out for each species. Species with less than 45 records from the study period were omitted from the phenology calculation. The protocol for determining flight season corresponds with the one used in the European Atlas [5]. The flight season for each species was determined based on the records and since the earliest and latest sightings often refer to unusual events, the flight season was defined as the first and last decade in which one to 99% of the records had been made. The main flight season, when there was a greater chance of observing the species, was also determined, being defined as the first and last decade in which 5% or more of the total number of records occur. We then compared these data with that for Turkey [4] and neighbouring countries presented in the European Atlas [5].

4. Results

During the study period, a total of 7877 visits were made to 703 sites resulting in 23,899 records with a count of 343,008 adults. The annual breakdown of these data is shown in Table 1 and Appendix A summarises the number of records and abundance per species per year for the study period and for completion, the earliest and latest annual sightings are given for each species.

Table 1. Some general results of the Cyprus Dragonfly monitoring schemes for the period 2013 to 2019.

	2013	2014	2015	2016	2017	2018	2019	2013–2019
Number of visits	865	1088	1042	1049	1062	1315	1456	7877
Number of sites	171	126	139	162	214	237	310	703
Number of records	2937	2893	3273	3233	2970	4089	4504	23,899
Count of adults	67,252	41,434	53,178	36,984	35,981	42,866	65,313	343,008

4.1. Status

The status of the dragonflies and damselflies on the Cyprus checklist based on the total number of sites from which each was observed during the study period is presented in Table 2. From the total of the 37 species ever observed on Cyprus, 15 can be considered at least as common ('common' and 'very common'), another 16 species are scarce and can be found at a rather reduced number of sites, three species are extremely rare and are limited to a handful of sites and three species were not recently observed. Thus 34 species were observed during the study period, and of these, there were a sufficient number of records to determine with confidence the flight season data for 30 of them, plus good indicative data for one other, *Caliaeschna microstigma*, for which we only have 47 records of adults on the wing.

Table 2. Status of the dragonflies and damselflies of Cyprus, based on the total number of sites (N_{max} = 703) where a species was observed during 2013–2019.

Species	Total	%	Status	Criteria
Sympetrum fonscolombii	282	40.1	very common	
Crocothemis erythraea	247	35.1	very common	
Trithemis annulata	231	32.9	very common	
Ischnura elegans	216	30.7	very common	
Sympetrum striolatum	202	28.7	very common	
Trithemis arteriosa	193	27.5	very common	
Orthetrum chrysostigma	178	25.3	very common	
Anax parthenope	152	21.6	very common	≥150 loc
Orthetrum brunneum	143	20.3	Common	
Orthetrum coerulescens	138	19.6	Common	
Calopteryx splendens	134	19.1	Common	
Sympecma fusca	124	17.6	Common	
Epallage fatime	117	16.6	Common	
Anax ephippiger	117	16.6	Common	
Pantala flavescens	109	15.5	Common	≥100 loc
Aeshna mixta	95	13.5	Rather scarce	
Onychogomphus forcipatus	93	13.2	Rather scarce	
Trithemis festiva	90	12.8	Rather scarce	
Chalcolestes parvidens	88	12.5	Rather scarce	
Orthetrum taeniolatum	76	10.8	Rather scarce	
Orthetrum sabina	73	10.4	Rather scarce	
Anax imperator	58	8.3	Rather scarce	≥50 loc

Table 2. *Cont.*

Species	Total	%	Status	Criteria
Anax immaculifrons	47	6.7	Scarce	
Selysiothemis nigra	41	5.8	Scarce	
Diplacodes lefebvrii	39	5.5	Scarce	
Orthetrum cancellatum	36	5.1	Scarce	
Erythromma lindenii	32	4.6	Scarce	
Caliaeschna microstigma	28	4.0	Scarce	
Ischnura intermedia	24	3.4	Scarce	
Erythromma viridulum	19	2.7	Scarce	
Lestes macrostigma	10	1.4	Scarce	≥10 loc
Sympetrum meridionale	4	0.6	extremely rare	
Aeshna isoceles	2	0.3	extremely rare	
Lestes barbarus	1	0.1	extremely rare	<10 loc
Ischnura pumilio	0	0.0	no longer present	
Aeshna affinis	0	0.0	no longer present	
Brachythemis impartita	0	0.0	no longer present	

4.2. Phenology

Dragonflies and damselflies were observed on the wing during every month of the year. The cumulative number of records and the number of species observed in each month during the period 2013–2019 is given in Figure 2. During January and February, the number of records and abundance was low but picked up in March and April, peaking in May and June through to August, followed by a gentle decline to the end of the year. A high percentage of the island's species were recorded from March to November, a consequence of many species having a long flight season.

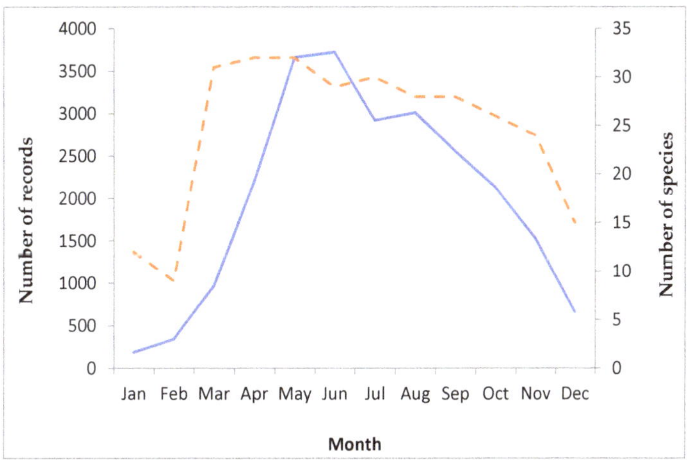

Figure 2. The total number of records (continuous blue line) and number of species recorded (dotted red line) for each month for the dragonflies and damselflies of Cyprus for the period 2013–2019.

Flight season per decade (10-day period) of adult Odonata in Cyprus for the period 2013–2019 is presented in Figure 3. Species with fewer than 45 records were not included in this analysis. A distinction was made between the flight season, which contains decades with up to 99% of the records, and the main flight season, being defined as the first and last decade in which 5% or more of the total number of records occurred.

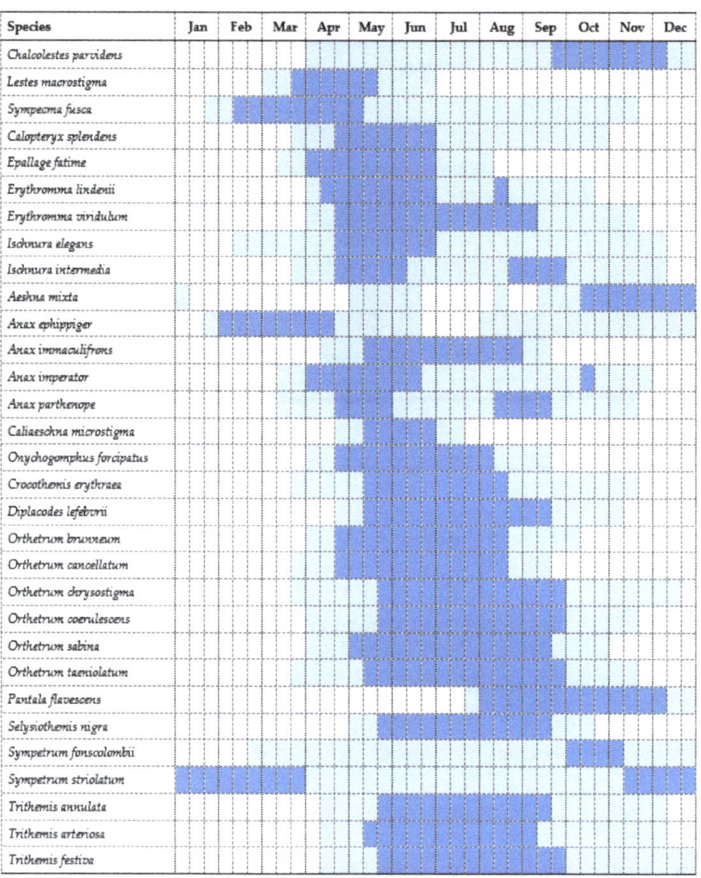

Figure 3. The flight season of the Odonata present in Cyprus based on records for the period 2013 to 2019. Flight season (light shading) and main flight season (dark shading).

5. Discussion

One of the characteristics of islands is that they often have a low number of species compared to neighbouring countries but a high rate of endemic species [11]. With only 37 species ever observed on Cyprus, the island has a relatively poor dragonfly diversity compared with neighbouring countries such as Turkey with 105 species [6], Syria with 67 species [6], Greece with 76 species [5] but is rather similar to Crete with 35 species [5] and Egypt with 39 species [12]. Although the nearby Mediterranean island of Crete has a similar size, it holds two endemic species [5], while no endemic species are present in Cyprus. The broad separation of Crete from mainland Greece (160 km) compared to the shorter distance from Cyprus to Turkey and Syria (70 km) may be a factor. It may result in a higher probability of migration and gene exchanges with mainland populations compared to Crete. Migration and gene exchanges lower the likelihood of local adaptation and species radiation. However, in some orders, Cyprus has similar or higher levels of endemism than Crete, for instance, six endemic butterflies compared to four in Crete. It has been suggested [8] that the absence of endemism in odonates might be a result of the unstable nature of the habitats and the need for recolonisation preventing the development of insular characteristics. Just a few species present in Cyprus have a rather restricted range, e.g., *Ischnura intermedia* [13], *Epallage fatime* [6], *Caliaeschna microstigma* [14] and occur also in West Asia. Nearly all dragonfly species found in Cyprus are widely dispersed either in

Europe, Asia, or Africa. Several of them are well known migrants such as *Anax ephippiger*, *Pantala flavescens* [15] and *Sympetrum fonscolombii*.

5.1. Status

Six of the eight species in the 'very common' category were recognised as being thus for a long time by comparison with 1994 [7] and 2003/2004 [8] data. The presence of *Orthetrum chrysostigma* and *Trithemis arteriosa* in this category, however, indicates a recent strong expansion on the island. The seven species in the common category include the first two obligate lotic species, *Calopteryx splendens* and *Epallage fatime*, whose distribution is restricted to the streams of the Troodos and western Cyprus, where *Trithemis festiva* is also locally common. Although also in the common category, the two most prominent migrants, *Anax ephippiger* and *Pantala flavescens* show strong yearly fluctuations in numbers (Appendix A). Although very common, and more so than *Anax parthenope* on many Mediterranean islands, *Anax imperator* is ranked as 'rather scarce' on Cyprus. This was already noted by Lopau and Adena [7] who attributed this to its late arrival on the island. However, two decades later it still has only managed to establish a modest presence. We, therefore, presume that the absence of many suitable reproduction habitats, e.g., permanent waters, has a stronger impact. *Orthetrum sabina*, although still rather scarce, has none the less expanded its range significantly since 1994. Earlier population sizes on Cyprus, and even more generally in Europe are small (<10 individuals) [7,8,16] but we have on several occasions observed populations of several hundred individuals. Although *Anax immaculifrons* has been found in less than 10% of the investigated sites, this species is rather widespread in the western part of Cyprus and is a regular visitor to some swimming pools in the Paphos area, where reproductive behaviour, including oviposition has been observed. *Caliaeschna microstigma* is unusual in being active in the late afternoon and early evening when only limited monitoring is carried out by the CDSG and also many of the localities where it occurs are not easily accessible. Consequently, it has almost certainly been under-recorded, which is supported by the large number of exuviae found, indicating that it is much more common than suggested by its scarce ranking. Of the species in the extremely rare category, *Sympetrum meridionale* was found to be the most elusive anisopteran with no established populations or localities where it could regularly be seen: there is just one record from April and two records from November. *Aeshna isoceles* was rediscovered in 2019 and was observed between April and July at two locations in a river valley on the Karpas Peninsula, where populations seemed to be well established [17]. Only a single *Lestes barbarus* individual was observed in August 2019 and this was assumed to be a migrant [17]. Finally, three species, *Ischnura pumilio*, *Aeshna affinis* and *Brachythemis impartita* were not observed during the study period and are assumed to be no longer present on the island. None of these three species ever had a strong presence on Cyprus. For *Brachythemis impartita*, there was only one set of records (1 male and 2 females) in 2006, and these were assumed to be accidental visitors. *Brachythemis impartita* has in recent decades extended its range into Europe and West Asia and to the northeast is found in the Levant and the southernmost part of Turkey [6]. Thus, we can expect that *B. impartita* may be able to colonise Cyprus. For *Ischnura pumilio*, there are only two records from Cyprus: one from 1893 and one from 1947. These also could reasonably be considered to be migrants with no established populations. It is known as a pioneer species with a preference for sparsely vegetated habitats. It seems unlikely that colonisation will occur from Turkey or Syria, where this habitat is also under severe pressure. *Aeshna affinis* was last recorded from Cyprus in 1994. It is a well-known disperser and was until two decades ago a rare observation in most of north-western Europe [5]. As Cyprus is at the south-eastern edge of the range of the species [6], it is not unlikely that the former observations are the result of one or several influxes on the island.

There is one published record of *Calopteryx virgo* from Cyprus, reported by Navas [18], who examined material collected by Mavromoustakis, an eminent local hymenopterist, in 1930 at an altitude above 1000 m asl in the Troodos. Since then, this species has appeared on every published checklist of the island's Odonata. Neither the CDSG nor any other observers have since been able to find a second specimen on the island. Lopau and Adena [7] already considered that this single record was a misidentification of a particularly blue *C. splendens* specimen. The ideal summer water temperatures for *C. virgo* are between 13 and 18 °C [19]. We measured water temperatures in streams in July and August and found that at altitudes above 1200 m, there were waters cool enough to support *C. virgo*. However, these occupy a very small geographic area and are on very precipitous mountain sides that are liable to have severe destructive flash floods during the winter months when rainfall on the Troodos can be torrential. Whereas other species such as *Epallage fatime* could recolonise such mountain streams by spreading upwards from the lower slopes, any species found only in the highest reaches would be particularly vulnerable to local extinction. We, therefore, concur with Lopau and Adena [7] and consider that the species never was found on Cyprus and remove *Calopteryx virgo* from the island's checklist.

5.2. Yearly Variation in Abundance

Monitoring programmes are typically set up to detect trends in the long term, such as the monitoring demanded by the European Union for the species mentioned on the Habitats Directive where a timeline of 24 years is stipulated [20]. It is obvious that our timeframe 2013–2019 is too short to detect statistically sound trends. Nevertheless, some clear differences in the observed number of adults over the years could be detected, which is much more marked than that of the number of records for each species (Appendix A). To correct for sampling effort, we calculated the mean numbers of adults being observed per visit and per year (Figure 4). These numbers correlate very well with the amount of rainfall level during the previous winter (Figure 4). By far, the highest abundance was observed in 2013. Although rainfall during the winter of 2012/2013 was somewhat above normal, this followed four previous winters of well above average rainfall. Reservoirs were full and there was permanent flow along much longer stretches of the rivers than observed in years of low rainfall. In contrast, rainfall during the 2013/2014 winter was well below average and the abundance of the island's odonate populations was immediately reduced. The well above average rainfall during the winter of 2014/2015 resulted in a noticeable rapid recovery of abundance but was reversed by the exceptionally low rainfall during the winters of the next three years (2015/2016, 2016/2017 and 2017/2018). Reservoir levels were low, many stretches of rivers had no flow during the entire period and significant overgrowth with reeds occurred. The winter of 2018/2019 was exceptionally wet, in fact the wettest since record keeping started in 1901. Reservoirs filled, dams overflowed and there was extensive flooding on most of the rivers. Extensive damage resulted; several of the sites that were being monitored were washed away, and, in many locations, the developing larvae may also have been affected. Nonetheless, there was an upswing in the adult abundance during the 2019 season.

This variability is typical of Cyprus' climate, which is characterised by repeating cycles of drought years such as seen 2016 to 2018, resulting in many local extinctions of odonate populations. Consequently, Cyprus' most successful odonate species are those that are habitat generalists and able to rapidly recolonise former habitats or new habitats as they are formed. However, there is also a climatic problem that is equally, if not more worrying than global warming for the flora and fauna of Cyprus, particularly for the odonates. From 1901 onwards, when climate variables started to be recorded, there was a clear decline in rainfall levels (Table 3). The average decline was almost one millimetre per year, a trend that is predicted to continue [9]. Seeing the impact of low rainfall levels, this does not bode well for the future of Cyprus' odonates and we may expect species that are habitat specific and less adept at recolonising to be hit first and hardest. Such clear declines in rainfall have already been observed in Mediterranean-type landscapes in south-western Australia over the last 40 years [21]. Besides climate change, the growing human impact during the past century has also been detrimental to many wetland habitats in Cyprus. The intensified exploitation of water resources and especially dam building, which is well recognised to have impoverished the valleys below the dams, has certainly affected the population of several species, especially those dependent upon streams and rivers or those species, which only occur in a limited number of sites.

Figure 4. Relation between annual rainfall (continuous blue line) and the mean adult count per visit per year (dotted line in red).

Table 3. Average annual rainfall levels for the four 30-year periods since 1901.

30-Year Period	1901/02–1930/31	1931/32–1960/61	1961/62–1990/91	1991/92–2020/21
Average annual rainfall mm	559	524	503	476

5.3. Phenology

With seven years of recording experience and a large number of records, we have a good dataset to determine the flight season for 31 of Cyprus' species with a high degree of confidence. Many have long flight seasons (Figure 3) and longer than that reported for their conspecifics in neighbouring countries [4–6].

Just three species, *Lestes macrostigma*, *Epallage fatime* and *Caliaeschna microstigma*, all univoltine, emerge early in the year and have short flight seasons (Figure 4). The duration of the flight season for the first two species is similar to Turkey, Greece and Bulgaria [4,22,23] but the emergence is around one month earlier, which can be explained by the year-round warmer climate in Cyprus. For *C. microstigma*, our flight season is shorter but falls well inside that for Turkey [24] and may be a consequence of our rather low number (47) of records.

At least five species are present year-round on Cyprus. One of these, *Sympecma fusca* overwinters as an adult across its range [25] and mating activity on Cyprus takes place from the beginning of February to the end of April. *Sympetrum striolatum* not only overwinters as an adult on Cyprus as previously mentioned [4,26], but it is the main breeding season. In Europe, the species has normally been observed only until December [27], but it overwinters as an adult in North Africa [28,29] and winter breeding is known to occur in northern Algeria [30]. On Cyprus, after emergence in spring, reproductive activity is delayed until mid-October and continues through to the end of March. For three other species, *Ischnura elegans*, *Sympetrum fonscolombii* and *Trithemis annulata*, the numbers of adults observed in January and February are very low, and most likely these species do not overwinter as adults or only very occasionally. For *I. elegans* and *T. annulate*, numbers start to pick up from March and April, respectively, and mating activity has then been observed right through the remainder of the flight season. However, although not visible on Figure 3, *S. fonscolombii*, shows a typical bivoltine lifecycle with mating occurring from the beginning of April to the end of June and then from mid-September to mid-November, when hundreds of females have been observed ovipositing in tandem over the complete area of many reservoirs and other lentic water bodies. It seems likely that *Anax imperator* and *A. parthenope* have a bivoltine lifecycle in Cyprus (Figure 3), but a dedicated survey of larvae and exuviae is needed to confirm this.

Apart from *S. striolatum*, two other species delay reproductive activity until autumn/winter. For *Chalcolestes parvidens*, there are a few records from January to April, but numbers start to build up from May. Reproductive activity, however, is delayed until the autumn/winter, starting from mid-September and peaking from mid-October to the end of November, much later than observed elsewhere [31]. This appears to be a response to Cyprus' warmer climate and a need to wait for cooler months to ensure the eggs enter a diapause. *Aeshna mixta* also delays reproductive activity to autumn/winter, emerging from March/April but then moving away from the breeding grounds with mating observed from mid-October to the end of December.

Cyprus' two main migrant species have very different flight seasons. There are records from every month except July for the migratory *Anax ephippiger*, although the main influx occurs from the beginning of February to the end of April when in some years vast swarms have been observed. The few individuals observed from August onwards are thought to be mainly offspring from the spring influx. In contrast, there are no records for *Pantala flavescens* from mid-January to the beginning of July, when it has then been observed to the end of the year. It is also confirmed to breed on Cyprus [15].

Orthetrum chrysostigma, *Crocothemis erythraea*, *Trithemis arteriosa* and *Trithemis festiva* also have long flight seasons with records from every month of the year except February or January in the case of *C. erythraea*. The first three species have their main distribution range in Africa where they can be observed year-round [32] just as in many parts of the Arabian Peninsula [6]. *Trithemis festiva* is a mainly Oriental species and its flight season on Cyprus is considerably longer than that reported for Greece and Turkey [33] and somewhat longer than that for West and Central Asia [6]. The range restricted and threatened *Ischnura intermedia* has at least two and possibly even three generations a year and records are only lacking from January and February [13].

The remaining species mostly have flight seasons that are longer, emerging earlier and staying on the wing longer than that reported for Greece and Turkey [4,5]. It is possible that the longer flight season reported for Cyprus may just be a consequence of more intensive year-round monitoring but could also be a result of the warmer climate enabling an earlier emergence and facilitating more generations per year.

Author Contributions: Conceptualisation, G.D.K. and D.J.S.; methodology, G.D.K. and D.J.S.; formal analysis, G.D.K. and D.J.S.; data curation and management, D.J.S. and R.L.S.; writing—original draft preparation, D.J.S. and G.D.K.; writing—review and editing, G.D.K., D.J.S. and R.L.S.; supervision, G.D.K. All authors have read and agreed to the published version of the manuscript.

Funding: This research received no external funding.

Institutional Review Board Statement: Not applicable.

Informed Consent Statement: Not applicable.

Data Availability Statement: All the data were obtained through volunteer-based monitoring and are available from the first author on reasonable request.

Acknowledgments: The authors are indebted to the volunteer members, past and present, of the CDSG, whose dedication, commitment and regular contributions to odonate monitoring made this paper possible: Ray Atkinson, Dave Bowler, Linda Crossley, Vicky Dobson, Pete Evans, Danne Johnson, Roger and Heather Kent, Dinos Konis, Jenny Lewis, Shane Marshall, Mary Michaelides, Allison Muller, Athina Papatheodoulou, Klaus Siedle, Matt Smith, Bill Stokes, Heather Stroud, Annie Walker, Dave and Jan Walker. Many thanks go to the many other observers who have contributed records to the CDSG database. The Paphos 3rd Age (P3A) is thanked for providing an umbrella under which the CDSG could operate and for providing financial support for equipment, software and books. Our thanks also go to the anonymous reviewers whose helpful comments improved our manuscript.

Conflicts of Interest: The authors declare no conflict of interest.

Appendix A

Table A1. Number of records and adult counts (records/counts of adults) of dragonflies and damselflies on Cyprus and earliest and latest annual sighting for each species stored in the CDSG database for the period 2013 to 2019. * Species for which the flight season extends into the early months of the following year.

Species	2013	2014	2015	2016	2017	2018	2019	Total Records	Total Ind.	Earliest Sighting	Latest Sighting
Chalcolestes parvidens	86/2189	71/354	122/1626	76/685	120/1780	111/2799	144/2716	730	12,149	20-iii-2014	02-ii-2015 *
Lestes barbarus	0/0	0/0	0/0	0/0	0/0	0/0	3/3	3	3	14-viii-2019	31-viii-2019
Lestes macrostigma	6/964	8/263	4/121	5/145	6/20	13/577	15/903	57	2993	04-iv-2019	16-vi-2017
Sympecma fusca	55/1937	86/3089	59/349	165/4121	35/175	101/2882	77/1414	578	13,967	year-round	year-round
Calopteryx splendens	178/4809	118/2140	173/3948	135/1653	121/2173	202/2548	227/2817	1154	20,088	07-iii-2014	21-xi-2019
Epallage fatime	110/2084	83/1387	87/2128	84/434	54/710	94/738	144/717	656	8198	07-iii-2013	20-viii-2013
Erythromma lindenii	13/53	19/361	14/216	17/197	19/197	23/133	13/45	118	1202	05-iv-2019	11-xi-2016
Erythromma viridulum	5/9	2/12	6/23	17/2205	9/907	19/284	16/276	74	3716	03-iv-2018	12-xi-2015
Ischnura elegans	299/10,197	379/5232	425/9197	436/5815	348/7653	467/6841	439/11,234	2793	56,169	year-round	year-round
Ischnura intermedia	49/364	35/242	78/1069	93/798	77/780	84/926	64/404	480	4583	05-iii-2017	05-xii-2019
Ischnura pumilio	0/0	0/0	0/0	0/0	0/0	0/0	0/0	0	0		
Aeshna affinis	0/0	0/0	0/0	0/0	0/0	0/0	0/0	0	0		
Aeshna isoceles	0/0	0/0	0/0	0/0	0/0	0/0	15/58	15	58	13-iv-2019	08-vii-2019
Aeshna mixta	45/212	63/207	85/315	35/66	82/163	79/194	75/240	465	1397	06-iii-2013	16-i-2016
Anax ephippiger	51/2508	20/294	10/37	15/104	34/142	26/268	54/1928	210	5281	26-i-2017	21-xii-2019
Anax immaculifrons	12/20	27/54	34/71	55/96	26/36	45/74	29/33	228	384	13-iv-2016	27-ix-2015
Anax imperator	37/78	10/21	7/12	12/17	16/18	16/21	11/13	109	180	06-iii-2013	28-xi-2014
Anax parthenope	133/482	154/771	131/454	139/375	141/551	182/444	149/581	1029	3658	04-ii-2014	21-xii-2018
Caliaeschna microstigma	9/14	9/10	2/2	8/20	8/11	14/77	11/10	61	144	24-iv-2014	13-vii-2013
Onychogomphus forcipatus	93/667	34/60	76/330	19/35	31/71	56/114	113/155	422	1432	20-iii-2013	16-x-2013
Brachythemis impartita	0/0	0/0	0/0	0/0	0/0	0/0	0/0	0	0		
Crocothemis erythraea	283/2393	296/2747	302/2389	301/1301	261/1326	391/2122	420/2230	2254	14,508	20-ii-2014	15-xiii-2014
Diplacodes lefebvrii	29/822	34/197	25/437	17/94	14/169	23/140	20/153	162	2012	21-iv-2016	17-xi-2018
Orthetrum brunneum	112/1122	105/404	107/540	95/265	118/465	148/524	120/302	805	3622	20-iii-2014	28-x-2017
Orthetrum cancellatum	21/234	18/133	19/814	24/663	28/380	32/366	31/197	173	2787	29-iii-2013	14-ix-2013
Orthetrum chrysostigma	193/2071	177/1599	248/3470	220/2062	196/1022	291/1341	355/2265	1680	13,830	02-iii-2018	15-i-2018 *
Orthetrum coerulescens	92/1186	130/1854	180/1706	182/1417	226/2171	279/2176	266/2434	1355	12,944	21-iii-2018	24-xi-2013
Orthetrum sabina	89/398	85/650	54/708	72/1299	78/749	74/955	89/793	541	5552	15-iii-2018	24-xi-2013
Orthetrum taeniolatum	38/506	31/94	47/163	32/71	13/21	31/76	48/333	240	1264	23-iii-2014	27-xi-2013
Pantala flavescens	1/1	13/24	2/5	9/13	13/17	45/74	146/233	229	367	09-vi-2018	11-i-2015 *
Selysiothemis nigra	23/1069	34/1006	31/2299	42/1269	35/828	32/300	34/547	231	7318	20-iv-2018	19-x-2018
Sympetrum fonscolombii	173/10,088	206/7825	254/11,290	185/4329	249/7306	238/7283	299/17,861	1604	65,982	year-round	year-round
Sympetrum meridionale	1/11	2/2	0/0	0/0	0/0	0/0	1/0	4	13	-	-
Sympetrum striolatum	162/4536	172/1536	129/590	196/1756	136/1204	258/1766	226/4009	1279	15,397	year-round	year-round
Trithemis annulata	294/12,382	275/6533	279/6213	286/4089	252/3839	321/5223	343/7248	2050	45,527	year-round	year-round
Trithemis arteriosa	72/716	60/137	68/224	124/331	103/343	176/639	268/1397	871	3787	23-iii-2018	06-i-2018 *
Trithemis festiva	172/3130	215/2196	137/2432	121/1259	218/754	239/961	1239/1764	1239	12,496	26-iii-2018	09-i-2016 *
Number of records	2937	2893	3273	3233	2970	4089	4504	23,899	-		
Total adult count	57,252	41,434	53,178	36,984	35,981	42,866	65,313	-	343,008		

References

1. Theobald, E.; Ettinger, A.; Burgess, H.; De Bey, L.; Schmidt, N.; Froehlich, H.; Wagner, C.; HilleRisLambers, J.; Tewksbury, J.; Harsch, M.A.; et al. Global change and local solutions. Tapping the unrealized potential of citizen science for biodiversity research. *Biol. Conserv.* **2015**, *181*, 236–244. [CrossRef]
2. Termaat, T.; van Strien, A.J.; van Grunsven, R.H.A.; De Knijf, G.; Bjelke, U.; Burbach, K.; Conze, K.-J.; Goffart, P.; Hepper, D.; Kalkman, V.J.; et al. Distribution trends of European dragonflies under climate change. *Divers. Distrib.* **2019**, *25*, 936–950. [CrossRef]
3. Bried, J.T.; Ries, L.; Smith, B.; Patten, M.; Abbott, J.; Ball-Damerow, J.; Cannings, R.; Cordero-Rivera, A.; Cordoba-Aguilar, A.; De Marco, P., Jr.; et al. Towards global volunteer monitoring of Odonate abundance. *BioScience* **2020**, *70*, 914–923. [CrossRef]
4. Kalkman, V.J.; van Pelt, G.J. The distribution and flight period of the dragonflies of Turkey. *Brachytron* **2006**, *10*, 83–153.
5. Boudot, J.-P.; Kalkman, V.J. (Eds.) *Atlas of the European Dragonflies and Damselflies*; KNNV Publishing: Zeist, The Netherlands, 2015; p. 381.
6. Boudot, J.-P.; Borisov, S.N.; De Knijf, G.; van Grunsven, R.; Schröter, A.; Kalkman, V.J. Atlas of the dragonflies and damselflies of West and Central Asia. *Brachytron* **2021**, 3–248.
7. Lopau, W.; Adena, J. *Die Libellenfauna von Cypern Naturkundliche Reiseberichte. Heft 19*; Private Publikation: Gnarrenburg, Germany, 2002; p. 73.
8. Flint, P. Observations of dragonflies (Odonata) from northern Cyprus. *Libellula* **2019**, *38*, 1–28.
9. Ministry of Agriculture, Rural Development and the Environment. Previously. Available online: https://www.cyprus.gov.cy/moa (accessed on 3 January 2020).
10. Water Development Department. Dams of Cyprus. Ministry of Agriculture, Natural Resources and Environment. 2009. Available online: https://www.cyprus.gov.cy/moa/wdd (accessed on 2 July 2020).
11. Kier, G.; Kreft, H.; Ming Lee, T.; Jetz, W.; Ibisch, P.L.; Nowick, C.; Mutke, J.; Barthlott, W. A global assessment of endemism and species richness across island and mainland regions. *Proc. Natl. Acad. Sci. USA* **2009**, *106*, 9322–9327. [CrossRef] [PubMed]
12. Boudot, J.-P.; Kalkman, V.J.; Azpilicueta Amorín, M.; Bogdanovic, T.; Cordero Rivera, A.; Degabriele, G.; Dommanget, J.-L.; Ferreira, S.; Garrigos, B.; Jovic, M.; et al. Atlas of the Odonata of the Mediterranean and North Africa. *Libellula* **2009**, *1* (Suppl. 9), 1–256.
13. De Knijf, G.; Sparrow, D.J.; Dimitriou, A.C.; Kent, R.; Kent, H.; Siedle, K.; Lewis, J.; Crossley, L. Distribution, ecology and status of a threatened species *Ischnura intermedia* (Insecta: Odonata), new for Europe. *Int. J. Odonatol.* **2016**, *19*, 257–274. [CrossRef]
14. Vilenica, M.; Kulijer, D.; Gligorovic, B.; Gligorovic, A.; De Knijf, G. Distribution, habitat requirements and vulnerability of *Caliaeschna microstigma* (Odonata: Aeshnidae) at the north-western edge of the species' range. *Odonatologica* **2021**, in press.
15. Sparrow, D.J.; De Knijf, G.; Smith, M.S.; Sparrow, R.; Michaelides, M.; Konis, D.; Siedle, K. The circumtropical *Pantala flavescens* is a regular visitor to Cyprus and reproducing on the island (Odonata: Libellulidae). *Odonatologica* **2020**, *49*, 289–311.
16. Kalkman, V.J. Orthetrum sabina (Schneider, 1845). In *Atlas of European Dragonflies and Damselflies*; Boudot, J.-P., Kalkman, V.J., Eds.; KNNV Publishing: Zeist, The Netherlands, 2015; pp. 283–285.
17. Sparrow, D.J.; Makris, C.; Sparrow, R.; Michaelides, M.; Konis, D.; De Knijf, G. First records of *Aeshna isoceles* and the rediscovery of *Lestes barbarus* on Cyprus (Odonata: Lestidae, Aeshnidae). *Not. Odonatol.* **2020**, *9*, 185–195.
18. Navas, R.P.L. Insect Orientalia. Serie 9. *Mem. Accad. Pontif. Nuovi Lincei* **1932**, *2*, 913–919.
19. Boudot, J.-P.; Prentice, S. Calopteryx virgo (Linnaeus, 1758). In *Atlas of European Dragonflies and Damselflies*; Boudot, J.-P., Kalkman, V.J., Eds.; KNNV Publishing: Zeist, The Netherlands, 2015; pp. 73–74.
20. DG Environment. *Reporting under Article 17 of the Habitats Directive: Explanatory Notes and Guidelines for the Period 2013–2018*; DG Environment: Brussels, Belgium, 2017; p. 188.
21. Care, N.; Chester, E.T.; Robson, B.J. Flow regime change alters shredder identity but not leaf litter decomposition in headwater streams affected by severe, permanent drying. *Freshw. Biol.* **2021**, *66*, 1813–1830. [CrossRef]
22. Boudot, J.-P.; Raab, R. Lestes macrostigma (Eversmann, 1836). In *Atlas of European Dragonflies and Damselflies*; Boudot, J.-P., Kalkman, V.J., Eds.; KNNV Publishing: Zeist, The Netherlands, 2015; pp. 58–60.
23. Kalkman, V.J.; Marinov, M.; Kutsarov, Y. Epallage fatime (Charpentier, 1840). In *Atlas of European Dragonflies and Damselflies*; Boudot, J.-P., Kalkman, V.J., Eds.; KNNV Publishing: Zeist, The Netherlands, 2015; pp. 78–79.
24. Kalkman, V.J.; Jovic, M. Caliaeschna microstigma (Schneider, 1845). In *Atlas of European Dragonflies and Damselflies*; Boudot, J.-P., Kalkman, V.J., Eds.; KNNV Publishing: Zeist, The Netherlands, 2015; pp. 184–185.
25. Jödicke, R. *Die Binsenjungfern und Winterlibellen Europas/Lestidae*; Westarp Wissenschaften/Die Neue Brehm-Bücherei (Bd. 631): Magdeburg, Germany, 1997; p. 277.
26. De Knijf, G.; Demolder, H. Early spring observations of Odonata from Cyprus. *Libellula* **2013**, *32*, 59–74.
27. Kalkman, V.J.; Sacha, D.; David, S. Sympetrum striolatum (Charpentier, 1840). In *Atlas of European Dragonflies and Damselflies*; Boudot, J.-P., Kalkman, V.J., Eds.; KNNV Publishing: Zeist, The Netherlands, 2015; pp. 309–311.
28. Samraoui, B.; Bouzid, R.; Boulahbal, R.; Corbet, P.S. Postponed reproductive maturation in upland refuges maintains life-cycle continuity during the hot, dry season in Algerian dragonflies (Anisoptera). *Int. J. Odonatol.* **1998**, *1*, 118–135. [CrossRef]
29. Samraoui, B.; Corbet, P.S. The Odonata of Numidia, Northeastern Algeria. Part II. Seasonal ecology. *Int. J. Odonatol.* **2000**, *3*, 27–39. [CrossRef]

30. Dijkstra, K.-D.B.; Schröter, A.; Lewington, R. *Field Guide to the Dragonflies of Britain and Europe*, 2nd ed.; Bloomsbury Nature Guides: London, UK, 2020; p. 336.
31. Boudot, J.-P.; Dyatlova, E. Chalcolestes parvidens (Artobolevskij, 1929). In *Atlas of European Dragonflies and Damselflies*; Boudot, J.-P., Kalkman, V.J., Eds.; KNNV Publishing: Zeist, The Netherlands, 2015; pp. 52–53.
32. Tarboton, W.; Tarboton, M. *A Guide to the Dragonflies & Damselflies of South Africa*; Struik Nature: Cape Town, South Africa, 2015; p. 224.
33. Boudot, J.-P. Trithemis festiva (Rambur, 1842). In *Atlas of European Dragonflies and Damselflies*; Boudot, J.-P., Kalkman, V.J., Eds.; KNNV Publishing: Zeist, The Netherlands, 2015; pp. 317–318.

 MDPI

Article

Aquatic Macrophyte Vegetation Promotes Taxonomic and Functional Diversity of Odonata Assemblages in Intermittent Karst Rivers in the Mediterranean

Marina Vilenica [1,*], Fran Rebrina [2], Renata Matoničkin Kepčija [2], Vedran Šegota [2], Mario Rumišek [2], Lea Ružanović [2] and Andreja Brigić [2]

1 Department in Petrinja, Faculty of Teacher Education, University of Zagreb, 44250 Petrinja, Croatia
2 Department of Biology, Faculty of Science, University of Zagreb, 10000 Zagreb, Croatia; fran.rebrina@biol.pmf.hr (F.R.); renata.matonickin.kepcija@biol.pmf.hr (R.M.K.); vedran.segota@biol.pmf.hr (V.Š.); rumisek.mario@windowslive.com (M.R.); lea.ruzanovic@gmail.com (L.R.); andreja.brigic@biol.pmf.hr (A.B.)
* Correspondence: marina.vilenica@ufzg.hr

Abstract: Assemblages of adult Odonata were studied in four intermittent karst rivers encompassing macrophyte-rich (MRH) and macrophyte-poor habitats (MPH) in southern Europe, where temporary lotic habitats are the predominant freshwater type but are still understudied. With a total of 25 recorded species, the studied habitats support species-rich Odonata assemblages, as already shown for intermittent rivers in the Mediterranean. Aquatic macrophyte abundance, conductivity, and water velocity are the most significant determinants of Odonata assemblages in the studied IRES. MRH promote higher Odonata abundance and the taxonomic and functional diversity of their assemblages compared to the MPH. Odonata assemblages in MRH are characterized by higher values of body size and a higher share of species preferring lentic and temporary hydrological conditions. Moreover, their assemblages are characterized by various patterns of nymphal development and drought resilience strategies. In contrast, MPH are preferred by lotic species, with nymphal development all year round and with no specific drought-resisting strategies. Our results contribute to the knowledge of diversity and ecological requirements of dragonflies and damselflies in IRES habitats, which could provide scientific background for future conservation activities and bioassessment protocols of such habitats and their biota.

Keywords: flow intermittence; environmental variables; aquatic macrophytes; karst; dragonflies; damselflies

Citation: Vilenica, M.; Rebrina, F.; Matoničkin Kepčija, R.; Šegota, V.; Rumišek, M.; Ružanović, L.; Brigić, A. Aquatic Macrophyte Vegetation Promotes Taxonomic and Functional Diversity of Odonata Assemblages in Intermittent Karst Rivers in the Mediterranean. *Diversity* **2022**, *14*, 31. https://doi.org/10.3390/d14010031

Academic Editor: Michael Wink

Received: 29 November 2021
Accepted: 27 December 2021
Published: 4 January 2022

1. Introduction

Approximately half of the running waters worldwide do not have continuous flow of surface water throughout the year and thus are categorized as temporary or non-perennial [1,2]. In the Mediterranean region of Europe, such rivers and streams are the predominant type of freshwater lotic habitats, due to dry climatic conditions, climate change, and land-use development. Non-perennial habitats are characterized by a wide range of hydrological regimes and can be categorized as intermittent (cease to flow seasonally or occasionally), ephemeral (flow only due to precipitation or snowmelt events), or episodic (flow primarily after heavy rainfall events) [3,4]. Here, we focus on intermittent rivers and streams (IRES), hydrologically highly dynamic and complex freshwater ecosystems that periodically cease to flow and run dry. Within such systems, three different flow categories can be distinguished: lotic (flowing), lentic phase (isolated pools), and dry riverbed, with the latter two being present during the dry periods [5,6].

Due to increasing anthropogenic pressures (e.g., river regulation, water abstraction, and pollution) and the global climate change, the Mediterranean basin is one of the most vulnerable regions in the world [7].The flow regimes of the IRES are rapidly changing, and the extent and intensity of dry periods in the IRES are expected to increase in the forthcoming future [6,8], which will also lead to serious water availability problems in the Mediterranean area [7]. Additionally, large parts of this area are densely populated, increasing the demand for irrigation and drinking water. The negative consequences of water abstraction and regulation are reflected in river hydrology modifications, with intensified drought effects [9]. Over the past few decades, water abstraction and impoundment have even caused many previously perennial rivers to become intermittent [10,11]. This trend is expected to continue in the near future [12], which will surely lead to irreversible changes in biological communities [13].

As IRES cover more than half of the global river network, it is essential to understand their contribution to biotic diversity at both local and landscape scales [5,6]. During the past decade, many studies have investigated and highlighted the importance of flow permanence for the composition and structure of aquatic macroinvertebrate assemblages [14–16]. Nevertheless, there are still large gaps regarding the environmental drivers that shape their diversity and composition in intermittent lotic habitats. Consequently, intermittent rivers and streams are still not included in biomonitoring programs in the majority of EU countries [8]. Yet, it is worth mentioning that in Croatia, there are two intermittent river/stream types in the National River Typology [17] with defined ecological status classes. In order to provide a scientific background for the development of widely applicable bioassessment methodology of IRES, it is essential to conduct further studies on the effect of flow intermittency on all aspects of stream ecology.

Odonata are an amphibious insect order (with aquatic nymphs and terrestrial adults) widely used as ecological indicators of freshwater ecosystem health [18–21]. Many studies showed that their assemblages are highly influenced by physicochemical water conditions [22–26] but even more importantly by habitat's morphology and structure (e.g., bottom substrate and structure of aquatic vegetation) [27–33]. Many Odonata species have life-history adaptations that enable them to occupy temporary habitats, such as desiccation-tolerant eggs or fast larval growth [34–36]. Nymphs of some large Odonata species can also use damp sediment beneath the stones for aestivation [37]. For those species whose drought-resisting abilities are low, perennial lotic habitats and pools [37] as well as artificial reservoirs [38] in the vicinity of IRES were shown to be suitable refuge sites during dry periods.

To improve our knowledge about the aquatic insect communities in IRES, we studied the assemblages of adult Odonata in karst intermittent rivers in the Dinaric Western Balkans ecoregion [39]. The Dinaric Alps extend over approximately 60,000 km^2 and are the largest continuous karst landscape in Europe [40]. Karst is a set of morphological, hydrological, and hydrogeological terrain features built of water-soluble rock. The Dinaric Western Balkans area is characterized by an extremely complex hydrological network [41] and extraordinary diversity of biota [42], yet it is still greatly understudied. Although Odonata are considered to be among the well-studied aquatic insect orders [43], their ecological requirements in karst rivers and streams are very poorly known [22,44], especially in intermittent habitats. Therefore, the main objectives of this study were (i) to compare Odonata assemblages (species richness, diversity, and abundance) in two focal habitat types: macrophyte rich and macrophyte poor, in the Mediterranean intermittent karst rivers; (ii) to examine the functional diversity of Odonata assemblages and detect changes in functional traits; and (iii) to determine the main environmental drivers that shape these assemblages. We hypothesize that the structure and abundance of aquatic macrophyte vegetation are the main environmental drivers shaping Odonata assemblages in the studied karst intermittent rivers, where macrophyte-rich habitats support the higher taxonomic and functional diversity of Odonata.

2. Materials and Methods

2.1. Study Area

The study was conducted in the Dinaric Western Balkan ecoregion (ER5) of Croatia [39]. Our study encompassed four Mediterranean karst intermittent rivers belonging to the Adriatic Sea basin: the Krčić, Čikola, Miljašić Jaruga, and Gudiča Rivers (Figure 1). The Krčić River catchment covers 157 km^2. Its source is located at the foot of the Dinara Mountain, near the town of Knin. The river flows for 10.5 km and ends with a 40 m high waterfall, which contributes to the forming of the Krka River [45]. The average annual discharge values for the period 1982–1990 for the Krčić River were 3.93 m^3/s [46]. The catchment area of the Čikola River is approximately 300 km^2. Its spring is located near Mirlović Polje village. The river runs for 39 km, ending as a tributary of the Krka River near Nos Kalik village [47,48]. The mean annual discharge for the Drniš hydrological station during 2003–2017 was 5.0 m^3/s [48]. The Miljašić Jaruga River is a part of the Bokanjac-Poličnik catchment area of 244.51 km^2 [49]. It springs near Suhovare village and flows for 25 km to its mouth in the Adriatic Sea near the town of Nin [50]. The mean annual discharge for the Boljkovac-Miljašić Jaruga hydrological station for the period 1961–2009 was 0.85 m^3/s [51]. The Bribišnica River belongs to the Prokljan Lake catchment area, which amounts to 596.22 km^2. It springs on the west side of the Bribirska Glavica hill. Near the Lađevci bridge, the river becomes the torrential Gudiča River, which flows for seven more kilometers and runs into the Prokljan Lake [49]. The Gudiča River generally carries on less than 1 m^3/s of water [52]. Throughout the text, this river is referred to as the Gudiča River.

Our study was conducted at a total of 12 study sites (three sites per river) (Figure 1). At each site, we analyzed the vegetation structure, Odonata assemblages, and measured physicochemical water parameters.

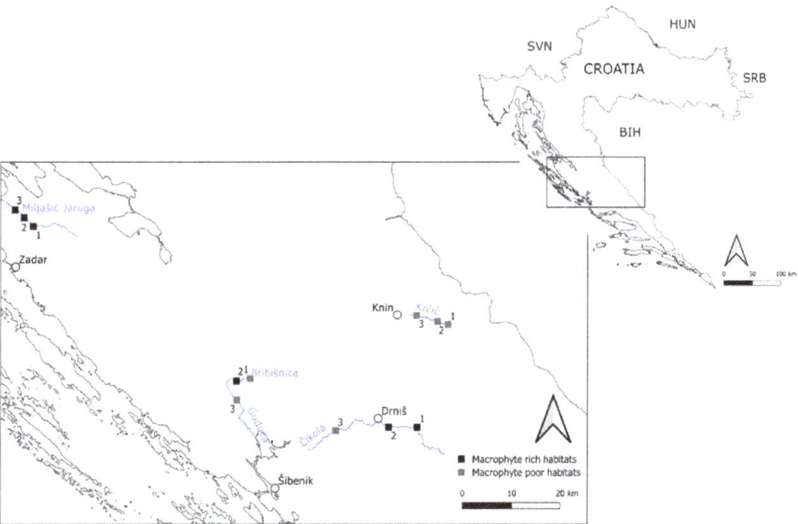

Figure 1. Geographical position of the four studied intermittent karst rivers located in the Croatian part of the ER5. Legend: SVN—Slovenia, BIH—Bosnia and Herzegovina, SRB—Serbia, HUN—Hungary.

2.2. Vegetation Analysis

During our third sampling event (30 June/1 July 2021), at each study site, we conducted a macrophyte vegetation survey (aquatic vascular plants and bryophytes) that included the assessment of species coverage and abundance. The sampling plot size was approximately 100 m^2. The assessment of macrophyte species coverage and abundance was performed using the expanded, nine-degree Braun–Blanquet scale (+ = up to 5 individuals;

1 = up to 50 individuals; 2m = more than 50 individuals; 2a = coverage between 5 and 15%; 2b = coverage between 15 and 25%; 3 = coverage between 25 and 50%; 4 = coverage between 50 and 75%; 5 = coverage over 75%) [53–55]. The assessment included both aquatic vascular plants and aquatic bryophytes, for which the cumulative plant coverage of each group was calculated. Additionally, a cumulative plant coverage was calculated separately for lower (<30 cm) and higher (>30 cm) aquatic vascular plants.

2.3. Environmental Variables

In April 2021, when all four rivers were flowing (i.e., in the lotic phase), the following environmental parameters were measured at each study site: water temperature, dissolved oxygen concentration and saturation (using the oximeter WTW Oxi 330/SET), conductivity (using the conductivity meter WTW LF 330), pH (using the pH-meter WTW pH 330), water width and depth (using a hand meter/measuring tape), and water velocity (using the SonTek Flow Tracker). At each site, the parameters were measured at three equally spaced points in a transect from the shoreline to the center of the river, perpendicular to the river flow. Additionally, at those same points at each sampling site, triplicate 1 L water samples were taken for the laboratory analysis of the chemical parameters (alkalinity, chemical oxygen demand, concentrations of nitrites, nitrates, and orthophosphates) using Standard Analytical Procedure [56]. At the visited study sites, substrates were composed mostly of fine sediment (silt, mud), lithal (stones, gravel), and aquatic vegetation.

2.4. Odonata Sampling

Odonata adults were investigated between the end of May and beginning of July 2021, every two weeks, during a total of three sampling events (30 May/1 June, 14 June/15 June, 30 June/1 July 2021). At each river, three sites were visited at each sampling event (Figure 2); the first site was the closest to the river source, while the third was the most distant. At each study site, Odonata were investigated along a 200 m transect during the period of 45 min (until no additional species were detected). Species flying or perching within ≈5 m of the transect route were documented and counted (high abundances of damselflies were estimated instantaneously). The surveys were conducted on sunny days, between 10 a.m. and 4 p.m. Adults were mostly observed visually, identified by eye, or using close focusing binoculars. Some species were sampled using an entomological net (e.g., those from the genus *Sympetrum*). Collected individuals were identified in the field, photographed, and released. Surveys of all sites were conducted on foot by the same observer (M. V.). Taxonomy follows Ref. [57].

2.5. Data Analyses

Among a total of 12 study sites, six macrophyte-rich (MRH) and six macrophyte-poor habitats (MPH) were identified based on their aquatic macrophyte species richness and abundance (>10% habitat covered with macrophytes in MRH, <10% in MPH). Species richness and abundance data were tested for normality using the Shapiro–Wilk test in SPSS Statistics ver. 27.0 [58]. Then, aquatic macrophytes (total) and vascular plants (low and high) were analyzed with respect to species richness and abundance in two focal habitats: MPH vs. MRH, using Mann–Whitney U-tests in SPSS Statistics ver. 27.0 [58]. Differences in physicochemical water parameters between the MPH and MRH were tested using generalized linear mixed models (GLMMs). In all the constructed models, we included sites (level 1) nested within macrophyte vegetation type (level 2) with sampling events as repeated measures. For all the physicochemical parameters, we applied the gamma distribution with log link function. The macrophyte vegetation type was used as a fixed effect in all models. To account for the variation introduced by potential differences among sampling sites and events, sites and sampling events were included in all models as random effects, with first-order autoregressive (AR1) covariance type, which was assumed for repeated measures over time [59]. Pairwise contrasts of estimated means were applied using a least significant difference (LSD) *post hoc* test. We constructed a full model for each target physicochemical variable, as recommended by [60]. The above-mentioned analyses were performed in SPSS Statistics ver. 27.0 [58].

Figure 2. Examples of the sites where Odonata were studied at four intermittent karst rivers (Croatia): sites characterized by poorly developed aquatic vegetation: (**a**) the Krčić River (Site 3), (**b**) the Guduča River (Site 1), (**c**) the Čikola River (Site 3); sites with well-developed aquatic vegetation: (**d**) the Čikola River (Site 2), (**e**) the Guduča River (Site 3), (**f**) the Miljašić Jaruga River (Site 1).

Assemblage metrics: diversity (Shannon diversity index, H′, Simpson diversity index, $1 - \lambda$), species richness (S) and abundance (N), were calculated for Odonata assemblages at each study site in each of the two habitat types (MPH and MRH). In community ecology, it is common to use several diversity indices differing by their sensitivity to rare or common species, i.e., the most commonly used Shannon diversity index is disproportionately sensitive to the rare species, while the Simpson diversity index is disproportionately sensitive to the most common species (see in [61]). Prior to the analysis, assemblage data were tested for normality using the Shapiro–Wilk test in SPSS Statistics ver. 27.0 [58]. The similarity of Odonata assemblages between the two habitats was examined using hierarchical cluster analysis (HCA) based on the Bray–Curtis similarity matrix in Primer 6.0 [62]. Species data were log (x + 1) transformed prior to the HCA. Study sites with no Odonata records were excluded from the HCA. To evaluate the differences in Odonata assemblage metrics between the two habitats, a set of generalized linear mixed models (GLMMs) was constructed. In the construction of all models, we used the same approach as for physicochemical parameters. For species richness and abundance, Poisson distribution was applied, while for diversity indices, gamma distribution was used, with the log link function. The significance of the models was tested using the least significant difference (LSD) *post hoc* test.

To quantify the functional diversity of Odonata assemblages, a total of 17 functional traits from four groups were used: (i) body size, (ii) nymphal development (all year, mainly in spring, mainly in summer, mainly in autumn, mainly in winter, unknown), (iii) hydrological preference (eupotamon—main channel and connected side arms; parapotamon—side arms connected only at the downstream end at mean water levels; plesiopotamon—no connectivity with the main channel at the mean water level, including lakes, where coverage by macrophytes does not exceed 20%; palaeopotamon—no connectivity with the main channel at mean water levels, including lakes and pools, where coverage by macrophytes exceeds 20%; temporary water bodies—temporary pools, where the water level is primarily dependent on ground water levels) and (iv) drought resilience form (no resilience strategy against droughts, egg diapause—resisting in the egg stage; nymph diapause—resisting in the nymphal stage; adult diapause—resisting in the adult stage; unknown resilience strategy) (retracted from Refs. [57,63,64]) (Appendix A).

The functional diversity of Odonata assemblages was quantified using the Rao quadratic diversity (RaoQ) coefficient, which is a measure of trait convergence or divergence patterns compared to random expectation. Community weighted mean (CWM) values were calculated for each functional trait in Odonata assemblages to quantify shifts in mean trait values within the assemblages, resulting from environmental selection for certain functional trait categories [65]. RaoQ and CWM values were calculated in CANOCO version 5.11 package [66]. Prior to the analysis, functional data were tested for normality using the Shapiro–Wilk test in SPSS Statistics ver. 27.0 [58]. Differences in the RaoQ coefficient and trait CWM values between the two habitats were tested using a generalized linear mixed model (GLMM). We used a gamma distribution for each variable with log link function. We used the same approach as for physicochemical parameters and assemblage metrics to construct the models and test their significance.

The relationship between Odonata assemblages and environmental variables was tested using canonical correspondence analysis (CCA). Odonata represented by fewer than 20 individuals were omitted from the CCA, and a total of 17 species was used in the analysis. To assess the influence of environmental factors on the spatial distribution of CWM values of functional traits in Odonata assemblages, redundancy analysis (RDA) was used. All the recorded species were included in the RDA. A total of six statistically significant environmental variables (water temperature, velocity, hardness, conductivity, abundance of vascular macrophytes, and bryophytes) were included in the CCA and RDA. Prior to the RDA, Odonata abundances were centered and standardized by the average functional traits, while they were log (x + 1) transformed prior to the CCA. To test the relationship between trait or species composition and environmental variables, a Monte Carlo test using 499 permutations ($p < 0.05$) was performed. These analyses were performed in the CANOCO version 5.11 package [66].

3. Results

3.1. Vegetation Analysis

Vascular macrophyte species richness and abundance were significantly higher in MRH (Table 1) compared to the MPH. The same pattern was observed for species richness and the abundance of low and high vascular plants (Table 1). In MRH, tall (e.g., *Phragmites australis* (Cav.) Steud., *Scirpus lacustris* L., *Cyperus longus* L., *Typha angustifolia* L.) and low vascular plants (e.g., *Lythrum salicaria* L., *Mentha aquatica* L., *Alisma plantago-aquatica* L., *Agrostis stolonifera* L.) are intermixed in mosaic assemblages (Figure 2d–f).

On the other hand, aquatic bryophytes were more species rich and abundant at MPH (Table 1), with bryophyte species such as *Cinclidotus aquaticus* (Hedw.) Bruch et Schimp., *C. fontinaloides* (Hedw.) P. Beauv., *Cratoneuron filicinum* (Hedw.) Spruce, *Fissidens crassipes* Wilson ex Bruch et Schimp., and *Rhynchostegium riparioides* (Hedw.) Cardot predominating. Those habitats are characterized by low abundance and low number (solely six taxa overall) of macrophyte vascular species (Figure 2a–c).

Table 1. Vegetation analysis of macrophyte-poor and macrophyte-rich habitats of four intermittent karst rivers (Croatia), with mean values ± standard error per habitat type (n = 6, for each habitat type). Different letters indicate a significant difference between the habitats (Mann–Whitney U-test, $p < 0.01$). Legend: LM—low macrophytes, HM—high macrophytes.

	Habitat Type	
	Macrophyte Poor	**Macrophyte Rich**
Vascular Plants		
Dominant vascular plants	*Mentha longifolia* (L.) L., *Oenanthe fistulosa* L.	*Phragmites australis* (Cav.) Steud., *Scirpus lacustris* L., *Cyperus longus* L.
Species richness (total, mean ± SE)	1.00 ± 0.45 [b]	10.50 ± 1.61 [a]
Abundance (total, mean ± SE)	3.33 ± 1.69 [b]	39.00 ± 5.72 [a]
Species richness of LM (mean ± SE)	1.00 ± 0.45 [b]	7.50 ± 1.09 [a]
Abundance of LM (mean ± SE)	3.33 ± 1.69 [b]	25.00 ± 3.52 [a]
Species richness of HM (mean ± SE)	0.00 [b]	2.83 ± 0.70 [a]
Abundance of HM (mean ± SE)	0.00 [b]	14.00 ± 2.68 [a]
Bryophytes		
Dominant bryophytes	*Cinclidotus aquaticus* (Hedw.) Bruch et Schimp., *C. fontinaloides* (Hedw.) P. Beauv., *Rhynchostegium riparioides* (Hedw.) Cardot, *Cratoneuron filicinum* (Hedw.) Spruce	*Cinclidotus fontinaloides* (Hedw.) P. Beauv., *Calliergonella cuspidata* (Hedw.) Loseke, *Fontinalis antipyretica* Hedw.
Species richness (total, mean ± SE)	3.33 ± 0.33 [a]	1.50 ± 0.81 [a]
Abundance (mean ± SE)	14.17 ± 1.25 [a]	3.83 ± 2.17 [b]

3.2. Environmental Variables

Alkalinity, water hardness, conductivity, water temperature, and velocity differed significantly between the MPH and MRH (Table 2). MPH were characterized by significantly lower water temperature, hardness, conductivity and alkalinity, and significantly higher water velocity compared to the MRH (Table 2). The other measured parameters did not significantly differ between the two habitat types (MPH and MRH) (Table 2).

Table 2. Physicochemical properties of water measured in macrophyte-poor and macrophyte-rich habitats of four intermittent karst rivers (Croatia), with mean values ± standard error per habitat type (n = 18, for each habitat type). GLMM (full model) output shows differences in physicochemical water properties between the habitats. For all the physicochemical parameters, the gamma distribution with log link function was applied. Macrophyte vegetation was used as a fixed effect, while sites were included as a random effect. Statistically significant effects obtained from the least significant difference *post hoc* test ($p < 0.05$) are reported in bold. Legend: F—F statistic, d.f. 1—degrees of freedom, d.f. 2—denominator degrees of freedom, MPH—macrophyte-poor habitats, MRH—macrophyte-rich habitats.

Environmental Variables	MPH (Mean ± SE)	MRH (Mean ± SE)	F	*p*	d.f. 1	d.f. 2
Alkalinity (mg $CaCO_3$/L)	126.94 ± 3.75	157.50 ± 2.63	23.589	**0.000**	1	34
Water hardness (mg $CaCO_3$/L)	224.48 ± 14.18	277.88 ± 10.78	10.201	**0.003**	1	34
Conductivity (µS/cm)	414.61 ± 20.04	547.72 ± 16.80	10.224	**0.003**	1	34
Water temperature (°C)	10.84 ± 0.62	14.09 ± 0.38	6.706	**0.014**	1	34
Water velocity (m/s)	0.58 ± 0.03	0.31 ± 0.06	7.667	**0.009**	1	34

Table 2. *Cont.*

Environmental Variables	MPH (Mean ± SE)	MRH (Mean ± SE)	F	*p*	d.f. 1	d.f. 2
Oxygen saturation (%)	98.89 ± 1.22	107.81 ± 2.53	2.790	0.104	1	34
Nitrates (mg N/L)	0.25 ± 0.04	0.18 ± 0.02	1.085	0.305	1	34
Water depth (cm)	27.22 ± 4.15	33.61 ± 4.42	0.985	0.328	1	34
Nitrites (mg N/L)	0.02 ± 0.01	0.01 ± 0.00	0.481	0.495	1	34
Chemical oxygen demand (mg/L)	3.61 ± 0.23	3.76 ± 0.37	0.315	0.578	1	34
o-phosphates (mg N/L)	0.02 ± 0.00	0.02 ± 0.00	0.208	0.652	1	34
Dissolved oxygen (mg/L)	10.80 ± 0.19	11.04 ± 0.25	0.176	0.678	1	34
pH	7.98 ± 0.08	7.96 ± 0.03	0	0.988	1	34

3.3. Odonata Species Occurrence

In total, 25 Odonata species were recorded (Table 3). Overall, the most numerous species was *Platycnemis pennipes* (Pallas, 1771), which was also most frequently recorded at MRH. *Calopteryx virgo* (Linnaeus, 1758) was the most numerous species at MPH (Table 3). Species recorded in low numbers (less than 20 individuals) were *Sympetrum fonscolombii* (Selys, 1840), *S. sanguineum* (Müller, 1764), *S. meridionale* (Selys, 1841), *Cordulegaster heros* Theischinger, 1979, *Aeshna affinis* Vander Linden, 1820, *Crocothemis erythraea* (Brullé, 1832), *Orthetrum cancellatum* (Linnaeus, 1758), and *Somatochlora meridionalis* Nielsen, 1935 (Table 3).

Table 3. Odonata recorded in macrophyte-poor and macrophyte-rich habitats of four intermittent karst rivers (Croatia). Species are represented by the total number of individuals (N) and frequency (%). Species codes are those used in the CCA.

		Habitat Type				Total	
		Macrophyte Poor		Macrophyte Rich			
Species Name	Species Code	N	%	N	%	N	%
Calopteryx splendens (Harris, 1782)	*Ca spl*	25	3.85	233	3.75	258	3.76
Calopteryx virgo (Linnaeus, 1758)	*Ca vir*	564	86.80	70	1.13	634	9.24
Chalcolestes viridis (Vander Linden, 1825)	*Ch vir*			675	10.86	675	9.84
Sympecma fusca (Vander Linden, 1820)	*Sy fus*			222	3.58	222	3.23
Ischnura elegans (Van der Linden, 1820)	*Is ele*			370	5.96	370	5.38
Coenagrion puella (Linnaeus, 1758)	*Co pue*			392	6.31	392	5.71
Erythromma lindenii (Selys, 1840)	*Er lin*			218	3.51	218	3.18
Platycnemis pennipes (Pallas, 1771)	*Pl pen*	20	3.08	3500	56.33	3520	51.28
Aeshna affinis Vander Linden, 1820		1	0.15	4	0.07	5	0.07
Aeshna isoceles (Müller, 1767)	*Ae iso*			31	0.50	31	0.45
Anax imperator (Selys, 1839)	*An imp*			30	0.48	30	0.44
Brachytron pratense (Müller, 1764)	*Br pra*			34	0.55	34	0.50
Onychogomphus forcipatus (Linnaeus, 1758)	*On for*	36	5.54	51	0.82	87	1.28
Cordulegaster heros Theischinger, 1979		3	0.46			3	0.04
Somatochlora meridionalis Nielsen, 1935		1	0.15	18	0.29	19	0.28
Libellula depressa Linnaeus, 1758	*Li dep*			25	0.40	25	0.36
Libellula fulva Müller, 1764	*Li ful*			45	0.72	45	0.66
Orthetrum cancellatum (Linnaeus, 1758)				17	0.27	17	0.25
Orthetrum coerulescens (Fabricius, 1798)	*Or coe*			102	1.64	102	1.49
Orthetrum brunneum (Fonscolombe, 1837)	*Or bru*			48	0.77	48	0.70
Sympetrum sanguineum (Müller, 1764)				4	0.06	4	0.06
Sympetrum fonscolombii (Selys, 1840)				4	0.06	4	0.06
Sympetrum striolatum (Charpentier, 1840)	*Sy str*			105	1.69	105	1.53
Sympetrum meridionale (Selys, 1841)				1	0.02	1	0.01
Crocothemis erythraea (Brullé, 1832)				14	0.23	14	0.20
Species richness (S)		**7**		**24**		**25**	
Abundance (N)		**650**		**6213**		**6863**	

3.4. Odonata Assemblages and Their Functional Diversity

Odonata species richness and diversity were five to seven times significantly higher in MRH than in the MPH (Table 4, Figure 3a,c,d). Furthermore, Odonata abundance was over nine times significantly higher in MRH than in the MPH (Table 4, Figure 3b). Moreover, the results of the cluster analysis revealed clear separation of MPH and MRH, with low similarity of their respective Odonata assemblages, accounting for less than 5% (Figure 4).

Table 4. GLMM (full model) output showing differences in Odonata assemblage metrics, functional diversity, and community weighted mean (CWM) values of functional traits between macrophyte-poor and macrophyte-rich habitats of four intermittent karst rivers (Croatia). Poisson distribution was applied for species richness and abundance, while for diversity indices and functional traits, gamma distribution was used with log link function. Vegetation was used as a fixed effect, with sites and months as random effects. Statistically significant effects obtained from the least significant difference *post hoc* test ($p < 0.05$) are reported in bold. Legend: F—F statistic, d.f. 1—degrees of freedom, d.f. 2—denominator degrees of freedom.

Assemblage Parameter		F	p	d.f. 1	d.f. 2
Species richness (S)		45.756	**0.000**	1	34
Abundance (N)		58.940	**0.000**	1	34
Shannon diversity (H′)		29.200	**0.000**	1	23
Simpson diversity (1 − λ)		19.700	**0.000**	1	23
Functional Parameter					
Functional diversity (RaoQ)		28.563	**0.000**	1	27
CWM body size		8.149	**0.008**	1	27
CWM hydrological preferences	eupotamon	182.582	**0.000**	1	27
	parapotamon	13.839	**0.001**	1	21
	plesiopotamon	91.216	**0.000**	1	22
	palaeopotamon	19.457	**0.000**	1	22
	temporary water bodies	7.247	**0.014**	1	19
CWM nymphal development	spring	20.020	**0.000**	1	17
	summer	128.804	**0.000**	1	17
	autumn	-	-	-	-
	winter	-	-	-	-
	all year	188.086	**0.000**	1	27
CWM drought resilience form	no drought resilience	138.067	**0.000**	1	26
	egg diapause	11.703	**0.003**	1	17
	nymph diapause	1.994	0.178	1	15
	adult diapause	-	-	-	-
	unknown resilience type	-	-	-	-

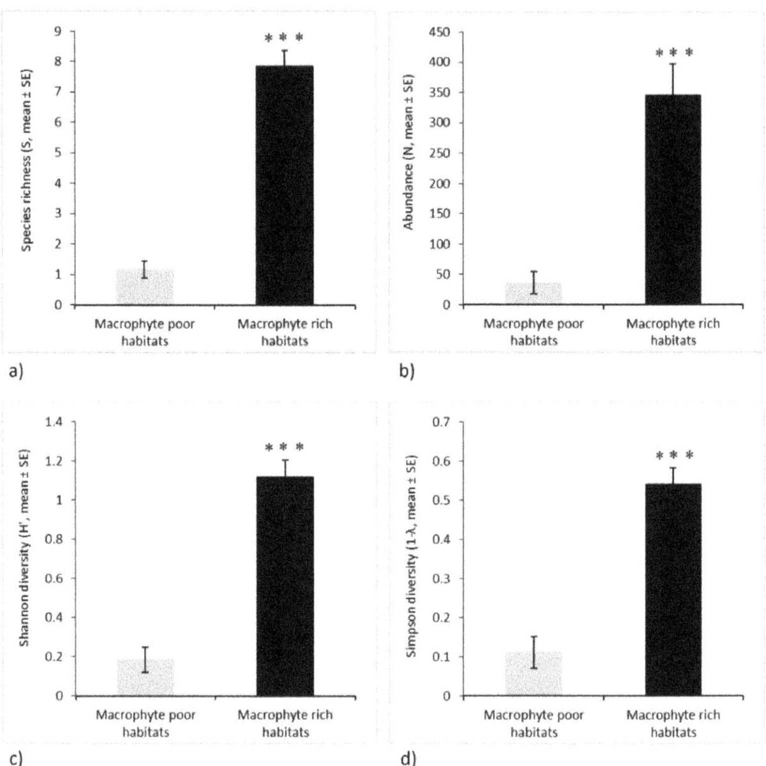

Figure 3. Odonata assemblages in macrophyte-poor and macrophyte-rich habitats of four intermittent karst rivers (Croatia). Each assemblage metric is shown by mean and standard error: (**a**) species richness (S), (**b**) abundance (N), (**c**) Shannon diversity index (H′), (**d**) Simpson diversity index (1 − λ). Asterisk indicates a significant difference between the habitats (GLMM including sites as fixed factors and months as repeated measures, least significant difference *post hoc* test, $p < 0.001$). Poisson distribution was applied for species richness and abundance, while for diversity indices, gamma distribution was used with log link function.

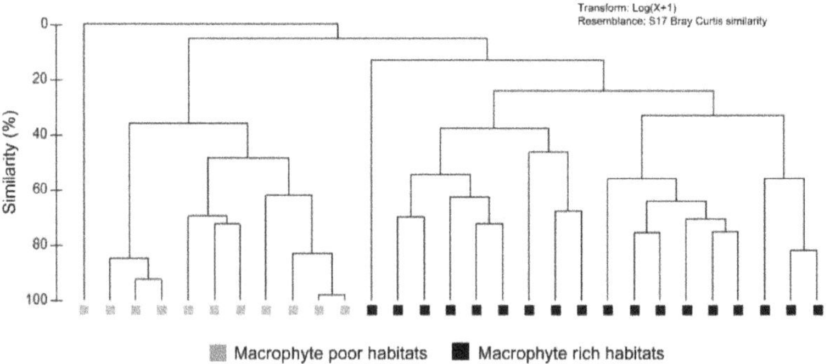

Figure 4. Cluster analysis of Odonata assemblages in MPH and MRH of four intermittent karst rivers (Croatia), based on Bray–Curtis similarity coefficient and species' log (x + 1)-transformed abundances. Study sites with no Odonata records were excluded from the analysis.

Functional diversity (RaoQ) was also significantly higher at MRH (Table 4, Figure 5a). Odonata assemblages inhabiting MRH are characterized by significantly higher CWM values of body size compared to the assemblages at MPH. In such habitats, other groups of functional traits were shown to be more diverse compared to those at MPH (Table 4, Figure 5). At MRH, a similar share of species with nymphal development all year, in spring and summer were recorded, while in MPH, species with nymphal development all year dominate (Figure 5). At MRH, we recorded a higher share of species with a preference for lentic hydrological conditions (plesiopotamon, palaeopotamon) and temporary water bodies, while those with a preference for eupotamon (lotic hydrological conditions) dominate at MPH. Finally, at MRH, we recorded a higher number of species with certain strategies to drought resilience (especially with egg diapause), while at MPH, mainly species with no resilience strategy against droughts were recorded (Table 4, Figure 5).

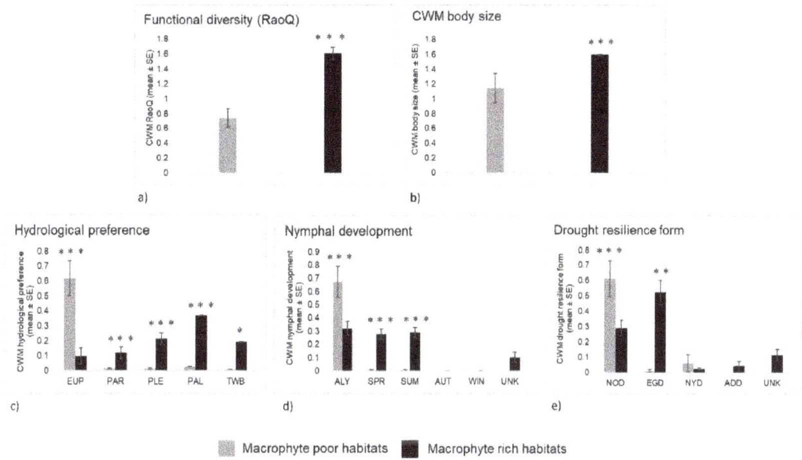

Figure 5. Odonata functional traits in macrophyte-poor and macrophyte-rich habitats of four intermittent karst rivers (Croatia) shown as mean and standard error: (**a**) functional diversity (RaoQ), (**b**) CWM body size, (**c**) hydrological preference, (**d**) nymphal development, (**e**) drought resilience form. Asterisk indicates a significant difference between the habitats (*** = $p < 0.001$), (** = $p < 0.01$), (* = $p < 0.05$). Legend: Hydrological preference: EUP—eupotamon, PAR—parapotamon, PLE—plesiopotamon, PAL—palaeopotamon, TWB—temporary water bodies. Nymphal development: ALY—nymphal development all year, SPR—nymphal development mainly in spring, SUM—nymphal development mainly in summer, AUT—nymphal development mainly in autumn, WIN—nymphal development mainly in winter, UNK—unknown nymphal development. Drought resistance forms: NOD—no resilience strategy against droughts; EGD—egg diapause, NYD—nymph diapause, ADD—adult diapause, UNK—unknown resilience strategy.

3.5. Odonata Species and Functional Traits Related to Environmental Variables

Odonata assemblages clearly differed with respect to two habitat types, as shown by statistically significant results of the CCA (explanatory variables accounted for 52.86%; F ratio = 3.9, $p = 0.002$). The first two ordination axes (eigenvalues 0.616 and 0.333) explained 40.62% of the variation (Figure 6). Axis 1 was related to conductivity (R = −0.672), water velocity (R = 0.616), and vascular macrophytes (R = −0.433), and axis 2 was related to bryophytes (R = 0.684) (Figure 6). These results indicate a strong separation between the MPH and MRH, with the latter mostly positioned to the left of zero, which is a pattern governed by higher conductivity and an abundance of vascular macrophytes, while all macrophyte-poor habitats were positioned to the right, in correlation with higher water velocity. Lentic species, such as *Anax imperator* (Selys, 1839), *Brachytron pratense* (Müller, 1764), and *Chalcolestes viridis* (Vander Linden, 1825) were positively associated with MRH, while the lotic ones such as *Onychogomphus forcipatus* (Linnaeus, 1758), *Calopteryx splendens* (Harris, 1782), and *C. virgo* were abundant in MPH (Figure 6).

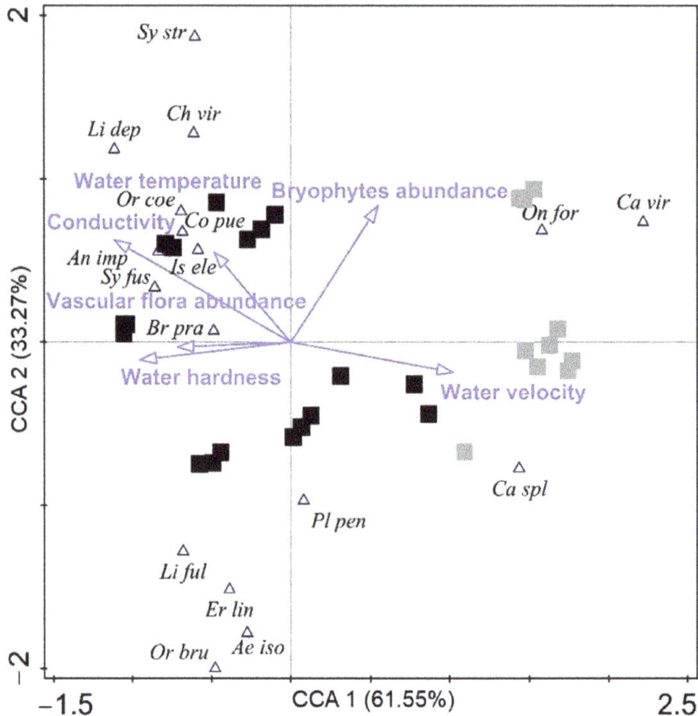

Figure 6. CCA ordination triplot on standardized and log (x + 1) transformed data of 17 Odonata species and six environmental variables recorded at four intermittent karst rivers (Croatia). Taxa codes (triangles) are presented in Table 3. Black squares represent macrophyte-rich habitats, and grey squares represent macrophyte-poor habitats.

Statistically significant results of RDA showed that Odonata functional traits differed with respect to the studied habitat types (explanatory variables accounted for 61.39%; F ratio = 5.8, p = 0.002). The first two ordination axes (eigenvalues 0.457 and 0.101) explained 55.75% of the variation (Figure 7). The first ordination axis is correlated mainly with macrophytes (vascular macrophytes (R = −0.834) and bryophyte abundance (R = 0.853), and the second axis is correlated mainly with water velocity (R = −0.382), once again indicating a strong separation of MRH (with abundant vascular macrophytes) and MPH (with higher water velocity).

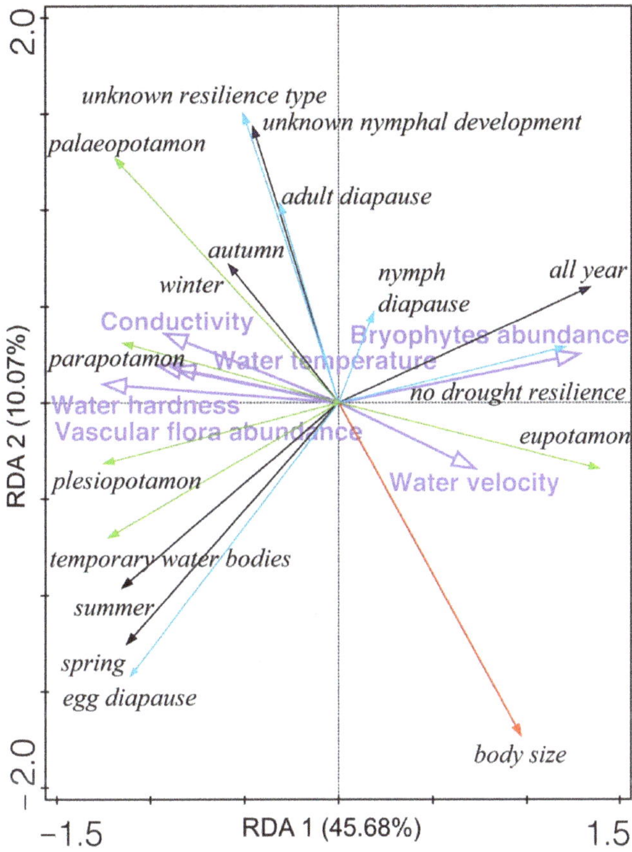

Figure 7. Redundancy analysis (RDA) ordination biplot showing the relationships between Odonata functional traits (CWM body size—red arrow, CWM hydrological preferences—green arrows, CWM nymphal development—black arrows, CWM drought resilience form—blue arrows) and six significant environmental variables (purple arrows) at four intermittent karst rivers (Croatia).

4. Discussion

With a total of 25 recorded species (37% of Croatian, 15% of European Odonata fauna) [67,68], our results indicate rather high Odonata species richness in the studied intermittent karst rivers, similar to previous studies on IRES in the Mediterranean area [69–71]. Although ephemeral and intermittent Mediterranean streams generally have different and less diverse macroinvertebrate communities compared to the perennial ones [16,70,72,73], previous studies observed a shift in community structure with changing hydrology. Lotic diversity tends to decrease with increasing flow intermittence, but in such habitats, there is often a compensation with an increase in lentic diversity, including Odonata [6,70]. This is not surprising, as higher numbers of European Odonata prefer lentic habitats or lotic ones with low water velocity, which are characterized by higher habitat heterogeneity [57,74]. One of the species recorded in our study, *C. heros*, is of international conservation concern. The species is endemic to Central and Southeastern Europe [57,75], and it is one of the eight near-threatened (NT) European Odonata [75]. It is indicative of pristine lotic habitats and thus is listed in Annexes II and IV of the EU Habitat Directive and in Annex I of the Bern Convention [75], implying that its habitats should not be altered [76]. The occurrence of *C. heros* in the studied IRES indicates the potential conservation value of such habitats.

The current study shows that Odonata assemblages in the studied intermittent rivers are highly influenced by habitat features, such as the structure and abundance of aquatic macrophytes, and physicochemical water parameters, particularly conductivity and water velocity. Physicochemical water properties and aquatic vegetation composition are strongly mutually influenced [77,78]. Our results corroborate the results of previous studies showing that the diversity of vascular macrophyte assemblages increases with decreasing water velocity, while bryophytes are generally associated with more turbulent water flow [78,79]. High abundances of vascular macrophytes lead to the increase in flow resistance and sedimentation of organic debris, which results in reduced water velocity, increased conductivity, and produces lentic conditions in a particular habitat [80,81]. In IRES, the occurrence of lentic conditions within a stream or river is enhanced by the absence of high flow periods that could limit the growth of vascular macrophytes [78,82].

Our study shows that abundant and diverse macrophyte vegetation promotes increased abundance, species richness, taxonomic, and functional diversity of Odonata assemblages in intermittent karst rivers. This corroborates the results of previous studies frequently demonstrating close relationship between Odonata and aquatic vegetation [31–33,83–87]. Odonata require aquatic vegetation to complete key stages of their life cycle, using it as shelter and hunting ground (both as nymphs and adults) [34,88,89], for emergence [90], perching, thermoregulation, and oviposition in the adult stage [34,91,92]. The presence of water and emergent macrophyte vegetation are the most important visual cues for adult habitat selection [93,94], yet they respond more to the structural variety of macrophytes than to macrophyte species composition [34,95].

Odonata assemblages in macrophyte-rich habitats showed higher values of body size, which is probably due to the preference of the recorded Anisoptera (e.g., *Aeshna isoceles* (Müller, 1767), *A. imperator*, *B. pratense*) toward such habitats. This preference is reflected in their higher species richness and abundance in macrophyte-rich habitats compared to macrophyte poor ones. Due to the lower water velocity and higher abundance of vascular macrophytes, such habitats had numerous lentic sections along the river course, and consequently a higher share of species with the preference for lentic hydrological conditions (such as *Sympecma fusca* (Vander Linden, 1820), *A. isoceles*, and *B. pratense*) and the species frequently occurring in temporary water bodies (such as *C. viridis*, *S. meridionale*, and *Libellula depressa* Linnaeus, 1758) [63,64]. During the fieldwork, we also observed high abundances of teneral individuals of many of the recorded lentic Zygoptera species (e.g., *C. viridis*, *S. fusca*), as well as the reproductive behavior of many of the lentic Anisoptera and Zygoptera, e.g., copulation and oviposition. Therefore, those species are very likely to complete their life cycle in macrophyte-rich intermittent rivers. Nevertheless, we strongly recommend future studies to be focused on systematic nymph-focused research. On the other hand, macrophyte-poor habitats were characterized by more turbulent water flow, and their assemblages consisted predominantly of lotic species (such as *C. virgo*, *O. forcipatus*, and *C. heros*) [63,64].

In macrophyte-rich habitats, we recorded a higher number of species with certain drought resilience strategies, especially egg diapause (such as in *C. viridis*, *A. affinis*, and most of the recorded *Sympetrum* species). In addition, Odonata assemblages in such habitats consisted of species whose nymphs develop all year round or specifically in spring or summer. After the oviposition into the waterbody, in order to survive harsh environmental conditions such as droughts, eggs may go through a diapause, or they begin to develop immediately into the aquatic nymphs. In intermittent habitats, rapid growth is crucial, as the nymphs must develop rapidly to emerge into aerial adults before the habitat dries out. Such nymphs generally have faster development that occurs within weeks after oviposition [96]. Therefore, drought resilience strategies and nymphal development are most likely closely related traits in the studied Odonata assemblages. Due to relatively high drought resilience, Odonata are often amongst the dominant and relatively diverse taxa in the Mediterranean intermittent rivers and streams [70,72,97]. In contrast, macrophyte-poor habitats in our study were characterized mainly by the species with all year-round nymphal development and with no drought resilience strategies [63,64]. However, for some of those species, drought-survival strategies and mechanisms may be insufficiently known (e.g., *C. splendens, C. heros*) [69,98], or their occurrence in the studied IRES could be the result of the adult search for food resources (e.g., *S. meridionalis, O. forcipatus*) [69,99]. Thus, their occurrence in IRES should be confirmed with future, nymph-focused studies.

5. Conclusions

The current study revealed high Odonata species richness in karst IRES ecosystems. One of the recorded species is of international conservation concern, indicating the potential conservation value of IRES habitats. Habitats with well-developed aquatic macrophytes promote higher abundance as well as the taxonomic and functional diversity of Odonata assemblages compared to the habitats with poorly developed macrophytes. In addition to aquatic vegetation, physicochemical water properties, particularly conductivity and water velocity, are shown to be amongst the most significant determinants of Odonata assemblages in the studied IRES. To define adequate conservation measures for habitats and the species they support, it is crucial to understand species diversity patterns related to the quality of their environment [100]. Therefore, the current study represents an interesting contribution to our knowledge of Odonata diversity and their ecological requirements in intermittent karst rivers in the Mediterranean. These results also provide some new insights that could be useful for sampling protocol development and the bioassessment of IRES.

Author Contributions: Conceptualization, M.V. and A.B.; methodology, M.V. and A.B.; formal analysis, A.B., M.V. and V.Š.; investigation, M.V.; resources, A.B. and R.M.K.; data curation, M.V., A.B., F.R., M.R., R.M.K., V.Š. and L.R.; writing—original draft preparation, M.V. and A.B.; writing—review and editing, M.V., A.B., F.R., M.R., R.M.K., V.Š. and L.R; visualization, M.V., A.B. and L.R.; supervision, M.V. and A.B.; project administration, M.V. and A.B.; funding acquisition, A.B. All authors have read and agreed to the published version of the manuscript.

Funding: This research was funded by the Croatian Scientific Foundation, DinDRY to A.B., grant number UIP-2020-02-5385.

Institutional Review Board Statement: The study was approved by The Ministry of Economy and Sustainable Development of the Republic of Croatia with a permission for investigating in the protected area and for sampling protected taxa (permission number 640-01/21-01/7).

Data Availability Statement: Data are available from the corresponding author upon request. The data are not publicly available [due to the authors' policy of saving unpublished data for future publications].

Acknowledgments: We kindly thank Vladimir Bartovsky and Marija Starčević for help with the sampling. Zlatko Mihaljević and employers of the Public Institution Nature of the Šibenik-Knin County for their help with study site selection. Ivka Štefanić and Vesna Gulin are thanked for their help with the laboratory analysis of the water samples. Finally, three anonymous reviewers are thanked for their useful comments and suggestions that helped to improve this paper.

Conflicts of Interest: The authors declare no conflict of interest.

Appendix A

Table A1. Four groups of functional traits characteristic for the recorded Odonata species (body size according to Ref. [57]), other functional traits adapted from Refs. [63,64]). Legend: eupotamon = main channel and connected side arms; parapotamon = side arms connected only at the downstream end at mean water levels, plesiopotamon (including lakes) = no connectivity with the main channel at mean water levels; coverage by macrophytes does not exceed 20%, palaeopotamon (incl. pools, ponds) = no connectivity with the main channel at mean water levels; coverage by macrophytes exceeds 20%.

Species Name	Functional Traits			
	Average Body Size (mm)	**Hydrological Preference**	**Nymphal Development**	**Drought Resilience Form**
Calopteryx splendens (Harris, 1782)	46.5	Lotic (eupotamon, parapotamon)	All year	No drought resilience
Calopteryx virgo (Linnaeus, 1758)	47.0	Lotic (eupotamon)	All year	No drought resilience
Chalcolestes viridis (Vander Linden, 1825)	43.5	Predominantly lentic (plesiopotamon, palaeopotamon), but can occur in lotic habitats. Recorded from temporary waterbodies.	Spring, summer	Egg diapause
Sympecma fusca (Vander Linden, 1820)	36.5	Predominantly lentic (plesiopotamon, palaeopotamon), but can occur in lotic habitats. Recorded from temporary waterbodies.	Mainly in summer	Adult diapause
Ischnura elegans (Van der Linden, 1820)	32.0	Eurytopic. Recorded from temporary waterbodies	All year	No drought resilience
Coenagrion puella (Linnaeus, 1758)	34.0	Predominantly lentic (plesiopotamon, palaeopotamon), but can occur in lotic habitats. Recorded from temporary waterbodies.	All year	No drought resilience
Erythromma lindenii (Selys, 1840)	33.0	Predominantly lentic (plesiopotamon, palaeopotamon), but can occur in lotic habitats	Unknown	Unknown
Platycnemis pennipes (Pallas, 1771)	36.0	Predominantly lotic (eupotamon, parapotamon) but can occur in lentic habitats	All year	No drought resilience
Aeshna affinis Vander Linden, 1820	61.5	Lentic (palaeopotamon). Recorded from temporary waterbodies	Spring, summer	Egg diapause
Aeshna isoceles (Müller, 1767)	64.0	Lentic (plesiopotamon, palaeopotamon)	All year	No drought resilience
Anax imperator (Selys, 1839)	75.0	Predominantly lentic (plesiopotamon, palaeopotamon), but can occur in lotic habitats	All year	No drought resilience
Brachytron pratense (Müller, 1764)	58.5	Predominantly lentic (plesiopotamon, palaeopotamon), but can occur in lotic habitats	All year	No drought resilience
Onychogomphus forcipatus (Linnaeus, 1758)	48.0	Lotic (eupotamon, parapotamon)	All year	No drought resilience
Cordulegaster heros Theischinger, 1979	80.5	Lotic (eupotamon)	All year	No drought resilience
Somatochlora meridionalis Nielsen, 1935	52.5	Lotic (eupotamon, parapotamon)	All year	No drought resilience

Table A1. *Cont.*

Species Name	Functional Traits			
	Average Body Size (mm)	Hydrological Preference	Nymphal Development	Drought Resilience Form
Libellula depressa Linnaeus, 1758	43.5	Predominantly lentic (plesiopotamon, palaeopotamon), but can occur in lotic habitats. Recorded from temporary waterbodies.	All year	Nymph diapause
Libellula fulva Müller, 1764	43.5	Predominantly lotic (eupotamon, parapotamon) but can occur in lentic habitats	All year	Nymph diapause
Orthetrum cancellatum (Linnaeus, 1758)	47.0	Predominantly lotic (eupotamon, parapotamon) but can occur in lentic habitats	All year	No drought resilience
Orthetrum coerulescens (Fabricius, 1798)	40.5	Predominantly lotic (eupotamon, parapotamon) but can occur in lentic habitats. Recorded from temporary waterbodies	All year	Nymph diapause
Orthetrum brunneum (Fonscolombe, 1837)	45.0	Lotic (eupotamon, parapotamon)	All year	Unknown
Sympetrum sanguineum (Müller, 1764)	36.5	Predominantly lentic (plesiopotamon, palaeopotamon), but can occur in lotic habitats. Recorded from temporary waterbodies.	Spring, summer	Egg diapause
Sympetrum fonscolombii (Selys, 1840)	36.5	Eurytopic. Recorded from temporary waterbodies	All year	Unknown
Sympetrum striolatum (Charpentier, 1840)	39.5	Eurytopic. Recorded from temporary waterbodies	All year	Egg diapause
Sympetrum meridionale (Selys, 1841)	37.5	Lentic (palaeopotamon). Recorded from temporary waterbodies	Spring, summer	Egg diapause
Crocothemis erythraea (Brullé, 1832)	40.5	Predominantly lentic (plesiopotamon, palaeopotamon), but can occur in lotic habitats	All year	No drought resilience

References

1. Lytle, D.A.; Poff, N.L. Adaptation to natural flow regimes. *Trends Ecol. Evol.* **2004**, *19*, 94–100. [CrossRef] [PubMed]
2. Datry, T.; Bonada, N.; Boulton, A. *Intermittent Rivers and Ephemeral Streams: Ecology and Management*, 1st ed.; Academic Press: Cambridge, MA, USA, 2017.
3. McDonough, O.T.; Hosen, J.D.; Palmer, M.A. Temporary streams: The hydrology, geography and ecology of non-perennially flowing waters. In *River Ecosystems: Dynamics, Management and Conservation*, 1st ed.; Elliot, H.S., Martin, L.E., Eds.; Nova Science Publ. Inc.: New York, NY, USA, 2011; pp. 259–289.
4. Arthington, A.H.; Bernardo, J.M.; Ilhèu, M. Temporary rivers: Linking ecohydrology, ecological quality and reconciliation ecology. *River Res. Appl.* **2014**, *30*, 1209–1215. [CrossRef]
5. Tockner, K.; Robinson, C.T.; Uehlinger, U. *Rivers of Europe*, 1st ed.; Academic Press: Cambridge, MA, USA, 2009.
6. Larned, S.T.; Datry, T.; Arscott, D.B.; Tockner, K. Emerging concepts in temporary-river ecology. *Freshw. Biol.* **2010**, *55*, 717–738. [CrossRef]
7. Vörösmarty, C.; McIntyre, P.; Gessner, M.; Dudgeon, D.; Prusevich, A.; Green, P.; Glidden, S.; Bunn, S.E.; Sullivan, C.A.; Reidy Liermann, A.; et al. Global threats to human water security and river biodiversity. *Nature* **2010**, *468*, 334. [CrossRef]
8. Stubbington, R.; Chadd, R.; Cid, N.; Csabai, Z.; Miliša, M.; Morais, M.; Munné, A.; Pařil, P.; Pešić, V.; Tziortzis, I.; et al. Biomonitoring of intermittent rivers and ephemeral streams in Europe: Current practice and priorities to enhance ecological status assessments. *Sci. Total Environ.* **2018**, *618*, 1096–1113. [CrossRef] [PubMed]
9. Boix, D.; García-Berthou, E.; Gascón, S.; Benejam, L.L.; Tornés, E.; Sala, J.; Benito, J.; Munné, A.; Solà, C.; Sabater, S. Response of community structure to sustained drought in Mediterranean rivers. *J. Hydrol.* **2010**, *383*, 135–146. [CrossRef]
10. Gleick, P.H. Global freshwater resources: Soft-path solutions for the 21st century. *Science* **2003**, *302*, 1524–1528. [CrossRef] [PubMed]
11. Shanafield, M.; Bourke, S.A.; Zimmer, M.A.; Costigan, K.H. An overview of the hydrology of non-perennial rivers and streams. *WIREs Water* **2021**, *8*, 1–25. [CrossRef]
12. Döll, P.; Schmied, H.M. How is the impact of climate change on river flow regimes related to the impact on mean annual runoff? A global-scale analysis. *Environ. Res. Lett.* **2012**, *7*, 14–37. [CrossRef]
13. Datry, T.; Larned, S.T.; Tockner, T. Intermittent rivers: A challenge for freshwater ecology. *BioScience* **2014**, *64*, 229–235. [CrossRef]
14. del Rosario, R.B.; Resh, V.H. Invertebrates in intermittent and perennial streams: Is the hyporheic zone a refuge from drying? *J. N. Am. Benthol. Soc.* **2000**, *19*, 680–696. [CrossRef]
15. Bonada, N.; Zamora-Munoz, C.; Rieradevall, M.; Prat, N. Ecological profiles of caddisfly larvae in Mediterranean streams: Implications for bioassessment methods. *Environ. Pollut.* **2004**, *132*, 509–521. [CrossRef]
16. Argyroudi, A.; Chatzinikolaou, Y.; Poirazidis, K.; Lazaridou, M. Do intermittent and ephemeral Mediterranean rivers belong to the same river type? *Aquat. Ecol.* **2009**, *43*, 465–476.
17. Government of the Republic of Croatia (GRC). Directive on Water Classification. Available online: https://www.irb.hr/ (accessed on 28 November 2021).
18. Golfieri, B.; Hardersen, S.; Maiolini, B.; Surian, N. Odonates as indicators of the ecological integrity of the river corridor: Development and application of the Odonate River Indec (ORI) in northern Italy. *Ecol. Indic.* **2016**, *61*, 234–247. [CrossRef]
19. Martín, R.; Maynou, X. Dragonflies (Insecta: Odonata) as indicators of habitat quality in Mediterranean streams and rivers in the province of Barcelona (Catalonia, Iberian Peninsula). *Int. J. Odonatol.* **2016**, *19*, 107–124. [CrossRef]
20. Oliveira-Junior, J.M.B.; Shimano, Y.; Gardner, T.A.; Hughes, R.M.; De Marco, P.; Juen, L. Neotropical dragonflies (Insecta: Odonata) as indicators of ecological condition of small streams in the eastern Amazon. *Austral Ecol.* **2015**, *40*, 733–744. [CrossRef]
21. O'Malley, Z.G.; Compson, Z.G.; Orlofske, J.M.; Baird, D.J.; Curry, R.A.; Monk, W.A. Riparian and in-channel habitat properties linked to dragonfly emergence. *Sci. Rep.* **2020**, *10*, 17665. [CrossRef]
22. Vilenica, M. Ecological traits of dragonfly (Odonata) assemblages along an oligotrophic Dinaric karst hydrosystem. *Ann. Limnol. -Int. J. Lim.* **2017**, *53*, 377–389. [CrossRef]
23. Vilenica, M.; Kerovec, M.; Pozojević, I.; Mihaljević, Z. Odonata assemblages in anthropogenically impacted lotic habitats. *J. Limnol.* **2020**, *80*, 1–9. [CrossRef]
24. Brito, J.; Calvão, L.; Cunha, E.; Maioli, L.; Barbirato, M.; Rolim, S.; Juen, L. Environmental variables affect the diversity of adult damselflies (Odonata: Zygoptera) in western Amazonia. *Int. J. Odonatol.* **2021**, *24*, 108–121. [CrossRef]
25. Adu, B.W.; Amusan, B.O.; Oke, T.O. Assessment of the water quality and Odonata assemblages in three waterbodies in Ilara-Mokin, south-western Nigeria. *Int. J. Odonatol.* **2019**, *22*, 101–114. [CrossRef]
26. Oliveira-Junior, J.M.B.; Juen, L. Structuring of dragonfly communities (insecta: Odonata) in Eastern Amazon: Effects of environmental and spatial factors in preserved and altered streams. *Insects* **2019**, *10*, 322. [CrossRef]
27. Brito, J.S.; Michelan, T.S.; Juen, L. Aquatic macrophytes are important substrates for Libellulidae (Odonata) larvae and adults. *Limnology* **2021**, *22*, 139–149. [CrossRef]
28. McAbendroth, L.; Ramsay, P.M.; Foggo, A.; Rundle, S.D.; Bilton, D.T. Does macrophyte fractal complexity drive invertebrate diversity, biomass and body size distributions? *Oikos* **2005**, *111*, 279–290. [CrossRef]
29. Silva, C.V.; Henry, R. Aquatic invertebrates assemblages associated with two floating macrophytes species of contrasting root systems in a tropical wetland. *Limnology* **2020**, *21*, 107–118. [CrossRef]

30. Vilenica, M.; Alegro, A.; Koletić, N.; Mihaljević, Z. New evidence of *Lindenia tetraphylla* (Vander Linden, 1825) (Odonata, Gomphidae) reproduction at the north-western border of its distribution. *Nat. Croat.* **2016**, *25*, 287–294. [CrossRef]
31. Calvão, L.B.; Juen, L.; Oliveira-Junior, J.M.B.; Batista, J.D.; De, M.P.J. Land use modifies Odonata diversity in streams of the Brazilian Cerrado. *J. Insect Conserv.* **2018**, *22*, 675–685. [CrossRef]
32. Huikkonen, I.; Helle, I.; Elo, M. Heterogenic aquatic vegetation promotes abundance and species richness of Odonata (Insecta) in constructed agricultural wetlands. *Insect Conserv. Divers.* **2019**, *13*, 374–383. [CrossRef]
33. Vilenica, M.; Pozojević, I.; Vučković, N.; Mihaljević, Z. How suitable are man-made water bodies as habitats for Odonata? *Knowl. Manag. Aquat. Ecosyst.* **2020**, *421*, 1–10. [CrossRef]
34. Corbet, P.S. *Dragonflies: Behavior and Ecology of Odonata*; Harley Books: Colchester, UK, 1999.
35. Stoks, R.; McPeek, M.A. Predators and life histories shape *Lestes* damselfly assemblages along a freshwater habitat gradient. *Ecology* **2003**, *84*, 1576–1587. [CrossRef]
36. Magnusson, A.K.; Williams, D.D. The roles of natural temporal and spatial variation versus biotic influences in shaping the physicochemical environment of intermittent ponds: A case study. *Arch. Hydrobiol.* **2006**, *165*, 537–556. [CrossRef]
37. Chester, E.T.; Robson, B.J. Drought refuges, spatial scale and recolonisation by invertebrates in non-perennial streams. *Freshw. Biol.* **2011**, *56*, 2094–2104. [CrossRef]
38. Deacon, C.; Samways, M.J.; Pryke, J.S. Aquatic insects decline in abundance and occupy low-quality artificial habitats to survive hydrological droughts. *Freshw. Biol.* **2019**, *64*, 1643–1654. [CrossRef]
39. Illies, J. *Limnofauna Europaea*, 2nd ed.; Gustav Fischer Verlag: Stuttgart, Germany, 1978.
40. Mihevc, A.; Zupan-Hajna, N.; Prelovšek, M. Case study from the Dinaric karst of Slovenia. In *Introduction to the Dinaric Karst*, 1st ed.; Mihevc, A., Prelovšek, M., Zupan Hajna, N., Eds.; Karst Research Institute: Postojna, Slovenia, 2010; pp. 49–66.
41. Bonacci, O.; Željković, I.; Galić, A. Karst rivers' particularity: An example from Dinaric karst (Croatia/Bosnia and Herzegovina). *Environ. Earth Sci.* **2013**, *70*, 963–974. [CrossRef]
42. Previšić, A.; Walton, C.; Kučinić, M.; Mitrikeski, P.T.; Kerovec, M. Pleistocene divergence of Dinaric Drusus endemics (Trichoptera, Limnephilidae) in multiple microrefugia within the Balkan Peninsula. *Mol. Ecol.* **2009**, *18*, 634–647. [CrossRef] [PubMed]
43. Suhling, F.; Sahlén, G.; Gorb, S.; Kalkman, V.J.; Dijkstra, K.D.B.; van Tol, J. Order odonata. In *Thorp and Covich's Freshwater Invertebrates*, 4th ed.; Thorp, J., Rogers, D.C., Eds.; Academic Press: Cambridge, MA, USA, 2015; pp. 893–932. [CrossRef]
44. Pešić, V.; Gligorović, B.; Savić, A.; Buczyński, P. Ecological patterns of Odonata assemblages in karst springs in central Montenegro. *Knowl. Manag. Aquat.* **2017**, *418*, 1–20. [CrossRef]
45. Bonacci, O. Hydrological investigations of Dinaric Karst at the Krčić catchment and the River Krka springs (Yugoslavia). *J. Hydrol.* **1985**, *82*, 317–326. [CrossRef]
46. Bonacci, O.; Jukić, D.; Ljubenkov, I. Definition of catchment area in karst: Case of the rivers Krčić and Krka, Croatia. *Hydrol. Sci. J.* **2006**, *51*, 682–699. [CrossRef]
47. Bonacci, O.; Terzić, J.; Roje-Bonacci, T. Hidrološka analiza krške rijeke Čikole. *Hrvat. Vode* **2018**, *26*, 281–292.
48. Bonacci, O.; Terzić, J.; Roje-Bonacci, T.; Frangen, T. An intermittent karst river: The case of the Čikola River (Dinaric Karst, Croatia). *Water* **2019**, *11*, 2415. [CrossRef]
49. Brkić, Ž.; Biondić, R.; Pavičić, A.; Slišković, I.; Marković, T.; Terzić, J.; Dukarić, F.; Dolić, M. *Određivanje Cjelina Podzemnih Voda na Jadranskom Slivu prema Kriterijima Okvirne Direktive o Vodama EU*; Hrvatski Geološki Institute: Zagreb, Croatia, 2006.
50. Magaš, D. Geographical factors of formation and development of the neolithic settlement Crno Vrilo. In *Crno Vrilo*, 1st ed.; Marijanović, B., Ed.; University of Zadar: Zadar, Croatia, 2009; pp. 7–23.
51. Lukač Reberski, J.; Rubinić, J.; Terzić, J.; Radišić, M. Climate change impacts on groundwater resources in the coastal karstic Adriatic area: A case study from the Dinaric karst. *Nat. Resour. Res.* **2020**, *29*, 1975–1988. [CrossRef]
52. Cukrov, N.; Frančišković-Bilinski, S.; Mikac, N.; Roje, V. Natural and anthropogenic influences recorded in sediments from the Krka river estuary (Eastern Adriatic coast), evaluated by statistical methods. *Fresnius Environ. Bull.* **2008**, *17*, 855–863.
53. Barkman, J.J.; Doing, H.; Segal, S. Kritische bemerkungen und vorschläge zur quantitativen vegetationsanalyse. *Acta Bot. Neerl.* **1964**, *13*, 394–419. [CrossRef]
54. Braun-Blanquet, J. *Pflanzensoziologie, Grundzüge der Vegetationskunde*, 3rd ed.; Springer: Berlin/Heidelberg, Germany, 1964.
55. Dierschke, H. *Pflanzensoziologie: Grundlagen und Methoden*, 1st ed.; UTB: Stuttgart, Germany, 1994.
56. APHA. *Standard Methods for the Examination of Water and Wastewater*, 18th ed.; American Public Health Association: Washington, DC, USA, 1992.
57. Dijkstra, K.D.B.; Lewington, R. *Field Guide to the Dragonflies of Britain and Europe*, 1st ed.; British Wildlife Publishing: Gillingham, UK, 2006.
58. IBM Corp. *IBM SPSS Statistics for Windows, Version 27.0*; IBM: Armonk, NY, USA, 2020.
59. Field, A. *Discovering Statistics Using SPSS*, 3rd ed.; SAGE Publications: London, UK, 2009.
60. Thiele, J.; Markussen, B. Potential of GLMM in modelling invasive spread. *CAB Rev. Perspect. Agric. Vet. Sci. Nutr. Nat. Resour.* **2012**, *7*, 1–10. [CrossRef]
61. Jost, L. Partitioning diversity into independent alpha and beta components. *Ecology* **2007**, *88*, 2427–2439. [CrossRef] [PubMed]
62. Clarke, K.R.; Gorley, R.N. *Primer V6: User Manual/Tutorial*; Plymouth: Auckland, New Zealand, 2006.
63. Freshwaterecology. Available online: https://www.freshwaterecology.info/fwe_search.php?og=mzb (accessed on 3 November 2021).

64. Schmidt-Kloiber, A.; Hering, D. www.freshwaterecology.info—An online tool that unifies, standardises and codifies more than 20,000 European freshwater organisms and their ecological preferences. *Ecol. Indic.* **2015**, *53*, 271–282. [CrossRef]
65. Ricotta, C.; Moretti, M. CWM and Rao's quadratic diversity: A unified framework for functional ecology. *Oecologia* **2011**, *167*, 181–188. [CrossRef]
66. ter Braak, C.J.F.; Šmilauer, P. *Canoco Reference Manual and User's Guide: Software for Ordination, Version 5.0*; Microcomputer Power: Ithaca, NY, USA, 2012.
67. Kotarac, M.; Šalamun, A.; Vilenica, M. *EU Natura 2000 Integration Project: Field Research and Laboratory Processing for Collecting New Inventory Data for Taxonomic Groups: Actinopterygii and Cephalaspidomorphi, Amphibia and Reptilia, Aves, Chiroptera, Decapoda, Lepidoptera, Odonata, Plecoptera–Final Report for the Taxonomic Group Odonata*; Ministry of Environmental and Nature Protection: Zagreb, Croatia, 2016.
68. Boudot, J.P.; Kalkman, V.J. *Atlas of the European Dragonflies and Damselflies*; KNNV Uitgeverij: Zeist, The Netherlands, 2015.
69. Hardersen, S. Dragonfly (Odonata) communities at three lotic sites with different hydrological characteristics. *Ital. J. Zool.* **2008**, *75*, 271–283. [CrossRef]
70. Belmar, O.; Velasco, J.; Gutiérrez-Cánovas, C.; Mellado-Díaz, A.; Millán, A.; Wood, P.J. The influence of natural flow regimes on macroinvertebrate assemblages in a semiarid Mediterranean basin. *Ecohydrology* **2012**, *6*, 363–379. [CrossRef]
71. Yalles Satha, A.; Samraoui, B. Environmental factors influencing Odonata communities of three Mediterranean rivers: Kebir-East, Seybouse, and Rhummel wadis, northeastern Algeria. *Rev. Ecol.* **2017**, *72*, 314–329.
72. Bonada, N.; Rieradevall, M.; Prat, N.; Resh, V.H. Benthic macro invertebrate assemblages and macrohabitat connectivity in Mediterranean-climate streams of northern California. *J. N. Am. Benthol. Soc.* **2006**, *25*, 32–43. [CrossRef]
73. Bonada, N.; Rieradevall, M.; Prat, N. Macroinvertebrate community structure and biological traits related to flow permanence in a Mediterranean river network. *Hydrobiologia* **2007**, *589*, 91–106. [CrossRef]
74. Oliveira-Junior, J.M.B.; Dias-Silva, K.; Teodósio, M.A.; Juen, L. The response of neotropical dragonflies (insecta: Odonata) to local and regional abiotic factors in small streams of the amazon. *Insects* **2019**, *10*, 446. [CrossRef]
75. Kalkman, V.J.; Boudot, J.-P.; Bernard, R.; Conze, K.-J.; De Knijf, G.; Dyatlova, E.; Ferreira, S.; Jović, M.; Ott, J.; Riservato, E.; et al. *European Red List of Dragonflies*; Publications Office of the European Union: Luxembourg City, Luxembourg, 2010.
76. The IUCN Red List of Threatened Species. Available online: https://doi.org/10.2305/IUCN.UK.2010-1.RLTS.T158700A5263990.en (accessed on 16 November 2021).
77. Biudes, J.F.V.; Camargo, A.F.M. Studying the limitant factors to primary production of aquatic macrophytes in Brazil. (Estudos dos Fatores Limitantes à Produção Primária por Macrófitas Aquáticas no Brasil). *Oecol. Aust.* **2008**, *12*, 7–19. [CrossRef]
78. Buffagni, A.; Erba, S.; Armanini, D.G. The lentic–lotic character of Mediterranean rivers and its importance to aquatic invertebrate communities. *Aquat. Sci.* **2010**, *72*, 45–60. [CrossRef]
79. Kemp, J.L.; Harper, D.M.; Crosa, G.A. Use of 'functional habitats' to link ecology with morphology and hydrology in river rehabilitation. *Aquat. Conserv. Mar. Freshw. Ecosyst.* **1999**, *9*, 159–178. [CrossRef]
80. Pitlo, R.H.; Dawson, F.H. Flow-resistance of aquatic weeds. In *Aquatic Weeds, The Ecology and Management of Nuisance Aquatic Vegetation*, 1st ed.; Pieterse, A.H., Murphy, K.J., Eds.; Oxford University Press: New York, NY, USA, 1990; pp. 74–84.
81. Lemly, A.D.; Hilderbrand, R.H. Influence of large woody debris on stream insect communities and benthic detritus. *Hydrobiologia* **2000**, *421*, 179–185. [CrossRef]
82. Johnson, W.C. Tree recruitment and survival in rivers: Influence of hydrological processes. *Hydrol. Process* **2000**, *14*, 3051–3074. [CrossRef]
83. Corbet, P.; Brooks, S. *Dragonflies, Collins New Naturalist Library No 106*; Collins: London, UK, 2008.
84. Perron, M.A.C.; Pick, F.R. Stormwater ponds as habitat for Odonata in urban areas: The importance of obligate wetland plant species. *Biodivers. Conserv.* **2020**, *29*, 913–931. [CrossRef]
85. Perron, M.A.C.; Richmond, I.C.; Pick, F.R. Plants, water quality and land cover as drivers of Odonata assemblages in urban ponds. *Sci. Total Environ.* **2021**, *773*, 145467. [CrossRef]
86. Daso, J.M.; Arquisal, I.B.; Yuto, C.M.M.; Mondejar, E.P. Species diversity of Odonata in Bolyok Falls, Naawan, Misamis Oriental, Philippines. *Aquac. Aquar. Conserv. Legis.* **2021**, *14*, 664–671.
87. Holtmann, L.; Brüggeshemke, J.; Juchem, M.; Fartmann, T. Odonate assemblages of urban stormwater ponds: The conservation value depends on pond type. *J. Insect Conserv.* **2019**, *23*, 123–132. [CrossRef]
88. Johansson, F.; Brodin, T. Effects of fish predators and abiotic factors on dragonfly community structure. *J. Freshw. Ecol.* **2003**, *18*, 415–423. [CrossRef]
89. Kriska, G. Dragonflies and damselflies—Odonata. In *Freshwater Invertebrates in Central Europe*, 1st ed.; Springer: Vienna, Austria, 2013; pp. 194–209. [CrossRef]
90. Boda, R.; Bereczki, C.; Ortmann-Ajkai, A.; Mauchart, P.; Pernecker, B.; Csabai, Z. Emergence behaviour of the red listed Balkan Goldenring (*Cordulegaster heros* Theischinger, 1979) in Hungarian upstreams: Vegetation structure affects the last steps of the larvae. *J. Insect Conserv.* **2015**, *19*, 547–557. [CrossRef]
91. Wildermuth, H. Habitat selection and oviposition site recognition by the dragonfly *Aeshna juncea* (L.): An experimental approach in natural habitats (Anisoptera: Aeshnidae). *Odonatologica* **1993**, *22*, 27–44.
92. Purse, B.V.; Thompson, D.J. Oviposition site selection by *Coenagrion mercuriale* (Odonata: Coenagrionidae). *Int. J. Odonatol.* **2009**, *12*, 257–273. [CrossRef]

93. Wildermuth, H. Dragonflies recognize the water of rendezvous and oviposition sites by horizontally polarized light: A behavioural field test. *Sci. Nat.* **1998**, *85*, 297–302. [CrossRef]
94. Ward, L.; Mill, P.J. Habitat factors influencing the presence of adult *Calopteryx splendens* (Odonata: Zygoptera). *Eur. J. Entomol.* **2005**, *102*, 47–51. [CrossRef]
95. Steytler, N.S.; Samways, M.J. Biotope selection by adult male dragonflies (Odonata) at an artifical lake created for insect conservation in South Africa. *Biol. Conserv.* **1995**, *72*, 381–386. [CrossRef]
96. McPeek, M.A. Ecological factors limiting the distributions and abundances of Odonata. In *Dragonflies: Model Organisms for Ecological and Evolutionary Research*, 1st ed.; Córdoba-Aguilar, A., Ed.; Oxford University Press: Oxford, UK, 2008; pp. 51–62.
97. Sanchez-Montoya, M.D.; Punti, T.; Suarez, M.L.; Vidal-Abarca, M.D.; Rieradevall, M.; Poquet, J.M.; Zamora-Munoz, C.; Robles, S.; Alvarez, M.; Alba-Tercedor, J.; et al. Concordance between ecotypes and macroinvertebrate assemblages in Mediterranean streams. *Freshw. Biol.* **2007**, *52*, 2240–2255. [CrossRef]
98. Pernecker, B.; Mauchart, P.; Csabai, Z. What to do if streams go dry? Behaviour of Balkan Goldenring (*Cordulegaster heros*, Odonata) larvae in a simulated drought experiment in SW Hungary. *Ecol. Entomol.* **2020**, *45*, 1457–1465. [CrossRef]
99. Petrovičová, K.; Langraf, V.; David, S.; Krumpálová, Z.; Schlarmannová, J. Distinct Odonata assemblage variations in lentic reservoirs in Slovakia (Central Europe). *Biologia* **2021**. [CrossRef]
100. Veech, J.A.; Summerville, K.S.; Crist, T.O.; Gering, J.C. The additive partitioning of species diversity: Recent revival of an old idea. *Oikos* **2002**, *99*, 3–9. [CrossRef]

MDPI

Article

Plecoptera (Insecta) Diversity in Indiana: A Watershed-Based Analysis

Evan A. Newman [1], R. Edward DeWalt [2,*] and Scott A. Grubbs [3]

[1] Department of Entomology, University of Illinois, Urbana, IL 61820, USA; evanan2@illinois.edu
[2] Illinois Natural History Survey, University of Illinois, Champaign, IL 61820, USA
[3] Center for Biodiversity Studies, Department of Biology, Western Kentucky University, Bowling Green, KY 42101, USA; scott.grubbs@wku.edu
* Correspondence: dewalt@illinois.edu; Tel.: +1-217-649-7414

Abstract: Plecoptera, an environmentally sensitive order of aquatic insects commonly used in water quality monitoring is experiencing decline across the globe. This study addresses the landscape factors that impact the species richness of stoneflies using the US Geological Survey Hierarchical Unit Code 8 drainage scale (HUC8) in the state of Indiana. Over 6300 specimen records from regional museums, literature, and recent efforts were assigned to HUC8 drainages. A total of 93 species were recorded from the state. The three richest of 38 HUC8s were the Lower East Fork White (66 species), the Blue-Sinking (58), and the Lower White (51) drainages, all concentrated in the southern unglaciated part of the state. Richness was predicted using nine variables, reduced from 116 and subjected to AICc importance and hierarchical partitioning. AICc importance revealed four variables associated with Plecoptera species richness, topographic wetness index, HUC8 area, % soil hydrolgroup C/D, and % historic wetland ecosystem. Hierarchical partitioning indicated topographic wetness index, HUC8 area, and % cherty red clay surface geology as significantly important to predicting species richness. This analysis highlights the importance of hydrology and glacial history in species richness of Plecoptera. The accumulated data are primed to be used for monograph production, niche modeling, and conservation status assessment for an entire assemblage in a large geographic area.

Keywords: stoneflies; aquatic insects; USA; species richness; hierarchical unit codes

Citation: Newman, E.A.; DeWalt, R.E.; Grubbs, S.A. Plecoptera (Insecta) Diversity in Indiana: A Watershed-Based Analysis. *Diversity* **2021**, *13*, 672. https://doi.org/10.3390/d13120672

Academic Editors: Marina Vilenica, Zohar Yanai and Laurent Vuataz

Received: 3 November 2021
Accepted: 10 December 2021
Published: 15 December 2021

Publisher's Note: MDPI stays neutral with regard to jurisdictional claims in published maps and institutional affiliations.

1. Introduction

Stoneflies (Plecoptera) are aquatic insects that are species-rich in temperate, mountain streams [1–3]. Approximately 3900 extant, valid species are classified into 17 families worldwide [4,5]. In North America (including Mexico), the number of extant species is just over 770 [3]. Plecoptera species exhibit a range of sensitivities to water and habitat quality changes and this makes them useful as indicators of water quality [6,7].

Plecoptera species across much of the world are thought to be imperiled by human activity and climate change. To survive climate change through the end of the 21st century, stoneflies may be forced to undergo dramatic range shifts, as suggested by modeling of generic distributions in North America [8]. In the USA, Plecoptera are the third-most imperiled group of freshwater aquatic organisms [9]. In Illinois, 29% of 77 species known at the time were considered extirpated or extinct [10]. Extirpations and range loss have also been reported for Indiana [11], Michigan [12], and Ohio [13,14]. Similar imperilment of the stonefly fauna of the Czech Republic has been reported [15]. It is estimated that in Europe and North America up to 35% of stonefly species are in decline and many of these species appear to meet International Union for Conservation of Nature criteria for inclusion in the Red List of Threatened Species [16].

Despite demonstrated stonefly range loss and extinctions [10], the paucity of high quality stonefly specimen-level data hampers our ability to understand historic distributions, the effects of human disturbance through the 20th century, current distribution and relative imperilment, and predicted distribution changes. Accumulating such data is difficult. Much of the developed world had already degraded water and habitat quality prior to the 1950s, leaving large rivers without their characteristic stonefly fauna and intact assemblages being present only in small streams and at higher altitudes [10,15,17]. Older literature often present lists of species from known locations that include misidentifications and lack corrective voucher specimens. Many ecological works and water quality agencies appropriately apply methods using higher taxonomy [18], though the resulting data rarely meet species-level conservation objectives [19].

Criteria for inclusion in such a distribution data set include identifications as provided by taxonomic experts, precise location data, and a unique identifier (catalogue number) that links data to a particular specimen or specimens. Specimens providing this kind of information are found in museums or research collections. They often result from a long history of taxonomic research within a state or region. Such assessments have been conducted on Indiana stoneflies since before 1900, mostly as an adjunct to taxonomic studies [20–26]. The most recent publication in [11] reported 87 species and two recent works [27,28] added two new species and one existing species to the Indiana total.

Fortunately, nearly all cited authors deposited their specimens in regional museums so that specimens and data would be available for broader analyses in the future. Recent USA National Science Foundation and Fish and Wildlife Service grants to DeWalt have enriched these data with contemporary collections, building a >6330 record data set of Indiana stoneflies. These data are critical to establishing where species occurred prior to major degradation, providing context for current distributions and a means to estimate range losses of individual species. They are also important to determine which drainages and areas of the state are the richest in species and allow for analysis of factors useful in predicting richness within drainages. This data set will ultimately be used to develop a distributional atlas for the state and conservation status assessments for the entire assemblage in Indiana.

The objectives of this study were twofold: (1) to use the aforementioned accumulated species data set to assess species richness and its distribution across watersheds in Indiana, and (2) to investigate the relative importance of natural and human disturbance variables for explaining species richness within individual watersheds. We anticipated that our data, with many specimens collected prior to 1950, would reflect historic distributions and that species richness would be best predicted by natural variables, not human disturbance factors. We also predict that the southern more rugged areas of Indiana would contain the richest watersheds. These data will be used for other secondary objectives such as a monograph of species distributions, taxonomic investigations of potential new species, and for conservation status assessments of the entire assemblage in Indiana, several of which are ongoing. The Indiana data are a subset of nearly 40,000 records gathered from Ohio to Iowa, Minnesota to Michigan.

2. Materials and Methods

Present-day Indiana reflects two major glacial events [23,24]. The Illinoian glaciation, approximately 100,000 years ago at maximum extent, covered about 80% of Indiana. The Wisconsinan glaciation, maximum extent 18,000 years ago, covered approximately 60% (Figure 1, Table 1). These glacial events left three major landscapes within the state—the Wisconsinan (twice glaciated), the Illinoian (once glaciated), and the unglaciated south-central region. The Wisconsinan landscape is occupied by low-gradient streams and is deeply buried in glacial till. The older Illinoian landscape is eroded to abundant ravine streams and mature river valleys, and in the southwest along the Wabash River windblown loess ridges are common. Some larger river valleys of Illinoian age are filled with Wisconsinan-era outwash, forming large, meandering rivers. The unglaciated region is rugged with high-gradient streams, abundant groundwater, and exposed limestone bedrock.

Table 1. Summary of Indiana historic vegetation and glaciation prior to European settlement.

Historic Vegetation	Area km^2	% Area
Mesophytic forest	7885	8.4
Deciduous—beech—maple	46,600	49.7
Deciduous—oak—hickory	27,968	29.8
Dry prairie	2565	2.7
Open wetlands/wet prairie	8131	8.7
Wooded wetlands	570	0.6
Glaciation		
Wisconsinan (twice glaciated)	58,996	62.9
Illinoian (once glaciated)	23,028	24.6
Unglaciated	11,713	12.5
Total area	**93,719**	

The United States Geological Survey (USGS) hydrologic unit codes (HUCs) at the HUC8 scale [29] (Figure 1) were used as watershed replicates.

Prior to European settlement, Indiana supported six major vegetation communities—dry prairie, oak–hickory upland forest, beech–maple upland forest, beech–oak–maple–hickory mesophytic forest, wooded wetland, and non-wooded wetland [30–33] (Figure 1, Table 1). Forests dominated and prairies and wetlands occupied the northern third of Indiana. Currently, 62% of land use is agricultural (Table 2).

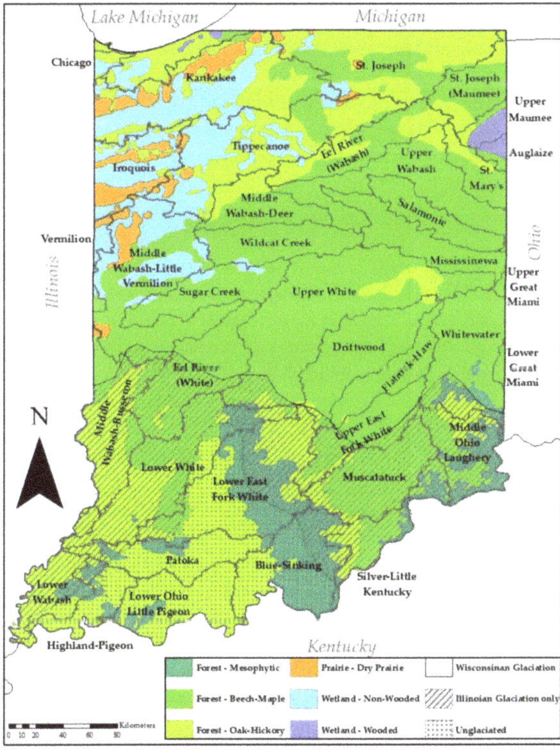

Figure 1. Map of Indiana vegetation prior to European settlement and glaciation. Adapted from Lindsey et al. [32].

Table 2. Summary of statewide land use of Indiana in 2016.

2016 Land Use	Area km^2	% Area
Agriculture	57,965	61.9
Forest	21,537	23.0
Built area	9756	10.4
Wooded wetlands	1996	2.1
Open water	1200	1.3
Herbaceous/shrub	815	0.9
Herbaceous wetlands	291	0.3
Barren	159	0.2
Total	**93,719**	

A large portion of the data used in this study resulted from examination of historical, borrowed specimens from many institutional and private collections, principal among these were the Illinois Natural History Survey Insect Collection (INHS), the Purdue University Entomological Research Collection (PERC), and the Western Kentucky University (WKUC). Sampling continued between 2000 and 2015 by DeWalt, Grubbs, and Donald W. Webb (deceased, INHS). Newman assumed lead of the project in 2016 and focused sampling on areas of the state where effort was sparse and rare species might be found. Throughout the century-long effort, sampling was not done at randomly selected locations, but was conducted at multiple locations within a full range of lotic habitats characteristic of the HUC8 being sampled (Figure 2). Resulting from this century of work is a highly detailed database of specimen and confirmed literature records. Historical and contemporary specimens were morphologically identified to the current state of the art. Recent literature used to identify species may be queried from the Plecoptera Species File Online [4]. Data from both larvae and adults were included where species-level identification was certain.

Border records were included in this analysis to increase the number of species within several Indiana peripheral drainages that were incompletely collected. These records met the following criteria for inclusion in the data set: the waterbody of the record formed a border with Indiana, or the locality of record was within 5 km of the state border and the same habitat existed in adjacent areas of Indiana. Border records were included from Illinois (110), Kentucky (3), Michigan (1), and Ohio (1).

Specimen data (locality labels, determination labels, and catalog numbers) were captured and normalized in a custom database. Most specimens were associated with their database record using a paper catalog number [34]. We georeferenced locations using an online mapping program [35] employing datum WGS-84. Where collectors provided coordinates, these were projected to verify the location and coordinates corrected accordingly. Precision of coordinates are provided as radius in meters: collector-provided = 10 m, localities with stream name and road crossing or small town name = 100–1000 m radius, localities with moderate population size to 50,000 people = 10,000 m, and Indiana county-level records = 100,000 m. State-level records were not mapped. County records were retained in analyses if drainage affiliation was certain.

Maps were exported from an ArcView 9.3 (ESRI) project file using a WGS-84 projection and overlaid on USGS HUC8 drainages. Each georeferenced record was thus assigned to a HUC8 drainage, allowing creation of a binary matrix of presence/absence of species by HUC8 drainage. Total species richness values were obtained from this matrix. Drainages with fewer than five recorded species were considered incompletely collected and were eliminated from analysis. Five species was the value for the Little Calumet drainage which was known to be well sampled [36]. Small drainages leaving or entering border states were trimmed to areas within Indiana.

Figure 2. Map of Plecoptera specimen localities in Indiana.

All statistical analyses were conducted in an R environment [37]. Linear regression models were used to examine the relationship of species richness to the number of unique localities and HUC8 drainage area. This was accomplished using the *lm* function. The completeness of species discovery in Indiana was analyzed by building a species presence–absence vs. unique site-date collection events matrix. This matrix was used to fit a species accumulation curve using the *specaccum* function in *vegan* using 100 permutations [38]. Data were further subjected to the *specpool* function in order to estimate the species richness of the study area.

A K-means cluster analysis was used to examine the similarity of species assemblages between different drainages. The number of clusters represented by the data was determined using the "elbow method", which indicated two clusters in the data. Jaccard distance between HUC8 species assemblages was calculated using *vegdist* function (*vegan* R package). This output was then subjected to hierarchical clustering using the *hclust* function (*stats* package). These data were plotted as a tree.

A natural and human disturbance variable set containing 116 environmental variables was assembled using three sources—USDA/NRCS Geospatial Gateway [39], USGS National Land Cover Database (NLCD) 2016 [40], and pre-European settlement vegetation from land survey data [32]. Variables fell into seven categories: climate, geology, hydrology, soils, topography, land cover, and historical ecosystem (vegetation).

Data from the USDA/NRCS and NLCD 2016 were in raster format while historical ecosystem data were formatted in shapefiles. All were treated similarly. Variables were extracted for each HUC8 drainage using ARCMap Spatial Analyst Tools, Zonal Statistics as Table to obtain a mean value for each HUC8. For datasets with several discrete values such as land cover, Spatial Analyst Tools, Tabulate Area was used and values were converted into percentages of coverage for each HUC8. Variable data were consolidated into a spreadsheet in Microsoft Excel (Microsoft, Redmond, WA, USA) for the first stage of variable reduction.

Multiple linear regression was used for variable set reduction followed by linear model-based variance partitioning to assess the effects of the environmental variables on Plecoptera species richness. Statistical methods for AIC based analyses were adapted from previous work [41].

To eliminate highly correlated variables, Pearson correlation coefficients were calculated. Pairs of variables were considered highly correlated if $r \geq 0.7$. In this case, one variable was removed from further analysis based on interpretation and experience of which variable was likely more important to stonefly species richness. This reduced the number of variables by 75, leaving 41. The remaining variables were examined for variance inflation factor (VIF) in multiple linear regression modeling (*vifstep* in R package *usdm*). Variables with a VIF > 10 were considered highly collinear and were dropped from further analysis [41]. This left 15 variables which were tested for their effect on species richness using relative weights and dominance analysis. This procedure examines independent variable contribution to variance in a multiple linear regression model [42]. This was accomplished using the package *yhat* [43] using the function *rlw*. The relative weight values were used to reduce the 15 variable set to nine, as this is the maximum number that can be used in the *hier.part* function (package *hier.part*) used later in the analysis. Six of seven categories were represented by the remaining variables: hydrology, soils, land cover, historic ecosystem, HUC8 area, and geology (Table 3).

A generalized linear model was fit for species richness based on the remaining nine predictors using a Poisson distribution. The dispersion parameter was calculated as 1.64, negating the need for an over-dispersion adjustment to the data [37,41].

Using the *dredge* function in the R package *MuMin* [44,45], all possible candidate models using the variables from the global model were ranked using Akaike information criteria (AICc). Score differences in AICc (ΔAICc) between the top-ranked model and all other models were used to select a group of models considered substantially supported [46]. Six models with a $\Delta AICc \leq 2$ were selected for further analysis. Model averaging was calculated for six well-supported models using R package *AICcmodavg*, using the *modavg* function. This method produced average coefficient estimates and 95% confidence intervals which were standardized to facilitate comparison of dissimilar variables [47]. Each predictor was assigned a relative importance value calculated by summing the model AICc weight of all models containing that particular variable. This analysis effectively eliminated % cherty red clay since it was not contained in any model where $\Delta AICc \leq 2$.

Further analysis was conducted using hierarchical variance partitioning, using the *hier.part* function in the *hier.part* R package [48]. Hierarchical variance partitioning estimates the percentage of variance explained by individual predictors in a linear model. This method compares all possible models in a multiple regression to obtain *I*, the measure of individual predictor effect on variance, and *J*, the contribution of a predictor when combined with each of the other predictors [49]. It provides another estimate of importance as confirmation of RIV importance, though they do not always agree. Data matrices of the top AICc importance predictors were randomized 1000 times with *rand.hp* to create distributions of *I* for each variable. Z-scores were computed with 95% confidence intervals for the value of *I* of each variable.

Table 3. Description, source, and statistics of nine variables included in the model.

Variable	Description	Source	Mean	Median	SD	SE	Min	Max
Drainage area km^2		[38]	2934	2404	1514	276	352	7044
Mean Horton overland flow	Soil infiltration exceeded by precipitation	[38]	6.12	6.20	1.40	0.26	2.47	8.14
Mean topographic wetness index (TWI)	Depth-to-water	[38]	13.31	13.35	0.58	0.11	12.14	14.27
% Wetlands 2016	Recent wetlands	[39]	2.60	1.22	3.32	0.61	0.19	13.38
% Developed land 2016	Built lands	[39]	9.89	7.02	6.55	1.02	4.48	35.26
% Historic wetland ecosystem	Historic wetlands	[33]	9.13	0	17.99	3.28	0	65.87
% Soils hydrogroup B	Medium coarse soils	[38]	32.92	31.72	12.51	2.28	9.16	58.44
% Soils hydrogroup C/D	Fine soils, slowly drained	[38]	3.57	1.63	5.30	0.97	0.01	21.57
% Cherty red clay	Chert in clay	[38]	4.37	0	14.22	2.60	0	69.97

3. Results

3.1. Species Richness and Patterns

A total of 6339 specimen records were amassed during this study. Unique sampling events totalled 2411 and were conducted at 1050 unique locations from 1879 to 2021 (Figure 2). As a result, we recorded 93 stonefly species from Indiana (Table S1). Eight of 10 North American families were represented with Perlidae (36 species), Perlodidae (17), Capniidae (14), and Taeniopterygidae (7) providing 80% of all species collected in the state. No endemics were found, but four species represent new state records—*Allocapnia pygmaea* (Burmeister, 1839), *Paracapnia angulata* Hanson, 1961, *Acroneuria lycorias* (Newman, 1839), and an undescribed species, *Perlesta* IN-5 (temporary name).

Linear regression indicated a significant, positive relationship between species richness and the number of unique localities sampled within HUC8 drainages (p = 1.85 × 10^{-09}, adjusted R^2 = 0.72) (Figure 3). Species richness demonstrated a significant, positive, but weaker relationship with HUC8 area (p-value: 0.005, adjusted R^2 = 0.23) (Figure 4). A species accumulation curve demonstrates that stoneflies were sampled to near completeness (Figure 5). The model estimated a terminal species richness of 95 (±5) species, just two more than observed values.

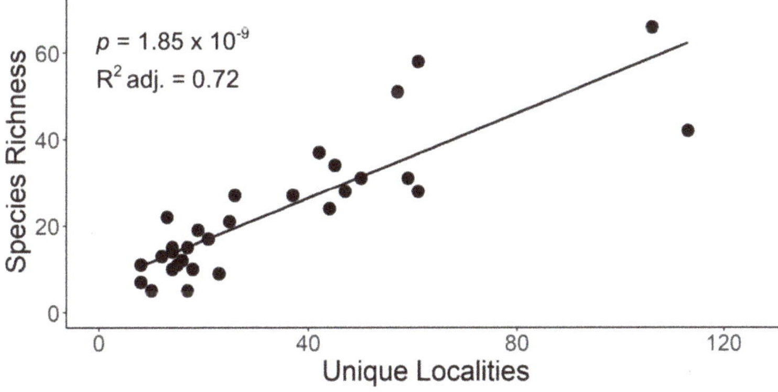

Figure 3. Scatterplot of stonefly species richness versus the unique localities within HUC8 watersheds of Indiana. Adjusted R^2, probability, and line of best fit provided for simple linear regression analysis.

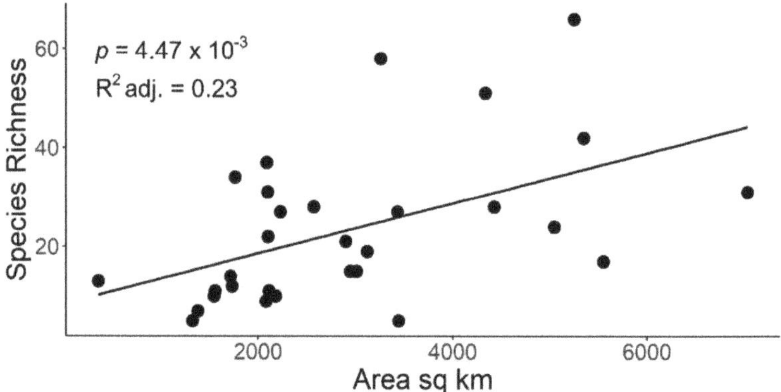

Figure 4. Scatterplot of stonefly species richness versus Indiana HUC8 watershed drainage area in km^2. Adjusted R^2, probability, and line of best fit provided for simple linear regression analysis.

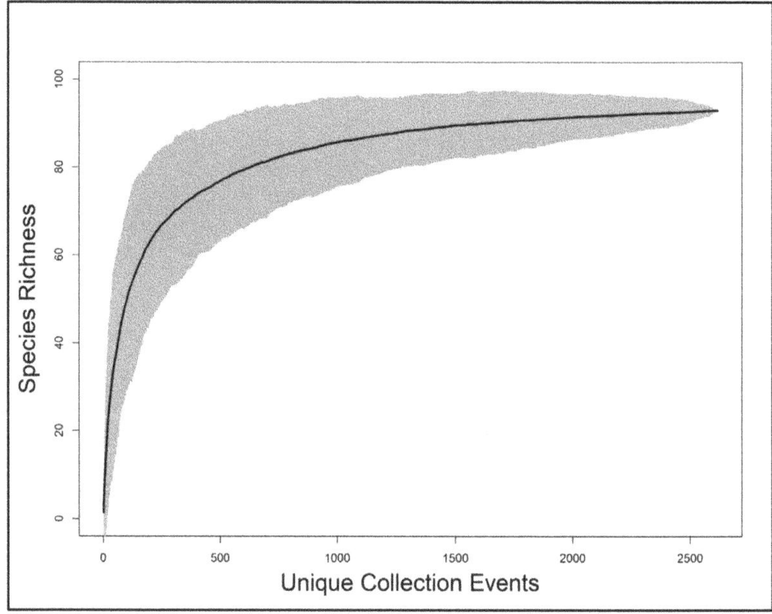

Figure 5. Species accumulation curve with 95% confidence interval (gray shade) for Indiana Plecoptera using unique collection events as replicates. Terminal estimation was 95 ± 5 species.

The mean number of species for the 30 HUC8 watersheds with five or more species was 23.3, with a median of 20.0. A heat map of species richness demonstrates that three hyperdiverse HUC8 drainages exist in Indiana (Figure 6, Table S2). These include the Lower East Fork White (66 species), the Blue-Sinking (58 species), and the Lower White (51 species), all in the southern third of the state. Other drainages with richness values above 30 species include the Middle Wabash–Little Vermilion (42 species), the Middle Ohio Laughery (37 species), Silver-L. Kentucky (34), Sugar Creek (31 species), and Upper White (31 species). These too occur in the southern half of the state. Eight watersheds were represented by four or fewer species, all were relatively small in drainage area, inadequately sampled, and unlikely to be diverse.

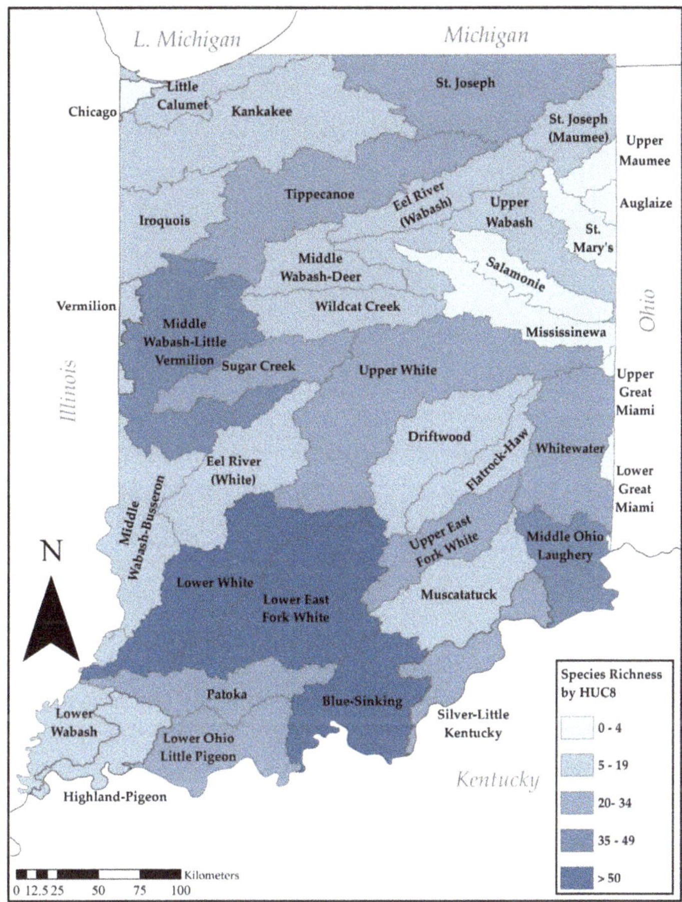

Figure 6. Heat map of Plecoptera species richness within HUC8 watersheds in Indiana.

A K-means cluster analysis of HUC8 species assemblages produced two clusters. Cluster 1 contains drainages from the unglaciated and once-glaciated southern landscapes of Indiana (Figure 7). Cluster 2 is composed of mostly northern drainages with a few more large, southern drainages filled with glacial outwash.

3.2. Relative Importance of Watershed Variables in Species Richness

Analysis by AICc provided six models for species richness with ΔAICc ≤ 2 (Table 4). Akaike weights of these six models ranged from 0.0575 to 0.1548 with a mean of 0.0852 $\pm$ 0.0148 SE. McFadden R^2 values suggest the models strongly explain variance in species richness and by a similar proportion among all models (McFadden R^2 = 0.52–0.54 for top-six models). AICc analysis identified four variables included in all top-six models—mean topographic wetness index (TWI), area km^2, % soils hydrogroup C/D, and % historic wetland ecosystem. Percent cherty red clay was the only variable not included in the ΔAICc ≤ 2 model set. All models contained between four and six variables. The best model, as determined by ΔAICc, was Model 1 with five variables (area, % soils hydrogroup CD, % development, % historic wetlands, and mean TWI), though all models were approximately equivalent.

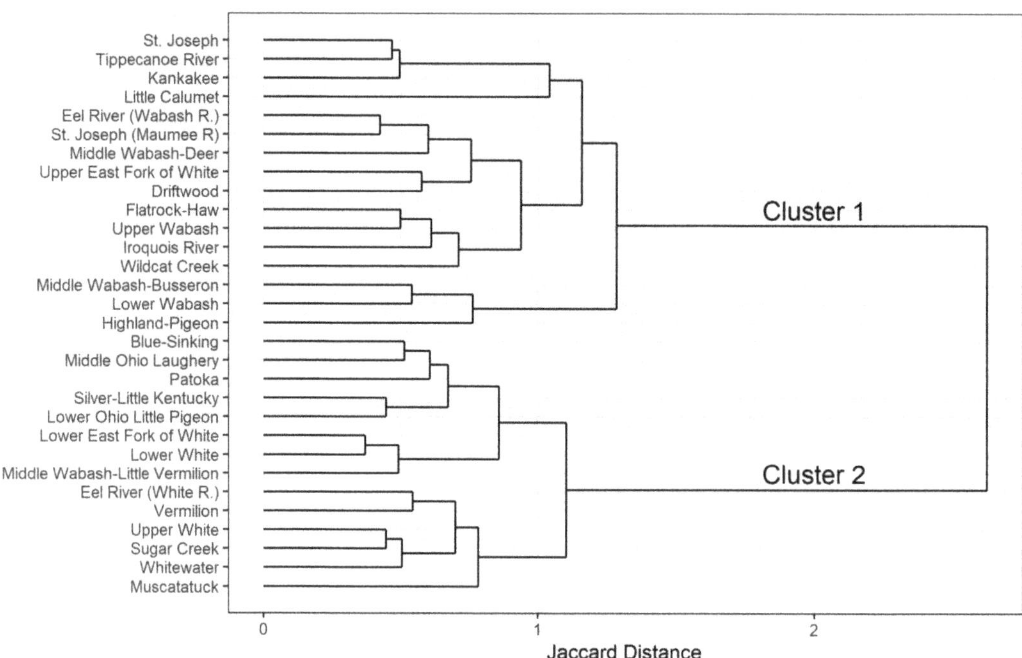

Figure 7. Hierarchical cluster tree based on K-means analysis of species by HUC8 drainage.

Table 4. Models with ΔAICc ≤ 2 and their variables (X), AICc score, ΔAICc, AICc weights, and McFadden R^2 values. Drainage area km^2 = Area, % soils hydrogroup B = % B, % soils hydrogroup CD = % CD, % cherty red clay = % Chert, %wetlands 2016 = % Wet, % developed 2016 = % Dev, % historical wetland ecosystem = % HWet, mean Horton overland flow = μ Hort, mean topographic wetland index = TWI.

Model	Area	% B	% CD	% Chert	% Wet	% Dev	% HWet	μHort	TWI	df	log Lik	AICc	ΔAICc	Akaike Weight	McFadden R^2
1	X		X			X	X		X	6	−97.8	211.2	0	0.1548	0.53
2	X	X	X			X	X		X	7	−96.6	212.2	1.00	0.0940	0.54
3	X	X	X				X		X	6	−98.5	212.7	1.50	0.0731	0.53
4	X		X				X		X	5	−100.2	212.8	1.62	0.0688	0.52
5	X		X			X	X	X	X	7	−96.9	213.0	1.79	0.0632	0.54
6	X		X		X	X	X		X	7	−97.0	213.2	1.98	0.0575	0.53
Global	X	X	X	X	X	X	X	X	X	10	−96.1	223.8	12.58	0.0003	0.54

Results of the AICc analysis were supported by standardized coefficients since none of the top-four variables contained zero within their 95% confidence intervals (Table 5). A positive predictor effect was found for three variables (area km^2, % soils hydrogroup B, and % wetlands 2016) and the five remaining variables had a negative effect on species richness (% developed 2016, mean Horton overland flow, % historic wetland ecosystem, mean TWI, and % soils hydrogroup C/D).

Table 5. Standard coefficients of variables included in ΔAICc $\leq$ 2 model set with 95% confidence intervals. % Cherty red clay not included in AICc models.

Variable	Mean	SE	Lower Conf. Limit	Upper Conf. Limit
% Developed 2016	−1.763	0.995	−3.714	0.187
% Historic wetland ecosystem	−0.751	0.280	−1.299	−0.203
Mean topographic wetness index	−0.567	0.082	−0.729	−0.406
Mean Horton overland flow	−0.045	0.035	−0.114	0.024
% Soils hydrogroup C/D	−0.036	0.013	−0.061	−0.011
Drainage area km^2	0.00022	0.00003	0.00012	0.00024
% Soils gydrogroup B	0.006	0.004	−0.001	0.013
% Wetlands 2016	2.115	1.731	−1.278	5.508

AICc relative importance values (RIV) ranged from 0.196 to 1.000 (Figure 8). Relative positions of the top five variables were (1) mean TWI (1.00), (2) area km^2 (1.00), (3) % soils hydrogroup C/D (0.91), (4) % historic wetland vegetation (0.88), and (5) % developed 2016 (0.62).

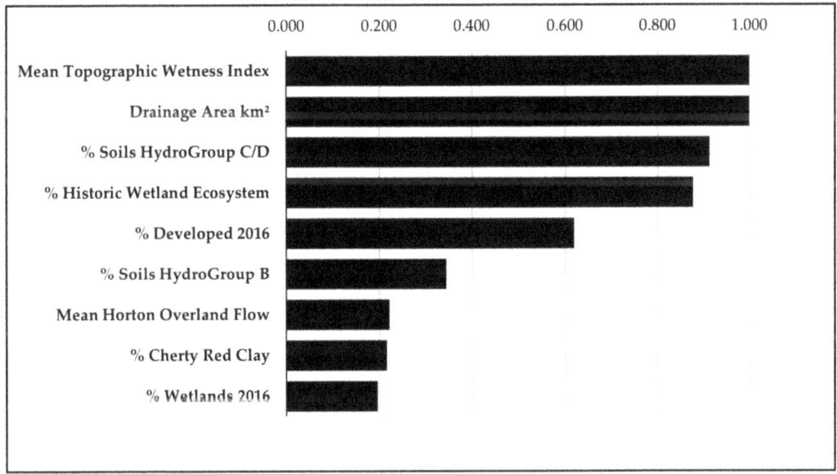

Figure 8. AICc Relative Importance Values (RIV) for nine variables included in the global model.

Relative positions of variables shifted when subjected to hierarchical partitioning. The relative positions of the top five highest scoring variables were (1) mean TWI (32.9%), (2) area km^2 (22.0%), (3) % cherty red clay (12.8%), (4) % Soils hydrogroup CD (9.5%), and % Soils hydrogroup B (7.9%) (Figure 9). Percent cherty red clay was not included in any of the top six AIC models but ranked in the top three by hierarchical partitioning. Percent soils hydrogroup B moved above % historic wetland ecoregion. The other three variables in aggregate accounted for less than 10% variation.

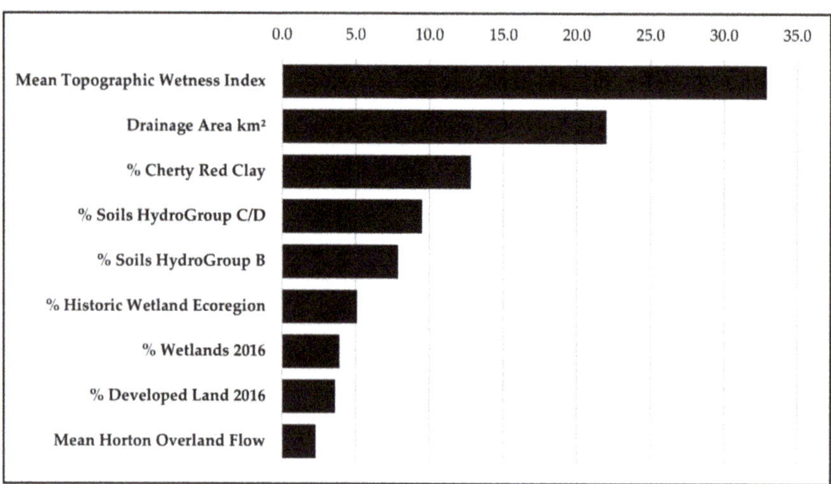

Figure 9. The %*I* contribution to variance in species richness determined by hierarchical partitioning.

Examining Z-scores indicated that mean TWI, area km^2, and % cherty red clay, each explained a significant percentage of variance, with Z-scores over the 95% confidence limit (Figure 10).

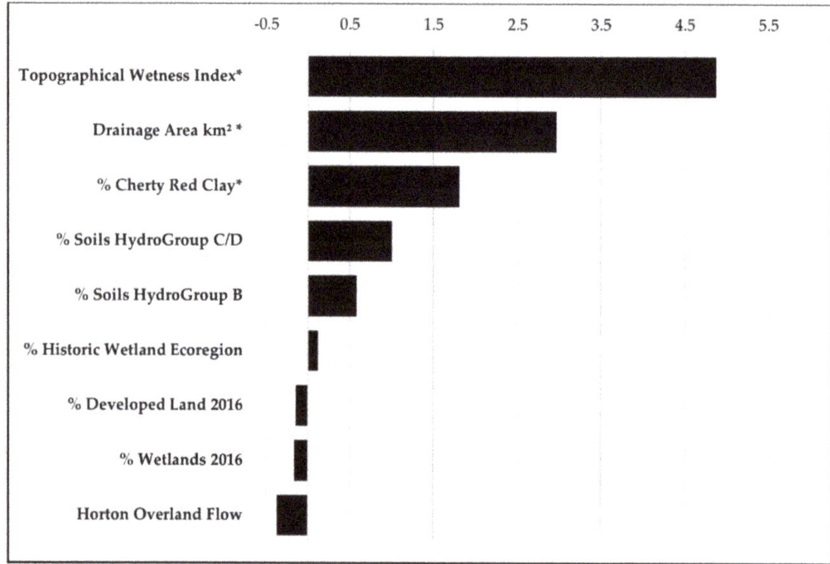

Figure 10. Z-scores of variables determined by hierarchical partitioning. * Denotes variables significant at 95% confidence interval.

4. Discussion

Large scale assessments of aquatic insect biodiversity are always difficult to conduct if species-level identification is needed. Sampling must be conducted instream for larvae and for adults in a wide variety of habitats and in multiple seasons; species succeed each other throughout the year [2]. In regions where water and habitat quality are degraded due to agricultural practices and human development many species will be missing or in distributions and densities too low for detection [10]. New sampling alone would not recover all species that were once present, nor would it produce natural ranges for species. Such is the case for Indiana, a midwestern USA state where at least 70% of land use is in agriculture and human development. Museum data are necessary to tally all species and to provide context for the species that still exist in an area. We have built a data set that allows us to recover the fauna of a highly disturbed landscape and ask questions about species richness of the entire assemblage of water and habitat quality sensitive species, how it is distributed across the landscape, and the relative importance of factors maintaining species richness.

4.1. Species Richness and Patterns

We recovered 93 species of stoneflies from Indiana, adding four species to the state tally [4]. Species richness was a positive function of unique sampling events and to a lesser degree by drainage area (Figures 3 and 4). Both relationships were expected, though some heavily sampled drainages produced relatively few species. Our sampling nearly exhausted all species possible in the region (Figure 5) with a prediction being 95 ± 5 species. The current value of 93 is 14 more than known for Illinois [11]. Conversely, the new tally for Indiana is 9 fewer than Ohio's 102 species [49] and 32 fewer than Kentucky [50]. Michigan has 68 species [12]. Regional tallies are known to be a function of both longitude and latitude, with decreases in species richness as both variables increase [13].

Indiana stonefly species richness is greatest in the southern part of the state, especially in HUC8s with large proportions of unglaciated area. The two richest HUC8s (Lower East Fork White, 66 and Blue-Sinking, 58) are largely unglaciated (Figure 6). The third-richest drainage (Lower White, 51 species) is an old, once-glaciated landscape, having been covered only by the Illinoisan glacial advance. These three drainages may have acted as a refuge during the Wisconsinan glaciation but were certainly first invaded from the Appalachian Mountains and plateaus east and south of the Ohio River [51].

A K-means cluster analysis of HUC8 assemblages suggests that there is substantial faunal turnover between northern and southern drainages (Figure 7). Cluster 1 drainages are mostly northern and of Wisconsinan glaciated landscapes. Cluster 2 is composed mainly of drainages from the southern half of Indiana in areas of Illinoian once-glaciated and unglaciated landscapes. The largest differences in the two clusters are within the winter- and spring-emerging Capniidae, Chloroperlidae, Leuctridae, Nemouridae, Perlodidae, and Taeniopterygidae. Cluster 1 contains 8 genera and 14 species, while Cluster 2 contains 21 genera and 52 species [4].

4.2. Relative Importance of Watershed Variables in Species Richness

AICc models 1–6 all explained 52–54% of the variation in species richness in drainages with only minor loss of information (Table 4). There is a pattern in the distribution of species richness within the HUC8 drainages. Four variables were always present in these six models: area km^2, % soils hydrogroup CD, % historic wetland ecosystem, and TWI, all but one is related to hydrology. Examination of the importance of variables by RIV, %*I* contribution, and Z-scores always rated TWI and area km^2 first and second. Ranking of other variables was inconsistent. One variable not found in the six models was % cherty red clay and it turned up the % I contribution of hierarchical partitioning and Z-scores in third position. Note that % developed 2016 appears to be unimportant in broad scale species richness of stoneflies in Indiana. All others are natural variables. This suggests that historical specimen data have helped to capture pre-European settlement stonefly assemblages.

TWI is a complex estimate of depth-to-water and had a negative relationship with Plecoptera species richness, i.e., areas with high TWI (marshes and wooded wetlands) supported lower species richness (as in the Little Calumet drainage). Conversely, low TWI values indicate areas that drain well, have higher slopes, coarser substrates, and higher dissolved oxygen values. These are conditions conducive to stonefly species richness (as in the high richness Lower East Fork White and Blue-Sinking drainages). Some studies used TWI as a predictive measure for the presence of wetlands [52]. Others found a negative relationship of TWI with understory vegetation species richness in an Alberta, Canada boreal forest, i.e., drier habitats had greater species richness than wetter ones [53]. Conversely, an assemblage of grassland-inhabiting passerine birds living in the floodplain of the Loire valley, France was positively associated with TWI [54].

HUC8 area km^2 was the second most important variable and a positive predictor of species richness. Larger HUC8s contain a diversity of habitat types (seeps, ravine streams, and medium and large streams), supporting greater species richness. In Ohio, no relationship between drainage area and species richness of stoneflies was found [13], but the analysis was based on the much larger HUC6-scale drainages which may have smoothed habitat differences between drainages. Several other large-scale analyses of aquatic insect distributions have been conducted, but none have analyzed the effect of drainage area on species richness: for stoneflies [55] and caddisflies [56] in the Ozark and Ouachita Mountains area and for caddisflies in Minnesota [57]. Also working in the Ozark/Ouachita/Interior Highlands area, drainage area accounted for approximately 35% of variance in species richness of native fish [58]. Their drainages approximated HUC8 to HUC12 scale.

The third-most important variable in AICc and fourth in hierarchical partitioning was % soils in HydroGroup C/D (I = 9.5%). Soils in hydrologic group C/D are notoriously slow-draining [59]. In Indiana, these soils are common in large river bottoms in the southern half of the state, but they make up a large proportion of soils in the Eel (Wabash), Mississinewa, Salamonie, Upper Wabash, St. Joseph (Maumee), St. Mary's, Maumee, and Auglaize drainages where five under-sampled drainages were eliminated from the analysis. This region makes up the Bluffton Till Plain [31,33]. The soils were largely formed by precursors of Lake Erie and deposited by minor re-advances of glaciers, leaving a series of concentric moraines that divide the HUC8s of this region. Others [12,50] reported similar results of low stonefly richness for adjacent drainages in northwest Ohio. This hydrologic soil group is a representation of permeability and a function of the least transmissive horizon of soil in a location [59]. They may be placed in a dual category if the water table is present within 60 cm of the surface, even if the soil makeup is conducive to faster draining. This is the case of our % hydrologic group C/D. These are sandy clay loams where the water table is naturally within 60 cm of the surface.

Percent historic wetland ecosystem is likely representing a similar impact, though from a different part of the state. The lower portion of the Maumee and its tributaries in Indiana were wetlands, part of the Great Black Swamp, a wooded wetland complex formed atop lake plains of ancient glacial Lake Maumee [60]. Additionally, other major wetland complexes were found in northwestern Indiana as part of the Grand Kankakee Marsh and drained by the Calumet, Kankakee, Iroquois, Tippecanoe, and Middle Wabash–Little Vermilion drainages [30]. This is also the region where most of the state's wet and dry prairies occurred, which were grouped together with other emergent wetlands in presettlement vegetation percentages [35]. These areas are rich in flat streams with low gradients and sandy soils. Few stoneflies live in such conditions [36].

Percent cherty red clay was not included in AICc importance since it was not included in any of the top six models where ΔAICc ≤ 2. However, it ranked third with a positive relationship to richness when the dataset was subjected to hierarchical partitioning and Z-scores. This category of surface geology is associated with unglaciated Indiana [30]. Cherty red clay is a surface geology in portions of six HUC8 drainages, but it only makes up >5% area in three (Blue-Sinking, 70.0%; Lower East Fork White, 39.7%; and Silver-Little Kentucky, 11.3%). Eroding from these soils are chert nodules, biologically formed inclusions found in limestone. The presence of chert indicates moderate and high slopes, coarse substrates, and groundwater influence, all important for supporting high stonefly species richness. This variable appears important in only a small area of Indiana. It may be a more important variable in unglaciated areas where limestone is more extensively distributed.

It is important to note that Plecoptera species richness differs greatly across the drainages of the state. The top four most species-rich HUC8s are the Lower East Fork White (66 species), the Blue-Sinking (58), the Lower White (51), and the Middle Wabash–Little Vermilion (42) and they greatly exceed the mean value of 23.3 species. It should also be noted that certain lower-scoring HUC8s greatly outperform their neighbors. The St. Joseph River, which drains to Lake Michigan, still supports a nearly complete assemblage of large perlid and perlodid stoneflies: *Acroneuria* (three species), *Agnetina* (one species), *Paragnetina* (one species), *Neoperla* (one species), *Isoperla* (three species). This number is approximately twice as many species as found in adjacent HUC8s to the east and south. Morainal systems in this drainage created a more varied topography, swifter current, glacial lakes, and abundant groundwater.

5. Conclusions

Our results highlight the importance of hydrology on species richness of Plecoptera. Three of the top four variables in AICc importance are directly tied to hydrology—TWI, % soils in hydrogroup C/D, and % historic wetlands. These factors are important in that they reflect available substrates, water availability, flow rate, and dissolved oxygen. Human disturbance variables were unimportant in explaining HUC8 diversity.

We could not directly demonstrate the importance of glacial history to Indiana Plecoptera species richness, but several of the hydrology variables are likely surrogates for glacial history. Percent cherty red clay may also be indicative of lack of glaciation. The cluster analysis strongly suggests a role for glacial history.

Land use over the past two centuries has had a significant impact on current species richness across the state [11]. This study tallies species present in museum records dating back to 1879. The older records place many species in drainages where they no longer occur [10,11]. Eight stonefly species have not been collected in Indiana since the early- to mid-20th century. These species primarily inhabited large rivers, especially the middle and lower Wabash and lower White rivers. The timing of these extirpations correlates with the change to mechanized agriculture and the use of chemical insecticides, herbicides, and fertilizers [9]. Currently, 62% of the land area of Indiana is in agriculture, and Wisconsinan glaciated areas share an outsized proportion of land devoted to agriculture. High stonefly species richness persists only in unglaciated and once-glaciated areas where topography is not conducive to widespread agriculture [10].

The work conducted here is of sufficient extent and intensity that additional analyses are possible. A distributional atlas of all the stoneflies species in Indiana is nearly ready for submission. Conservation status assessments with these data are also planned for early 2022. The latter will help inform conservation efforts of species, important drainages, and individual waterbodies. It has taken the efforts of many scientists over the past century to gather these data.

Indiana data and those of a seven state region are now being used to predict pre-European distribution of stoneflies throughout the wider Midwest, USA. This will be the precursor for predicting future distributions as influenced by climate and other variables.

Supplementary Materials: The following are available online at https://www.mdpi.com/article/10.3390/d13120672/s1, Table S1: Presence/absence data matrix for Indiana stoneflies (Plecoptera) and USGS HUC8 drainages they occur within. Genus included where the species was undeterminable but present in the HUC8, Table S2: USGS HUC8 watersheds, HUC8 code, species richness, number of specimen records, and drainage area.

Author Contributions: Conceptualization, E.A.N., R.E.D. and S.A.G.; methodology, E.A.N. and R.E.D.; formal analysis, E.A.N.; data curation, E.A.N.; writing—original draft preparation, E.A.N. and R.E.D.; writing—review and editing, E.A.N., R.E.D. and S.A.G.; supervision, R.E.D.; project administration, R.E.D.; funding acquisition, R.E.D. All authors have read and agreed to the published version of the manuscript.

Funding: The USA National Science Foundation, grant number CSBR 14-58285 funded a graduate research assistantship and tuition for E.A.N. NSF DEB 09-18805 ARRA and USA Fish and Wildlife Service DOI-USFWS X-1-R-1 funded accumulation of specimen data. The Indiana Department of Natural Resources and the Indianapolis Zoo grant number 0017556043 funded a status assessment of stoneflies. The University of Illinois, Department of Entomology and the Society for Freshwater Science Conservation grant competition provided summer travel grants to E.A.N. for this project. The International Society of Plecopterologists provided generous travel funds to E.A.N. to present at an international meeting.

Institutional Review Board Statement: Not applicable.

Informed Consent Statement: Not applicable.

Data Availability Statement: Most specimen data used in this study are available from the INHS Insect Collection database portal at http://inhsinsectcollection.speciesfile.org/InsectCollection.aspx (Accessed on 13 December 2021). These data are also replicated at the Global Biodiversity Information Facility portal at https://www.gbif.org/ (Accessed on 13 December 2021). Museum data from PERC are held in a private database that will soon become available through GBIF. We have provided all data to the Indiana Department of Natural Resources as a deliverable.

Acknowledgments: We thank Arwin Provonsha, former collection manager at the Purdue University Entomological Research Collection, who, over a decade, gave us access to and provided loans of stonefly specimens. Permits (NP20-05) to sample Indiana state properties were facilitated by Ronald P. Hellmich, Director, Division of Nature Preserves, Indiana Department of Natural Resources. We are grateful to Jason L. Robinson (INHS) for digitizing the pre-European settlement vegetation data.

Conflicts of Interest: The authors declare no conflict of interest. The funders had no role in the design of the study; in the collection, analyses, or interpretation of data; in the writing of the manuscript, or in the decision to publish the results.

References

1. Fochetti, R.; Tierno de Figueroa, J.M. Global diversity of stoneflies (Plecoptera; Insecta) in freshwater. *Hydrobiologia* **2008**, *595*, 365–377. [CrossRef]
2. DeWalt, R.E.; Kondratieff, B.C.; Sandberg, J.B. Order Plecoptera. In *Ecology and General Biology: Thorp and Covich's Freshwater Invertebrates*; Thorp, J., Rogers, D.C., Eds.; Academic Press: Cambridge, MA, USA, 2015; pp. 933–949. ISBN 9780123850263.
3. DeWalt, R.E.; Ower, G.D. Ecosystem services, global diversity, and rate of stonefly species descriptions (Insecta: Plecoptera). *Insects* **2019**, *10*, 99. [CrossRef] [PubMed]
4. DeWalt, R.E.; Maehr, M.D.; Hopkins, H.; Neu-Becker, U.; Stueber, G.; Plecoptera Species File Online. Version 5.0/5.0. 2021. Available online: http://Plecoptera.SpeciesFile.org (accessed on 25 October 2021).
5. South, E.J.; Skinner, R.K.; DeWalt, R.E.; Davis, M.A.; Johnson, K.P.; Teslenko, V.A.; Lee, J.J.; Malison, R.L.; Hwang, J.M.; Bae, Y.J.; et al. A new family of stoneflies (Insecta: Plecoptera), Kathroperlidae, fam. n., with a phylogenomic analysis of the Paraperlinae (Plecoptera: Chloroperlidae). *Insect Syst. Divers.* **2021**, *5*, 1. [CrossRef]
6. Barbour, M.T.; Gerritsen, J.; Snyder, B.D.; Stribling, J.B. *Rapid Bioassessment Protocols for Use in Streams and Wadeable Rivers: Periphyton, Benthic Macroinvertebrates and Fish*, 2nd ed.; U.S. Environmental Protection Agency Office of Water: Washington, DC, USA, 1999.
7. Lenat, D.R. A biotic index for the southeastern United States: Derivation and list of tolerance values, with criteria for assigning water-quality ratings. *J. N. Am. Benthol. Soc.* **1993**, *12*, 279–290. [CrossRef]
8. Shah, D.N.; Domisch, S.; Pauls, S.U.; Haase, P.; Jähnig, S.C. Current and future latitudinal gradients in stream macroinvertebrate richness across North America. *Freshw. Sci.* **2014**, *33*, 1136–1147. [CrossRef]
9. Master, L.L.; Stein, B.A.; Kutner, L.S.; Hammerson, G.A. Vanishing assets: Conservation status of U.S. species. In *Precious Heritage: The Status of Biodiversity in the United States*; Stein, B.A., Kutner, L.S., Adams, J.S., Eds.; The Oxford University Press: New York, NY, USA, 2000; pp. 93–118.
10. DeWalt, R.E.; Favret, C.; Webb, D.W. Just how imperiled are aquatic insects? A case study of stoneflies (Plecoptera) in Illinois. *Ann. Entomol. Soc. Am.* **2005**, *98*, 941–950. [CrossRef]
11. DeWalt, R.E.; Grubbs, S.A. Updates to the Stonefly Fauna of Illinois and Indiana. *Illiesia* **2011**, *7*, 31–50.
12. Grubbs, S.A.; Pessino, M.; DeWalt, R.E. Michigan Plecoptera (Stoneflies): Distribution patterns and an updated species list. *Illiesia* **2012**, *8*, 162–173.
13. DeWalt, R.E.; Cao, Y.; Tweddale, T.; Grubbs, S.A.; Hinz, L.; Pessino, M.; Robinson, J.L. Ohio USA stoneflies (Insecta, Plecoptera): Species richness estimation, distribution of functional niche traits, drainage affiliations, and relationships to other states. *ZooKeys* **2012**, *178*, 1–26. [CrossRef]

14. Grubbs, S.A.; Pessino, M.; DeWalt, R.E. Distribution patterns of Ohio stoneflies, with an emphasis on rare and uncommon species. *J. Insect Sci.* **2013**, *13*, 72. [CrossRef]
15. Bojková, J.; Komprdová, K.; Soldán, T.; Zahrádková, S. Species loss of stoneflies (Plecoptera) in the Czech Republic during the 20th century. *Freshw. Biol.* **2012**, *57*, 2550–2567. [CrossRef]
16. Sánchez-Bayo, F.; Wyckhuys, K.A.G. Worldwide decline of the entomofauna: A review of its drivers. *Biol. Conserv.* **2019**, *232*, 8–27. [CrossRef]
17. Zwick, P. Stream habitat fragmentation—A threat to biodiversity. *Biod. Conserv.* **1992**, *1*, 80–97. [CrossRef]
18. Pires, M.M.; Grech, M.G.; Stenert, C.; Maltchik, L.; Epele, L.B.; McLean, K.I.; Kneitel, J.M.; Bell, D.A.; Greig, H.S.; Gagne, C.R.; et al. Does taxonomic and numerical resolution affect the assessment of invertebrate community structure in New World freshwater wetlands? *Ecol. Indic.* **2021**, *125*, 107437. [CrossRef]
19. Lenat, D.R.; Resh, V.H. Taxonomy and stream ecology—The benefits of genus and species-level identifications. *J. N. Am. Benthol. Soc.* **2001**, *20*, 287–298. [CrossRef]
20. Needham, J.G.; Claassen, P.W. *A Monograph of the Plecoptera or Stoneflies of America North of Mexico*; The Thomas Say Foundation: Lafayette, IN, USA, 1925; Volume 11, 397p.
21. Frison, T.H. Fall and winter stoneflies, or Plecoptera, of Illinois. *Ill. Nat. Hist. Surv. Bull.* **1929**, *18*, 344–409. [CrossRef]
22. Frison, T.H. The stoneflies, or Plecoptera, of Illinois. *Ill. Nat. Hist. Surv. Bull.* **1935**, *20*, 281–467. [CrossRef]
23. Frison, T.H. Studies of North American Plecoptera with special reference to the fauna of Illinois. *Ill. Nat. Hist. Surv. Bull.* **1942**, *22*, 233–355.
24. Ricker, W.E. A first list of Indiana stoneflies (Plecoptera). *Proc. Indiana Acad. Sci.* **1945**, *54*, 225–230.
25. Bednarik, A.F.; McCafferty, W.P. A checklist of the stoneflies, or Plecoptera, of Indiana. *Great Lakes Entomol.* **1977**, *10*, 223–226.
26. Grubbs, S.A. Studies on Indiana stoneflies (Plecoptera), with an annotated and revised state checklist. *Proc. Entomol. Soc. Wash.* **2004**, *106*, 865–876.
27. Grubbs, S.A.; DeWalt, R.E. *Perlesta ephelida*, a new Nearctic stonefly species (Plecoptera: Perlidae). *ZooKeys* **2012**, *194*, 1–15. [CrossRef]
28. Grubbs, S.A.; DeWalt, R.E. *Perlesta armitagei* n. sp. (Plecoptera: Perlidae): More cryptic diversity in darkly pigmented *Perlesta* from the eastern Nearctic. *Zootaxa* **2018**, *4442*, 83–100. [CrossRef] [PubMed]
29. Hydrologic Unit Maps. Available online: https://water.usgs.gov/GIS/huc.html (accessed on 1 January 2021).
30. Malott, C.A. The physiography of Indiana. In *Handbook of Indiana Geology*; Indiana Department of Conservation Publication; William B. Burford, Contractor for State Printing and Binding: Indianapolis, IN, USA, 1922; Volume 21, pp. 59–256.
31. Gray, H.H. *Physiographic Divisions of Indiana*; Indiana Geological Survey Special Report 61; Indiana Geological and Water Survey: Bloomington, IN, USA, 2000; 15p.
32. Lindsey, A.A.; Crankshaw, W.B.; Qadir, S.A. Soil relations and distribution map of the vegetation of presettlement Indiana. *Bot. Gaz.* **1965**, *126*, 155–163. [CrossRef]
33. Homoya, M.A.; Abrell, D.B.; Aldrich, J.R.; Post, T.W. The Natural Regions of Indiana. *Proc. Indiana Acad. Sci.* **1985**, *94*, 245–268.
34. DeWalt, R.E.; Yoder, M.; Snyder, E.; Dmitriev, D.; Ower, G. Wet collections accession: A workflow based on a large stonefly (Insecta, Plecoptera) donation. *Biodivers. Data J.* **2018**, *6*, e30256. [CrossRef] [PubMed]
35. Acme Mapper 2.2. Available online: http://mapper.acme.com (accessed on 28 October 2021).
36. DeWalt, R.E.; South, E.J.; Robertson, D.R.; Marburger, J.E.; Smith, W.W.; Brinson, V. Mayflies, stoneflies, and caddisflies of streams and marshes of Indiana Dunes National Lakeshore, USA. *ZooKeys* **2016**, *556*, 43–63. [CrossRef] [PubMed]
37. R Core Team. *R: A Language and Environment for Statistical Computing*; R Foundation for Statistical Computing: Vienna, Austria, 2014; Available online: http://www.R-project.org/ (accessed on 1 January 2021).
38. Oksanen, J.; Blanchet, F.G.; Friendly, M.; Kindt, R.; Legendre, P.; McGlinn, D.; Minchin, P.R.; O'Hara, R.B.; Simpson, G.L.; Solymos, P.; et al. *Vegan: Community Ecology Package, R Package Version 2.5-7*; R Project for Statistical Computing: Vienna, Austria, 2020; Available online: https://CRAN.R-project.org/package=vegan (accessed on 1 January 2021).
39. Geospatial Data Gateway. Available online: https://datagateway.nrcs.usda.gov/ (accessed on 1 January 2021).
40. Wickham, J.; Stehman, S.V.; Sorenson, D.G.; Gass, L.; Dewitz, J.A. Thematic accuracy assessment of the NLCD 2016 land cover for the conterminous United States. *Remote Sens. Environ.* **2021**, *257*, 112357. [CrossRef]
41. South, E.J.; DeWalt, R.E.; Cao, Y. Relative importance of Conservation Reserve Programs to aquatic insect biodiversity in an agricultural watershed in the Midwest, USA. *Hydrobiologia* **2019**, *829*, 323–340. [CrossRef]
42. Quinn, G.P.; Keough, M.J. *Experimental Design and Data Analysis for Biologists*; Cambridge University Press: New York, NY, USA, 2002.
43. Budescu, D.V. Dominance analysis: A new approach to the problem of relative importance of predictors in multiple regression. *Psychol. Bull.* **1993**, *114*, 542–551. [CrossRef]
44. Nimon, K.; Oswald, F.; Roberts, J.K. *Yhat: Interpreting Regression Effects, R Package Version 2.0-0*; R Project for Statistical Computing: Vienna, Austria, 2021; Available online: https://CRAN.R-project.org/package=yhat (accessed on 15 January 2021).
45. Burnham, K.P.; Anderson, D.R. *Model Selection and Multimodel Inference: A Practical Information-Theoretic Approach*, 2nd ed.; Springer: New York, NY, USA, 2002.
46. Barton, K. *MuMIn: Multi-Model Inference*; R package Version 1.15.1; R Project for Statistical Computing: Vienna, Austria, 2015.

47. Burnham, K.P.; Anderson, D.R. Multimodel inference: Understanding AIC and BIC in model selection. *Sociol. Methods Res.* **2004**, *33*, 261–304. [CrossRef]
48. Chevan, A.; Sutherland, M. Hierarchical partitioning. *Am. Stat.* **1991**, *45*, 90–96. [CrossRef]
49. DeWalt, R.E.; Grubbs, S.A.; Armitage, B.; Baumann, R.W.; Clark, S.M.; Bolton, M.J. Atlas of Ohio Aquatic Insects: Volume II, Plecoptera. *Biodivers. Data J.* **2016**, *4*, e10723. [CrossRef]
50. Tarter, D.C.; Chaffee, D.L.; Grubbs, S.A.; DeWalt, R.E. New state records of Kentucky (USA) stoneflies (Plecoptera). *Illiesia* **2015**, *11*, 167–174.
51. Pessino, M.; Chabot, E.T.; Giordano, R.; DeWalt, R.E. Refugia and postglacial expansion of *Acroneuria frisoni* Stark & Brown (Plecoptera: Perlidae) in North America. *Freshw. Sci.* **2014**, *33*, 232–249. [CrossRef]
52. Ganser, C.; Wisely, S.M. Patterns of spatio-temporal distribution, abundance, and diversity in a mosquito community from the eastern Smoky Hills of Kansas. *J. Vector Ecol.* **2013**, *38*, 229–236. [CrossRef]
53. Echiverri, L.; MacDonald, S.E. Utilizing a topographic moisture index to characterize understory vegetation patterns in the boreal forest. *For. Ecol. Manag.* **2019**, *447*, 35–52. [CrossRef]
54. Besnard, A.G.; La Jeunesse, I.; Pays, O.; Secondi, J. Topographic wetness index predicts the occurrence of bird species in floodplains. *Divers. Distrib.* **2013**, *19*, 955–963. [CrossRef]
55. Poulton, B.C.; Stewart, K.W. The stoneflies of the Ozark and Ouachita Mountains (Plecoptera). *Mem. Am. Entomol. Soc.* **1991**, *38*, 1–116.
56. Moulton, S.R., II; Stewart, K.W. Caddisflies (Trichoptera) of the Interior Highlands of North America. *Mem. Am. Entomol. Inst.* **1996**, *56*, 1–313.
57. Houghton, D.C. Biological diversity of the Minnesota caddisflies (Insecta, Trichoptera). *ZooKeys* **2012**, *189*, 1–389. [CrossRef]
58. Matthews, W.J.; Robison, H.W. Influence of drainage connectivity, drainage area and regional species richness on fishes of the Interior Highlands in Arkansas. *Am. Midl. Nat.* **1998**, *139*, 1–19. [CrossRef]
59. United States Soil Conservation Service. Hydrologic Soil Groups. In *National Engineering Handbook*; Section 630, Chapter 7; United States Department of Agriculture: Washington, DC, USA, 2007.
60. Kaatz, M. The Black Swamp: A study in historical geography. *Ann. Assoc. Am. Geogr.* **1955**, *45*, 1–35. [CrossRef]

MDPI

Article

The Stoneflies (Insecta: Plecoptera) of Israel: Past, Present, Future ... ?

Zohar Yanai

The Steinhardt Museum of Natural History, Tel Aviv University, Tel Aviv 6997801, Israel; yanaizohar@gmail.com

Abstract: Of the more than 3900 described species worldwide, stoneflies (order Plecoptera) are represented in Israel, a semi-arid country, by as few as five species. As a group of highly sensitive aquatic insects, they are restricted to the northernmost watershed of the Sea of Galilee, where the most pristine streams in Israel are found. The Israeli stoneflies are not often collected in the field, and they have not been recorded in the literature in the last 30 years. In order to provide an up-to-date picture, I gathered the available historical records of the local fauna, as well as all verified data from the last decade, and compared the two datasets. Despite the unprecedented efforts that have recently been invested in studying freshwater macroinvertebrates in Israel, a sharp decrease in stonefly occurrence is evident. Whilst the populations of three species have dramatically declined (*Protonemura zernyi*, *Leuctra hippopus*, and *L. kopetdaghi*), the remaining two have not been collected at all in over four decades and are considered locally extinct (*Brachyptera galeata* and *Marthamea beraudi*). These findings highlight the joint impact of multiple stressors on the stream system in the Sea of Galilee Watershed—namely, stream pollution and water diversion on the local level and climate change on the global level. If the current trends continue, there is a great concern that the entire local stonefly fauna will become extinct, and many stream-dwelling taxa may follow soon after.

Keywords: anthropogenic impact; distribution; local extinction; museum study; Plecoptera; population decline

Citation: Yanai, Z. The Stoneflies (Insecta: Plecoptera) of Israel: Past, Present, Future ... ? *Diversity* **2022**, *14*, 80. https://doi.org/10.3390/d14020080

Academic Editors: Piero G. Giulianini and Michael Wink

Received: 11 January 2022
Accepted: 20 January 2022
Published: 24 January 2022

1. Introduction

The approximately 3900 valid species of stoneflies (order Plecoptera) are well distributed across temperate regions and in mountainous landscapes but are extremely rare in semi-arid and arid regions [1–4]. Similar to many other aquatic insects, their life cycle includes an aquatic nymph stage and an aerial adult stage. The nymphs develop in freshwater bodies, in particular streams, where they employ various feeding strategies, including detritivory, herbivory, and carnivory. They obtain dissolved oxygen by absorption through their body integument and specialised gills. They require well-oxygenised waters and, therefore, can usually be found only in unpolluted, running waters with low temperatures, compared with many other freshwater invertebrates. The adults are typically winged and emerge following several moulting events. Adults, many of which are short lived and do not feed, often remain in the vicinity of aquatic habitats, where they mate and oviposit in the water [1,2].

Stoneflies are not adapted to arid and semi-arid regions, and their nymphs require pristine conditions [1], a combination that naturally dictates their distribution. In Israel, they are confined to the Sea of Galilee Watershed in the very north of the country, where human disturbance is relatively moderate and the climatic conditions are humid-mediterranean. While stoneflies are abundant in Lebanon [5], they are completely absent from other neighbouring countries, such as Jordan, highlighting that this watershed is apparently the southernmost stonefly suitable habitat in the Levant (in accordance to "Nehring Line", see [6]).

Only five species are known from Israel. Evidence for the first two species is found in sporadic reports [7–9]. Bromley [10] was the first, and so far the only one, to provide a full checklist of the local species alongside information regarding their identification, distribution, and phenology. Her publication was based on material collected primarily in the 1970s, which was partially identified by P. Zwick. Part of this material is currently housed in the entomological collection in the Steinhardt Museum of Natural History at Tel Aviv University and has been examined as part of the preparation of the current study. Additional material should be housed in the old Inland Water Ecological Service (IES) collection at the Hebrew University of Jerusalem [10] but could not be located for the purposes of this study (E. Gavish-Regev, pers. comm.). Bromley [10] determined two distinct geographical clusters of stoneflies in Israel (Figure 1)—namely, the headwaters (main tributaries) of the Jordan River (three species) and streams in the central Golan Heights (two species). Following these publications, Israeli stoneflies have rarely been mentioned in the literature [11–15] and always as part of a wider invertebrate community. The paucity of scientific and grey literature regarding Israeli stoneflies does not mean that local research has stopped. In fact, numerous sampling expeditions in the streams of northern Israel have been carried out in the last decade. These expeditions included studies for several graduate theses focusing on stream macroinvertebrates; independent research projects; growing interest and demand by local authorities, such as the Nature and Parks Authority, to assess the ecological status of streams; above all, those carried out by the Israel Center for Aquatic Ecology. Although no accurate numbers of old or recent sampling expeditions are available, it is highly probable that the sampling effort over the last few years is as great, if not greater, than in the past.

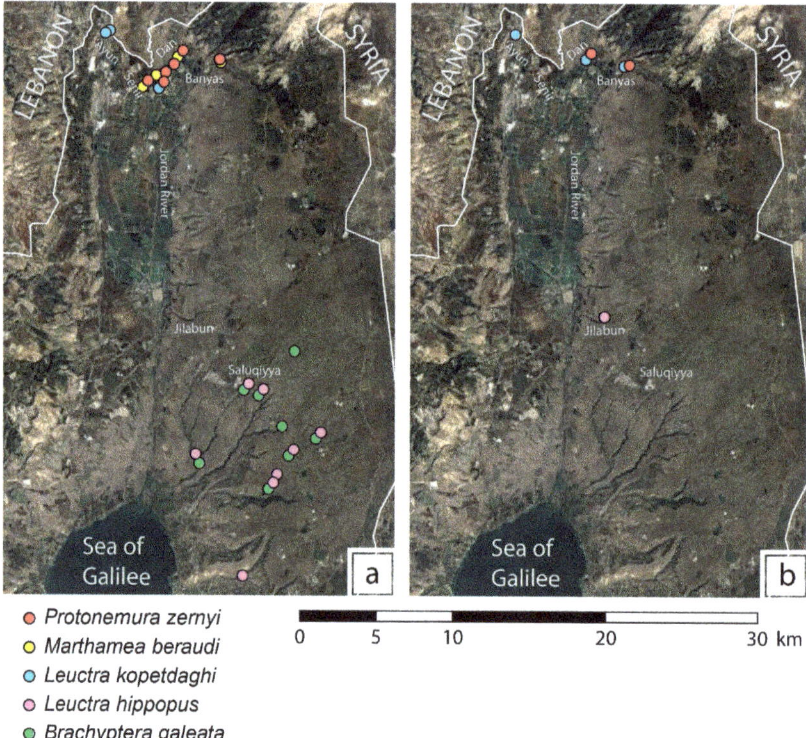

Figure 1. Geographical distribution of the five stonefly species in Israel. The two geographical clusters are the headwaters of the Jordan River in the north of the map and the Golan Heights in the south. All the streams mentioned in the main text are indicated by their names: (**a**) past distribution, i.e., prior to 1988 (reproduced based on [10]); (**b**) present known distribution, i.e., data points from 1989 and onwards.

As they are environmentally sensitive (most of them are narrow-ranged stenothermal species), stoneflies are usually considered to be good bioindicators of ecological changes [1,2,4]. Recently, a growing number of studies worldwide have compared old and current distribution patterns of stoneflies in order to assess small- and large-scale spatial trends (e.g., [16–18]) and, in particular, the effects of climate change. In this study, I applied a similar methodology to the Israeli fauna, by gathering the latest data on the occurrence of stoneflies in Israel, drawing an up-to-date picture of their distribution, and discussing their potential response to future environmental change.

2. Materials and Methods

In order to assemble a database with all of the old and the new records of Israeli stoneflies, I combined all of the available data from the literature and from museum collections. I examined all of the locatable stonefly material ever collected in Israel (48 museum entries with a total of 222 individuals). Most of this material is stored in the entomological collection of the Steinhardt Museum of Natural History at Tel Aviv University (SMNH). Additional collections I examined include those in the Israel Centre for Aquatic Ecology (ICAE) and the Museum of Zoology, Lausanne, Switzerland (MZL), which houses an important collection of aquatic insects from the Levant. As noted above, I attempted to check the collection of the Hebrew University in Jerusalem (HUJI), but 13 existing vials contain foreign material only. Other important entomological collections from the Levant (in Budapest, Prague, and Vienna) do not include Israeli stoneflies either (D. Murányi, pers. comm.). The lists below include all of the identified Israeli stoneflies that could be located. Additional, unidentified material found in the SMNH collection (ca. 15 pinned adults) was collected in the known sites in Israel between 1943 and 1984, and therefore, no information is lost by excluding them. The analysed material includes nymphs and adults (marked below as N and A, respectively). The majority of the material was determined by P. Zwick in the early 1980s. I re-examined the old material, to confirm the validity of my own identifications, using the identification key presented in [10] and then identified the more recent material (i.e., post-1980s). Material from MZL was first determined by J.-P. Reding. It is noteworthy that some collection details in the old material are missing from the original labels.

The material was collected over the years by several collectors, who, according to the details in the published literature and personal communication with the available collectors, employed various sampling techniques. Nymphs were usually collected from streams using aquatic hand nets of various mesh sizes. Adults were collected either by hand nets in the vicinity of streams or in light traps. In most cases, newer material was preserved in ethanol (at least 70%), and the old adult material was occasionally pinned. Sampling effort varied greatly over the years, with a few peaks representing the thorough fieldwork led by the IES and the Hebrew University team during the 1970s–1980s, and by scientists from Tel Aviv University in the last decade. The latter include former students of A. Gasith, and more recently, members of the ICAE, led by Y. Hershkovitz. Previously recorded localities are mostly found in nature reserves which are frequently sampled for various scientific and monitoring purposes. Despite the inconsistency in sampling effort, it is fair to assume that the presence of sustainable populations would have been identified, as current fieldwork in the northern streams of Israel is at least as intensive as it always has been. It is unlikely that a stable population, which was recorded in the past and persisted in the same sites, escaped all modern sampling efforts.

Building DNA libraries based on the barcoding segment of the mitochondrial cytochrome *c* oxidase (COI) gene is a common practice nowadays in order to keep track, identify, and compare sampled specimens. No such data are available for stoneflies of the five local species (neither from Israel nor from foreign populations). In order to bridge this gap, I extracted DNA and sequenced the 658-bp COI segment from fresh material of *P. zernyi* and of *L. kopetdaghi*. Extraction and amplification protocols are detailed in [19]. The resulting sequences (Table 1) were compared against available sequences in GenBank.

Table 1. New mitochondrial cytochrome *c* oxidase (COI) sequences, with closest available sequences in GenBank database (accessed on 29 November 2021).

Species	GenBank Accession Number	Closest Available Taxon
Protonemura zernyi	OL352236	*Protonemura meyeri* [KY262295]: 88% identity
Protonemura zernyi	OL352237	*Protonemura meyeri* [KY262295]: 88% identity
Leuctra kopetdaghi	OL352238	*Leuctra inermis* [HM376115]: 88% identity

A literature survey included the few papers published on Israeli stoneflies (mainly [10]), as well as two recently completed Master's theses that include stonefly records [14,15]. Approximate locality data for Figure 1 were retrieved from collected material and from the literature. Stream names and their spelling vary in the literature and on collection labels, and they are often replaced by a landmark such as a named waterfall. The following list of common spellings should aid the non-local reader: Ayun = Ayun = Iyyon = HaTanur; Banyas = Banias = Panyas = Hermon; Senir = Hatzbani = Hazbani; Jilabun = Gilbon; Qusbiye = Qusbiya = Salukiyya = Saluqiyya.

3. Results

3.1. *Brachyptera galeata Koponen, 1949*

In the past, this species was reported to be fairly abundant, albeit confined to three small streams in the Golan Heights [10]. However, this species has not been found in any of the numerous field surveys conducted in the central Golan Heights in the past four decades, suggesting that it has disappeared from Israel. The old material was collected by D. Furth from 'Qusbiye', a ruined Syrian village in the central Golan. This locality is not characterised by any reliable water body that may explain the occurrence of this population, but the collecting probably occurred near a small bridge nearby (D. Furth, pers. comm.)—facilities in this area have considerably changed over the years. Saluqiyya Springs, found less than 2km away, used to be the most important water source around Qusbiye (see below), and probably the closest stonefly suitable habitat to the village.

Genetic information: the COI segment has never been sequenced for this species.

Available material in collections: ISRAEL: SMNH378713, 2A, Qusbiye, 31.i.1978, Furth D.G. leg. • SMNH378714, 20N, Qusbiye, 3.ii.1981, Furth D.G. leg. • SMNH378715, 51N, Qusbiye, 3.ii.1981, Furth D.G. leg.

3.2. *Leuctra hippopus Kempny, 1899*

This is a common stonefly, with a wide Palearctic distribution, reaching its southern limit of distribution in Israel. It was previously recorded in the 1970s from a few springs and streams in the Golan Heights [10] (Figure 1a). Two *Leuctra* nymphs were collected in the Jilabun Stream in 2015, in moderately running water upon medium and large stones with plenty of aquatic vegetation and leaf litter. Although they are too small to be identified to the species level with certainty, they are assumed to belong to *L. hippopus* based on their distribution. Despite the fact that the Jilabun Stream has never been a locality record for the species, it is fairly close to the other *L. hippopus* localities and relatively far from the *L. kopetdaghi* localities (Figure 1). It is only owing to this unusual sample that the species is not considered extinct from Israel.

Genetic information: *L. hippopus* is the only Israeli stonefly species with available genetic data in GenBank and in BOLD databases, owing to sequences obtained from its central European populations. No sequences from Israel are available, and my attempts to extract DNA from the fresh small specimens have failed.

Available material in collections: ISRAEL: SMNH378716, 13A, Saluqiyya Springs, 31.i.1978, Furth D.G. leg. • SMNH378717, 3A, Saluqiyya Springs, 31.i.1978, Furth D.G. leg. • SMNH342801, 2N, Jilabun Stream, 29.x.2015, Yanai Z. leg.

3.3. *Leuctra kopetdaghi* Zhiltzova, 1972

This is a common stonefly in West and Central Asia, with Israel being its southernmost limit of distribution. The species was recorded in two important tributaries of the Jordan headwaters, the Ayun and Tal Streams, in the 1970s [10]. More recently it was recorded from Ayun in 2004 and 2010 [13], and also from the Dan Stream. Nymphs were collected around the year, providing no new information on phenology. They were found in the main course of a stream in moderate current velocity with a substrate of pebbles, gravel, and leaf litter [15].

Genetic information: a single specimen from the Dan Stream was successfully sequenced for its COI barcoding segment (Table 1). An 88% similarity to the closest match on GenBank confirms that the specimen probably belongs to the genus *Leuctra* but not to any species available in GenBank, i.e., it is very likely that this is the first genetic contribution of *L. kopetdaghi*.

Available material in collections: ISRAEL: SMNH378718, 2♀A, Ayun Stream (haTanur Waterfall), 21.ii.1974, Furth D. leg. • SMNH342800, 1N, Dan Stream (Tel Dan), 10.iii.2017, Gattolliat J.-L. & Truskanov N. leg. • ICAE1160, 3N, Dan Stream (Tel Dan), 7.v.2018, Weiss A. leg. • ICAE1164, 1N, Dan Stream (Tel Dan), 19.viii.2018, Weiss A. leg. • ICAE1165, 2N, Dan Stream (Tel Dan), 19.viii.2018, Weiss A. leg. • ICAE1169, 1N, Dan Stream (Tel Dan), 4.xi.2018, Weiss A. leg.

3.4. *Marthamea beraudi* (Navás, 1909)

This species was detected in the material from Israel for the first time by Illies [7], mentioned as '*Lerpa beraudi*', and later placed in its current generic position by Zwick [9]. *Marthamea beraudi* is currently known from Lebanon and from Israel, in the headwaters of the Litani and the Jordan Rivers, respectively [5,9,10]. It is assumed to favour potamal habitats [9], although in Israel, it was found in the crenal/rhithral sections of the Banyas, Dan, and Senir Streams [10,11] (Figure 1a). It was recently claimed to be found in the Dan Stream [14]; however, I failed to locate the relevant samples for validation, and I suspect that they were in fact misidentified *P. zernyi* (supported by the fact that these were reported to be rare and very young specimens [15]).

The material listed by Zwick [11] should be housed in the collection of Tel Aviv University (which was integrated into the SMNH collection), although a recent check revealed that part of this collection was missing.

Genetic information: the COI segment has never been sequenced for this species.

Available material in collections: ISRAEL: SMNH374414, 1A, Banyas Stream, 4.vi.1943, Bytinski-Salz H. leg. • SMNH374415, 1A, Banyas Stream, 4.vi.1943, Bytinski-Salz H. leg. • SMNH378719, 3N, 7A, Banyas Stream, 16.v.1968 • SMNH378730, 2A, Banyas Stream, 16.v.1968 • SMNH378720, 2N, Banyas Stream, 31.i.1970 • SMNH378726, 2N, Banyas Stream, 31.i.1970 • SMNH378724, 1N, Senir Stream, 17.v.1967 • SMNH378727, 1A, Senir Stream, 16.v.1968 • SMNH378723, 2N, Senir Stream, 16.i.1970 • SMNH378725, 2N, Senir Stream, 16.i.1970 • SMNH378721, 3N, Senir Stream, 31.i.1970 • SMNH378731, 3N, Senir Stream, 31.i.1970 • SMNH374416, 1A, Senir Stream, 9.v.1972, Kugler J. leg. • SMNH378729, 1N, Senir Stream, vii.1974, Freidberg A. leg. • SMNH378728, 1N, Dan Stream (Tel Dan), 2.i.1973, Freidberg A. leg. • SMNH378722, 1N, two labels: (Banyas Stream, 17.v.1967; Senir Stream, 16.v.1968).

3.5. *Protonemura zernyi* Aubert, 1964

This species is only known from Lebanon, where it was described based on a single male [5,20], and from Israel. Zwick [8] reported and described the female and the nymph based on two males, one female, and one female nymph, which were collected in Tel Dan, Israel on 15.iv.1971. These specimens are housed in an unknown collection. He also determined two adults and two nymphs collected in the early 1970s, which are found in the SMNH collection. However, they must be different specimens than those reported, based on their collection details. The species was found in the Dan Stream during special excursions in 1979 and in 1982–1984 [11,12]. More recently, *P. zernyi* has been occasionally collected in the Banyas and Dan Streams (e.g., [14,15]; Weiss et al. in prep.). Although it remains rare, compared with other aquatic invertebrates, it seems to maintain its rank as the most abundant stonefly in Israel (Figure 1). It was also reported from the Ayun Stream recently [14], a first record for the species from this stream, yet the relevant sample could not be located and was excluded in this study since it cannot be verified. In recent years, *P. zernyi* nymphs were mainly collected in brooks near the mainstream course in the Tel Dan Nature Reserve, where water runs moderately, but quiet waters are available near banks and between boulders. The sites are well shaded by riparian trees and accommodate aquatic herbaceous vegetation with leaf litter. The substrate is composed of pebbles and gravel. Nymphs were most successfully collected in the spring (March–May) but were recorded in smaller numbers in August and in November.

Genetic information: the barcoding segment of the mitochondrial COI gene was sequenced for the first time for *P. zernyi*. Two sequences are available in GenBank (Table 1), and both support the claim that this is an independent species that has never been sequenced before within *Protonemura*.

Available material in collections: ISRAEL: SMNH378732, 1♀A, Senir Stream, 13.ii.1970 • SMNH378733, 2N, 1A, Dan Stream (Tel Dan), 2.i.1973, Freidberg A. leg. • MZL-GBIFCH 00981383, 2N, Dan Stream (Tel Dan), 8.v.1990, Sartori M. leg. • SMNH342803, 1N, Dan Stream (Tel Dan), 5.vi.2015, Yanai Z. & Cohen S. leg. • ICAE750, 4N, Dan Stream (Tel Dan), 14.vi.2015, Hershkovitz Y. leg. • SMNH342802, 8N, Dan Stream (Tel Dan), 5.xi.2015, Yanai Z. leg. • ICAE784, 1N, Dan Stream (Tel Dan), 10.viii.2016, Hershkovitz Y. leg. • SMNH342797, 21N, Dan Stream (Tel Dan), 10.iii.2017, Gattolliat J.-L. & Truskanov N. leg. • SMNH342798, 4N, Dan Stream (Tel Dan), 10.iii.2017, Gattolliat J.-L. & Truskanov N. leg. • SMNH342799, 2N, Dan Stream (Tel Dan), 10.iii.2017, Gattolliat J.-L. & Truskanov N. leg. • MZL-GBIFCH00981380, 7N, Dan Stream (Tel Dan), 10.iii.2017, Yanai Z. & Gattolliat J.-L. leg. • MZL-GBIFCH00981381, 1♀N, 1♂N, Dan Stream (Tel Dan), 10.iii.2017, Yanai Z. & Gattolliat J.-L. leg. • MZL-GBIFCH00981382, 20N, Dan Stream (Tel Dan), 10.iii.2017, Yanai Z. & Gattolliat J.-L. leg. • ICAE1164, 1N, Dan Stream (Tel Dan), 19.viii.2018, Weiss A. leg. • SMNH342804, 1N, Dan Stream (Tel Dan), 6.viii.2020, Yanai Z. & Hershko A. leg. • ICAE1168, 8N, Dan Stream (Tel Dan), 4.xi.2018, Weiss A. leg. • ICAE1162, 1N, Banyas Stream, 22.v.2018, Weiss A. leg. • ICAE1170, 4N, Banyas Stream, 5.xi.2018, Weiss A. leg. • ICAE reference collection, 3N, 1A, Banyas Stream, 10.xi.2021, Hershkovitz Y. & Luz D. leg. • SMNH379008, 1N, 1A, Banyas Stream, 10.xi.2021, Hershkovitz Y. & Luz D. leg.

4. Discussion

Examination of the historical stonefly material proved to be complicated due to a combination of insufficient details in the literature, difficulties in locating some of the material, and poor labelling of many of the old vials. The rough resolution of information given in the old literature makes it even harder to identify exact sites and characterise the nature of the habitats of the local stoneflies. In many cases, in particular, in the old labels and in the literature (e.g., [10]), a stream is only mentioned by its name, and no coordinates or detailed data regarding the exact site are given. These streams can vary greatly, from their crenal headwaters to their rhithral and, sometimes, potamal segments [21]. When necessary (e.g., in Figure 1), I tried to indicate the most probable site for stoneflies based on my own knowledge and on other collected material. For example, most of the stoneflies from the Banyas and Dan Streams were probably found in the upper segments close to the Panyas Ruins and Tel Dan, respectively. Similarly, the assumption that Saluqiyya Springs are more accurate than the old records mentioned as 'Qusbiye' is based on acquaintance with this site. It is, therefore, challenging to accurately assess populations and habitats retroactively and to assess past and current threats to this group. Nevertheless, it is possible to draw conclusions regarding general trends in stonefly distributions.

Protonemura zernyi is currently the most common stonefly in Israel, followed by *L. kopetdaghi*. Both were found in the northern geographical cluster (Figure 1), suggesting that their populations survived the environmental changes that occurred in this region over the years (see below). Both species are known today from fewer localities than in the past, suggesting slight population declines, although this cannot be ascertained without reliable long-term population monitoring. *Protonemura zernyi* persists in half of its previously known localities, perhaps owing to the fact that these are crenal stream sections, which are relatively protected and exhibit stable conditions, such as temperature (<17 °C) and chemical composition [11,12], despite a slight decline in water discharge [22,23]. *Leuctra kopetdaghi* is still common in the same water system and has recently been found in the Dan Stream as well. A probable explanation is that streamflow in all streams in the watershed has declined over the years [22,24], and perhaps, the unique conditions that were once available in smaller tributaries are now only available in the Dan Stream and in the rehabilitated Ayun Stream. The water transfer from Dan to Ayun [13,25] may also be of crucial importance for the *L. kopetdaghi* population in the latter stream. The origins of the Jordan River are protected in nature reserves and are, therefore, relatively moderately disturbed and serve as refugia for threatened crenal species. Similar to stoneflies, other orders of sensitive freshwater insects are well established there. Tel Dan supports 35% and 24% of the Israeli diversity of mayflies (Ephemeroptera) and caddisflies (Trichoptera), respectively ([12,14,15]; unpublished data), highlighting the importance of Tel Dan as a conservation hotspot at the local scale.

In central Golan Heights, *L. hippopus* has not been found in any of its historical sites, but a very rare finding from the Jilabun Stream suggests that some *Leuctra* still exist in the Israeli Golan Heights. These nymphs were impossible to identify and may represent either an expansion northward by *L. hippopus* or southward by *L. kopetdaghi*. The former seems more likely, considering the location (Figure 1b) and the ecological features of the stream, which is a typical Golan stream [21]. The question of the specific identity of this population can be resolved by future samplings, through either collection of adults or more mature nymphs, or successful COI sequencing. In any case, this single observation in the Jilabun Stream probably represents either a sink population from a Lebanese source or a fragile relict population, rather than a stable, sustainable one. Compared with other local stoneflies, *L. hippopus* is probably the most tolerant to environmental changes. It is distributed across a very wide range in the Palearctic and, unlike many other species, has exhibited an increase in populations in disturbed (channelised, polluted) sites in the Czech Republic [17].

Available material in SMNH, as well as reliable published information, suggests that *M. beraudi* was last documented and collected in Israel in 1974 [9]. Almost 50 years later, with considerable sampling efforts in its habitats, it appears that *M. beraudi* is extinct in Israel. *Marthamea beraudi* requires a set of unique conditions that can rarely be found in Israel, including low water temperature and high current velocity [12]. Recent changes in rainfall and water discharge in the Dan and adjacent streams [22–24] may have rendered them uninhabitable for *M. beraudi*.

The case of *B. galeata* is similar, with 40 years of absence from Israel. In the Golan Heights, organic pollution due to cattle activity affects most water sources. Touristic pressure has increased rapidly, including trampling and swimming in the Golan streams. Saluqiyya springs, where the most important population of *B. galeata* was sampled by D. Furth in the 1970s, has been for decades the main source for a large plant for bottled water, which altered the structural environment and significantly decreased the amount of water downstream.

The stonefly fauna in Lebanon is much richer, with many endemic species [5]. It appears that Lebanese stonefly populations are more stable than those in Israel and may be subject to weaker anthropogenic pressures in some habitats. The Israeli populations, on the other hand, are on the margin of the species distribution range and limited to a very small area which is experiencing massive climatic and anthropogenic changes. It is reasonable to expect that, at least for a few species, Lebanese populations represent a source that supports sink populations in Israel. International collaboration, with long-term sampling and genetic analyses, may shed light on these migration patterns.

Sequencing the barcoding COI segment for *P. zernyi* and *L. kopetdaghi*, had a threefold purpose in this study—namely, (a) it is the best available, genetics-based method to confirm the generic position of these species. In addition, it allowed me to confirm that these species had not yet been sequenced for their COI, i.e., to confirm that these were not synonyms or misidentifications of any other species that could be found in GenBank; (b) it is a step towards the inclusion of stoneflies in the ongoing process of building a COI library for all of the local freshwater macroinvertebrates; (c) it contributes to the global knowledge of these two species, allowing wider systematic studies. These three goals have been achieved for *P. zernyi* and *L. kopetdaghi*, and their COI sequences are now available for future comparisons and analyses (Table 1). I am hopeful that we will soon obtain sequences for the remaining three species—if not from Israeli populations, then from nearby Lebanese ones, furthering the three abovementioned aims.

Despite the fact that the Israeli stonefly fauna is rather poor, a loss of two out of five known species in just a few decades is alarming. It corresponds to recently reported trends in much faunistically richer regions, which have also had severe declines in their stonefly faunas [16–18]. Whilst in some countries (e.g., Czech Republic, USA [16–18]) there is a long tradition of monitoring freshwater invertebrates using a standard, consistent methodology, such data are not always available. Nevertheless, even in countries with more sporadic data, such as Israel, analysing long-term changes in populations and distribution patterns using publications and collection-based methods is of scientific value. It is particularly important in understanding ongoing processes and setting future conservation and research priorities.

Stoneflies are widely used as bioindicators of stream health, as they have strict environmental requirements, thus highlighting their importance for management practices. The knowledge of the concrete environmental requirements of the local stoneflies is extremely limited due to a lack of ecological, long-term research. Whilst this gap should be bridged via field surveys, laboratory experiments, and modelling techniques, it is likely that generally Israeli stoneflies are affected by similar environmental factors as other stoneflies (e.g., [17,18]). Their habitats in Israel suffer from multiple stressors, including water pumping and overtake, damming and other hydrological barriers, and increasing pressure from recreation. Smaller streams in the Golan Heights (the southern stonefly cluster, Figure 1) are usually subject to organic pollution due to cattle presence, whilst the important Dan and Banyas Streams in the northern cluster regularly receive wastewaters from commercial fishponds. Furthermore, the area is affected by climate change, with predictions of a significant decrease in rainfall and an increase in temperature in the future [23,24]. For stenothermal species such as stoneflies, this means that current habitats may become entirely unsuitable. With the continuing degradation in habitat quality in northern Israel, I expect that stonefly species will keep seeking the few refugia left, either within Israel (such as *L. kopetdaghi*) or closer to their source populations in Lebanon (probably the cases of *B. galeata*, *L. hippopus*, and *M. beraudi*). The trend seen in Israeli stoneflies, often considered quick responding bioindicators, is expected to be seen in the future in other species, especially in light of climate change, which may push the entire zoogeographical boundary ('Nehring Line' [6]) northwards. It is, therefore, of crucial importance to minimise all anthropogenic disturbances to the stream systems in northern Israel, with special attention on removing hydrological barriers and reducing agricultural pollution (e.g., cattle grazing and fishpond wastewaters). As the ability to reverse climatic processes is limited, they can still be mitigated by ensuring sufficient quantities of water from natural sources, such as springs in the streams. Finally, proper monitoring, studying, and reporting of stream biota are essential for tracking trends and identifying threats before it is too late.

5. Conclusions

It is reasonable to state that two stonefly species (*P. zernyi*, *L. kopetdaghi*) have maintained their populations, despite a slight decline. One species (*L. hippopus*) has severely declined but potentially maintains a local relict population in the Golan Heights. The remaining two species (*B. galeata*, *M. beraudi*) are extinct in Israel, remaining endemic to Lebanon. Accumulating evidence of local anthropogenic disturbances and global changes appear to be related to the observed decline. Plecoptera species are sensitive aquatic invertebrates that cope with a given set of anthropogenic and climatic stressors. Monitoring and understanding their response can predict the responses of other groups and serve as a warning sign for irreversible future changes.

Funding: This research received no external funding.

Institutional Review Board Statement: Not applicable.

Data Availability Statement: Collecting details that were analysed in this study are given in the text. Genetic sequences are available on NCBI GenBank (www.ncbi.nlm.nih.gov/, accessed 19 January 2022). Museum specimens are available for study in accordance with the policy of the housing institutes. Additional data are available from the author upon request.

Acknowledgments: Such a complicated investigation after evasive material and vague data, is only possible with the kind assistance of the few lucky people who collected, handled, and reported the presence of Israeli stonefly specimens. I thank A. Chipman, C. Dimentman, E. Gavish-Regev (Jerusalem), D. Furth (Washington/Tel Aviv), J.-L. Gattolliat (Lausanne), Y. Hershkovitz, A.L.L. Friedman, A. Katz, A. Weiss (Tel Aviv), and D. Murányi (Eger) for their efforts in looking for fresh and old material, investigating enigmatic labels, and providing ecological clarifications for this study. I am grateful to D. Murányi, T. Marcus, N. Truskanov, three anonymous reviewers, and especially to H. Bromley, for their insightful comments and contributions to various versions of the manuscript.

Conflicts of Interest: The author declares no conflict of interest.

References

1. Fochetti, R.; Tierno de Figueroa, J.M. Global diversity of stoneflies (Plecoptera; Insecta) in freshwater. *Hydrobiologia* **2008**, *595*, 365–377. [CrossRef]
2. DeWalt, R.E.; Kondratie, B.C.; Sandberg, J.B. Order Plecoptera. In *Ecology and General Biology: Thorp and Covich's Freshwater Invertebrates*; Thorp, J., Rogers, D.C., Eds.; Academic Press: Cambridge, MA, USA, 2015; pp. 933–949, ISBN 9780123850263.
3. DeWalt, R.E.; Maehr, M.D.; Hopkins, H.; Neu-Becker, U.; Stueber, G. Plecoptera Species File Online. Version 5.0/5.0. 2022. Available online: http://Plecoptera.SpeciesFile.org (accessed on 17 January 2022).
4. DeWalt, R.E.; Ower, G.D. Ecosystem Services, Global Diversity, and Rate of Stonefly Species Descriptions (Insecta: Plecoptera). *Insects* **2019**, *10*, 99. [CrossRef] [PubMed]
5. Vinçon, G.; Dia, A.; Kovács, T.; Murányi, D. A new stonefly from Lebanon, *Perlodes thomasi* sp. n. (Plecoptera: Perlodidae). *Illiesia* **2013**, *9*, 18–27.
6. Por, D.F. An outline of the zoogeography of the Levant. *Zool. Scr.* **1975**, *4*, 5–20. [CrossRef]
7. Illies, J. Katalog der rezenten Plecoptera. *Tierreich* **1966**, *82*, 1–632. (In German)
8. Zwick, P. *Protonemura zernyi* Aubert (Insecta: Plecoptera), an addition to the fauna of Israel. *Isr. J. Zool.* **1972**, *21*, 49–51.
9. Zwick, P. *Marthamea beraudi* (Navás) and its European congeners (Plecoptera: Perlidae). *Ann. Limnol. Int. J. Limnol.* **1984**, *20*, 129–139. [CrossRef]
10. Bromley, H. A note on the Plecoptera of Israel. *Isr. J. Entomol.* **1988**, *22*, 1–12.
11. Por, F.D.; Bromley, H.J.; Dimentman, C.; Herbst, G.N.; Ortal, R. River Dan, headwater of the Jordan, an aquatic oasis of the Middle East. *Hydrobiologia* **1986**, *134*, 121–140. [CrossRef]
12. Degani, G.; Herbst, G.N.; Ortal, R.; Bromley, H.J.; Levanon, D.; Glazman, H.; Regev, Y. Faunal relationships to abiotic factors along the River Dan in northern Israel. *Hydrobiologia* **1992**, *246*, 69–82. [CrossRef]
13. Moran, O. Ayun Stream water allocation project: First year monitoring. In *A Report Submitted to the Israeli Nature and Parks Authority (INPA)*; Moran Consulting and Development: Pardes Hanna, Israel, 2011. (In Hebrew)
14. Uekotter, L. Diversity of Ephemeroptera, Plecoptera and Trichoptera in Large Streams of the Lake Kinneret Catchment. Master Thesis, Ruhr University Bochum, Bochum, Germany, University of Duisburg-Essen, Duisburg, Germany, 2016.
15. Weiss, A. Identifying the Eco-Hydrological Niches of Macroinvertebrates to Assess the Impact of Prolonged drought in the Upper Jordan River. Master Thesis, Tel Aviv University, Tel Aviv-Yafo, Israel, 2019.
16. DeWalt, R.E.; Favret, C.; Webb, D.W. Just How Imperiled Are Aquatic Insects? A Case Study of Stoneflies (Plecoptera) in Illinois. *Ann. Entomol. Soc. Am.* **2005**, *98*, 941–950. [CrossRef]
17. Bojková, J.; Komprdová, K.; Soldán, T.; Zahrádková, S. Species loss of stoneflies (Plecoptera) in the Czech Republic during the 20th century. *Freshw. Biol.* **2012**, *57*, 2550–2567. [CrossRef]

18. Bojková, J.; Rádková, V.; Soldán, T.; Zahrádková, S. Trends in species diversity of lotic stoneflies (Plecoptera) in the Czech Republic over five decades. *Insect Conserv. Divers.* **2014**, *7*, 252–262. [CrossRef]
19. Barber-James, H.M.; Zrelli, S.; Yanai, Z.; Sartori, M. A reassessment of the genus *Oligoneuriopsis* Crass, 1947 (Ephemeroptera, Oligoneuriidae, Oligoneuriellini). *ZooKeys* **2020**, *985*, 15–47. [CrossRef] [PubMed]
20. Aubert, J. Quelques Plécoptères du Muséum d'Histoire naturelle de Vienne. *Ann. Naturhistor. Mus. Wien.* **1964**, *67*, 287–301. (In French)
21. Hershkovitz, Y.; Hering, D.; Gal, G.; Pottgiesser, T.; Feld, C.K. Stream types of the Lake Kinneret (Sea of Galilee) watershed. *Int. J. River Basin Manag.* **2018**, *16*, 133–143. [CrossRef]
22. Givati, A.; Tal, A. The hydrological status in Lake Kinneret Watershed–observed and predicted trends based on hydroclimatic models. *Ecol. Environ.* **2017**, *8*, 1–8. (In Hebrew)
23. Samuels, R.; Hochman, A.; Baharad, A.; Givati, A.; Levi, Y.; Yosef, Y.; Saaroni, H.; Ziv, B.; Harpaz, T.; Alpert, P. Evaluation and projection of extreme precipitation indices in the Eastern Mediterranean based on CMIP5 multi-model ensemble. *Int. J. Clim.* **2018**, *38*, 2280–2297. [CrossRef]
24. Givati, A.; Thirel, G.; Rosenfeld, D.; Paz, D. Climate change impacts on streamflow at the upper Jordan River based on an ensemble of regional climate models. *J. Hydrol. Reg. Stud.* **2019**, *21*, 92–109. [CrossRef]
25. Akron, A.; Ghermandi, A.; Dayan, T.; Hershkovitz, Y. Interbasin water transfer for the rehabilitation of a transboundary Mediterranean stream: An economic analysis. *J. Environ. Manag.* **2017**, *202*, 276–286. [CrossRef] [PubMed]

Article

Species Composition of Aquatic (Nepomorpha) and Semiaquatic (Gerromorpha) Heteroptera (Insecta: Hemiptera) in Kaeng Krachan National Park, Phetchaburi Province, Thailand

Sajeemat Attawanno and Akekawat Vitheepradit *

Department of Entomology, Kasetsart University, Bangkok 10900, Thailand; sajeemat.ra@ku.th
* Correspondence: agrawv@ku.ac.th; Tel.: +66-8-5370-3213

Abstract: The species composition of aquatic (Nepomorpha) and semiaquatic (Gerromorpha) Heteroptera were examined from protected and unprotected study sites in three streams associated with Kaeng Krachan National Park. At each stream, both quantitative and qualitative sampling methods were used during seven collecting events (November 2018 to June 2020). A total of 11 families, representing 33 genera and 60 species, were collected in this study, with more Nepomorpha families but higher species richness in Gerromorpha. The species richness of both protected and unprotected sampling sites were lowest during the fifth sampling event. Nevertheless, there was no significant difference in richness between protected and unprotected sampling sites for any sampling event based on a paired *t*-test analysis. Based on an nMDS analysis, the patterns of species composition of aquatic and semiaquatic heteropterans were unclear among protected and unprotected sampling sites. The use of aquatic and semiaquatic Heteroptera as bioindicators for habitat quality is still uncertain. Additional physiochemical characters of the water and physical characters of the stream may lead to a clearer picture of the relationship between aquatic and semiaquatic Heteroptera and stream habitat quality.

Keywords: Heteroptera; aquatic; species compositions; Thailand

Citation: Attawanno, S.; Vitheepradit, A. Species Composition of Aquatic (Nepomorpha) and Semiaquatic (Gerromorpha) Heteroptera (Insecta: Hemiptera) in Kaeng Krachan National Park, Phetchaburi Province, Thailand. *Diversity* **2022**, *14*, 462. https://doi.org/10.3390/d14060462

Academic Editors: Marina Vilenica, Zohar Yanai, Laurent Vuataz and Michael Wink

Received: 30 April 2022
Accepted: 6 June 2022
Published: 9 June 2022

1. Introduction

Kaeng Krachan National Park is the largest national park in Thailand, located in Phetchaburi and Prachuap Khiri Khan provinces, covering an area of approximately 2900 km^2 [1]. The area was announced as a national park in 1981 and as a World Heritage Site in 2021 [2]. The national park is apart of the Tenasserim Mountain Range where the Himalayan, Indochinese, and Sumatran faunae meet [3]. The national park is one of the best in Southeast Asia in terms of the preservation of wildlife habitats, with outstanding wildlife management and protection [4]. The national park fauna is highly diverse with many endemic species [5]. Therefore, numerous research projects have been conducted within the national park [5]. Nevertheless, the diversity on aquatic insects in this biologically rich national park remains unexplored. With a high biodiversity and distinguished conservation management, Kaeng Krachan National Park is a perfect study area to examine the differences in composition patterns of aquatic and semiaquatic Heteroptera between protected and unprotected sampling sites. Furthermore, studies within the park offer an opportunity to discover undescribed species of aquatic and semiaquatic Heteroptera within the faunistically complex area.

Semiaquatic Heteroptera species are in infraorder Gerromorpha [6,7]. Gerromorpha consists of eight families, 161 genera, and nearly 2120 species worldwide [8,9]. Aquatic Heteroptera are classified under infraorder Nepomorpha, with 11 families, 137 genera, and approximately 2006 species distributed worldwide [9,10]. Most Gerromorpha species live on the water surface, floating plants, and vegetation on the margins of freshwater habitats, whereas Nepomorpha live under water, including in both lentic and lotic habitats [7].

In the last several decades, the aquatic and semiaquatic Heteroptera in Thailand have received great attention [11–24]. Numerous new species have been discovered from this rich country [25–28]. Nevertheless, the diversity of aquatic and semiaquatic Heteroptera at Kaeng Krachan National Park remains unexplored.

Owing to their high diversity and habitat specialization, aquatic and semiaquatic Heteroptera are excellent organisms for studies in evolutionary biology, ecology, and conservation biology [29]. Most ecological studies on aquatic and semiaquatic Heteroptera investigate the relationship of these two infraorders with biotic factors (e.g., aquatic vegetation, and riparian vegetation) [30]. For example, riparian vegetation in lotic habitats is positively correlated with the nepomorphan and gerromorphan species richness [31]. Additionally, the communities of Nepomorpha and Gerromorpha in streams are influenced by abiotic factors, including stream velocity, electric conductivity, pH, organic matter, and water temperature [32,33]. Recently, ecological studies have shown effects of negative environmental changes on aquatic and semiaquatic Heteroptera communities [34–37]. Ephemeroptera, Plecoptera, Trichoptera, and Odonata were considered to be more sensitive to water quality and habitat degradation than aquatic and semiaquatic Heteroptera [38]. Nonetheless, aquatic and semiaquatic Heteroptera have been used as bioindicators in various aquatic systems [39,40]. For example, the species compositions of Nepomorpha communities may reflect ecological integrity, habitat diversity, and water quality, as well as stress by pollutants [41,42]. Furthermore, changes in the habitat structure can alter the physicochemical characters of water, which subsequently negatively effects the taxonomic diversity of Gerromorpha [35,36,43]. Therefore, gerromorphans are potential candidates as bioindicators to monitor environmental changes, especially those of anthropogenic source [30,44]. In this research, the compositions of aquatic and semiaquatic Heteroptera were compared between protected and unprotected sampling sites within three streams, each of which originates within Kaeng Krachan National Park. Aquatic and semiaquatic Heteroptera are suitable taxa for this study because (1) they are well-known taxonomically, and (2) the group has potential as indicators for habitat quality.

2. Materials and Methods

2.1. Study Area

Kaeng Krachan National Park covers the rain forests of the Tenasserim Mountain Range in the west of Thailand. The national park is located in the Mae Klong Watershed. The Phetchaburi and Pranburi rivers are two major rivers that originate from the uplands inside the national park and run through Phetchaburi and Prachuap Khiri Khan provinces, respectively.

The criteria for the paired study sites were: (1) the protected section was located in the national park, (2) the unprotected section was located at least 1 km outside the national park, and was noticeably impacted by human activities (e.g., agriculture, urbanization), and (3) sites must have been similar in size and stream morphology (e.g., the presence of riffles, stream width) for both sections (Figures 1 and 2). Three protected and unprotected sampling sites (six sites total) were chosen on three streams. The paired stream sites are (protected and unprotected sampling sites): (1) Mae Kra Dung La Waterfall (MWF) and Mae Kra Dung La Stream (MS); Ban Krang (BK) and Huai Sat Yai (HSY); and Pa La-U Waterfall (PWF) and Huai Palaw (HPL) (Figures 1 and 2).

2.2. Measurement of Physical Characters of Sampling Sites

Global Positioning System (GPS) technology was used to record latitude, longitude, and elevation (WGS84 datum). At each sampling site, basic physical characters were measured: stream width, stream depth, and riparian width (Table 1). Physical characters were measured at three randomly chosen spots and then averaged. The types of substrate and presence of marginal vegetation were also recorded at each sampling site (Table 1).

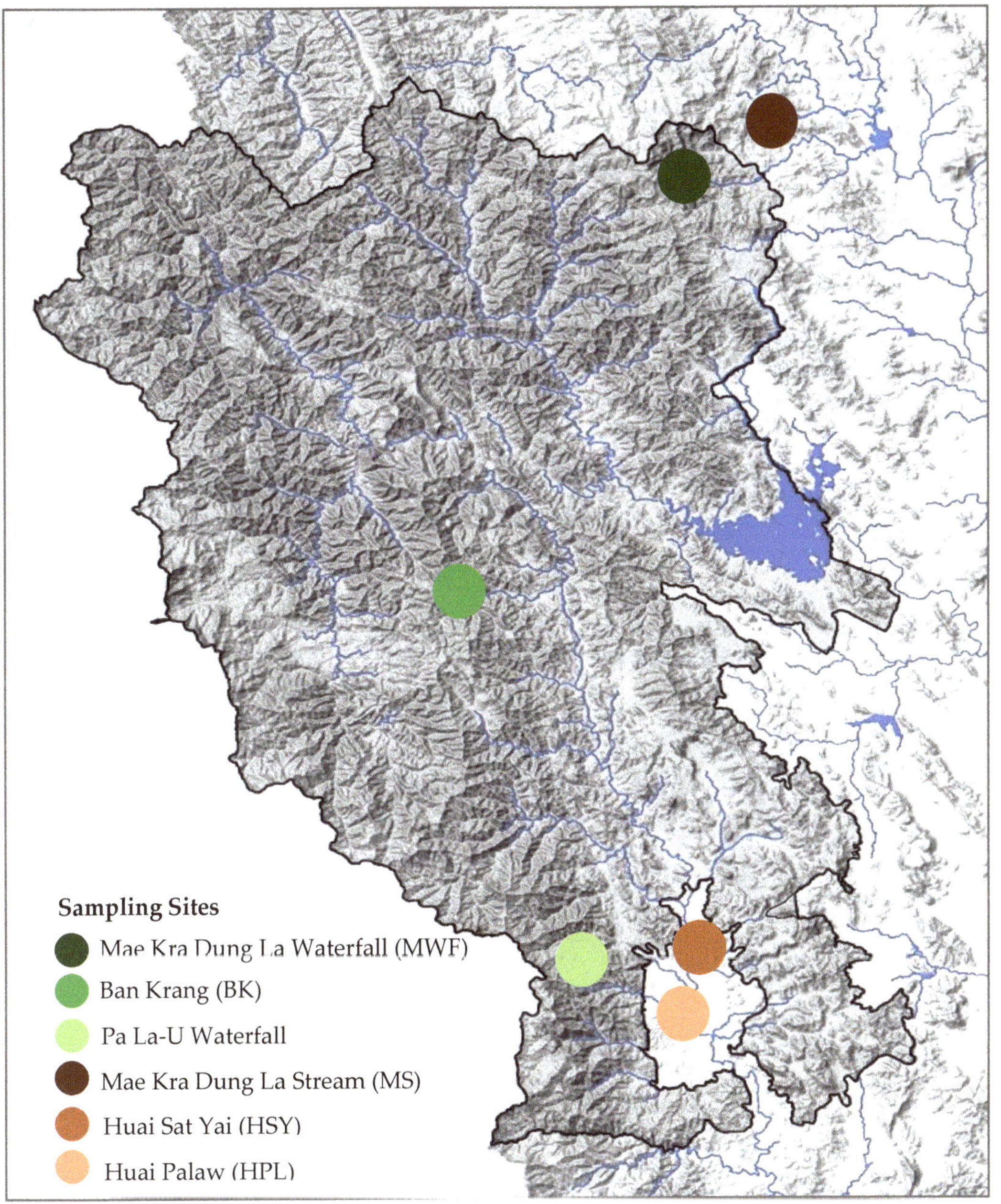

Figure 1. Map of Kaeng Krachan National Park showing three protected sampling sites located inside the national parks, including Mae Kra Dung La Waterfall (MWF); Ban Krang (BK); Pa La-U Waterfall (PWF), and three unprotected sampling sites located outside the national parks, including Mae Kra Dung La Stream (MS); Huai Sat Yai (HSY); Huai Palaw (HPL).

Figure 2. Protected sampling sites located inside the national park: (**a**) Mae Kra Dung La Waterfall (MWF); (**b**) Ban Krang (BK); (**c**) Pa La-U Waterfall (PWF), and unprotected sampling sites located outside the national park: (**d**) Mae Kra Dung La Stream (MS); (**e**) Huai Sat Yai (HSY); (**f**) Huai Palaw (HPL).

Table 1. Physical characters of sampling sites in the study. * = protected sampling sites and √ = found.

Sampling Sites	GPS Coordinates	Stream Width (m) Average (Min-Max)	Depth (m) Average (Min-Max)	Riparian Width (m)	Marginal Vegetation	Substrate Types in Stream			
						Boulder	Cobble	Gravel	Sand
MWF *	13°11.175′ N 099°32.472′ E (335 m)	5.78 (2.90–11.50)	0.16 (0.09–0.31)	>30	√	√	√		
BK *	12°48.144′ N 099°25.640′ E (386 m)	3.12 (2.00–4.80)	0.19 (0.05–0.37)	>30	√		√		
PWF *	12°45.250′ N 099°78.848′ E (232 m)	7.36 (1.40–4.23)	0.20 (0.04–0.45)	>30	√	√	√		
MS	13°12.182′ N 099°32.500′ E (350 m)	3.16 (1.56–4.45)	0.26 (0.07–0.44)	5–30			√		
HSY	12°30.832′ N 099°34.151′ E (176 m)	7.80 (3.25–12.00)	0.33 (0.10–0.70)	1–5				√	√
HPL	12°32.017′ N 099°29.940′ E (199 m)	9.37 (1.72–12.26)	0.21 (0.10–0.46)	1–5			√		

2.3. Sampling and Identification of Aquatic and Semiaquatic Heteroptera

To determine the composition patterns of aquatic and semiaquatic Heteroptera, six sampling sites were sampled at each site (two in each stream) using both quantitative and qualitative sampling methods. Each site was sampled seven times between November 2018 and June 2020. Three mesohabitats (i.e., gravel, margin, water surface) were identified and sampled. At each sampling site, three samples of each mesohabitat (3×3) were collected using a quantitative method and one sample of each mesohabitat (1×3) was collected using a qualitative method. Therefore, 12 samples were taken at each sampling site during the sampling events. Samples were collected using an aquatic D-net, although the specific sampling techniques differed among mesohabitats. For quantitative sampling, gravel sampling was conducted over a 2 m swath by kicking the substrate while holding the net downstream. Marginal stream vegetation and roots were swept with the D-net back and forth three times. The net was swept over the water surface three times to collect surface-dwelling insects (e.g., Gerridae, Veliidae).

For qualitative sampling, one sample of each mesohabitat was collected using a similar technique as during quantitative sampling. However, sampling continued until no recognizably new morphospecies were collected in three consecutive samples. All samples were sorted in the field using soft forceps to remove specimens, which were placed into container with 80% ethyl alcohol and labeled. Specimens were identified to species level and counted under a stereo microscope using various taxonomic keys [24,45,46].

2.4. Data Analysis

Taxonomic richness was tested for a normal distribution using the Shapiro–Wilk test. A pairwise t-test was performed to compare protected and unprotected sampling sites, significance was set at alpha = 0.05. The normality test and paired t-test were performed by Jamovi 2.3.9 [47]. A non-metric multidimensional scaling algorithm (nMDS) was used to reveal community patterns of aquatic and semiaquatic Heteroptera between protected and unprotected sampling sites of the seven collecting events based on abundance data. nMDS was performed using PC–ORD 5.0 [48].

3. Results

Protected and unprotected sampling sites overlapped in physical characteristics (Table 1). Protected sampling sites are located 232 to 386 m above sea level with an average stream width from 3.12 to 7.36 m and an average stream depth from 0.16 to 0.20 m. Likewise, the unprotected sampling sites are located 176 to 350 m above sea level with an average stream width from 3.16 to 9.37 m and an average stream depth from 0.21 to 0.33 m. Conversely, the riparian width, substrate types, and the presence of marginal vegetation between protected and unprotected sampling sites were dissimilar. The protected sampling sites have a wider riparian width (>30 m) and marginal vegetation with larger substrates (boulder and cobble), whereas the unprotected sampling sites have a narrower riparian width (1–5 m) and lack of marginal vegetation with smaller substrates (cobble, gravel, and sand).

Approximately 2000 specimens were collected in this study. Specimens were identified to the species level when taxonomic knowledge was available, or were assigned to morphospecies otherwise. Nevertheless, at least three possible undescribed species of Heteroptera have been discovered. Two possible undescribed species of *Metrocoris* were collected from Ban Krang (BK) and Pa La-U Waterfall (PWF), and a single possible undescribed specimen of *Ranatra* was collected from root mats at Pa La-U Waterfall (PWF) (Table 2).

Table 2. Taxa collected from protected sampling sites located inside the national park and uprotected sampling sites located outside the national park. * = protected sampling sites.

Species	Sites					
	MWF *	BK *	PWF *	MS	HSY	HPL
Nepomorpha						
Aphelocheiridae						
Aphelocheirus (M.) asiaticus (Hoberlandt & Stys)					x	x
Aphelocheirus (A.) grik Polhemus & Polhemus					x	x
Helotrephidae						
Helotrephes otoeis Nieser & Chen				x		x
Hydrotrephes jani Zettel	x					x
Tiphotrephes indicus (Distant)	x				x	x
Micronectidae						
Micronecta quadristrigata Breddin			x			x
Naucoridae						
Gestroiella limnocoroides Montandon					x	x
Gestroiella siamensis Polhemus, Polhemus & Sites					x	x
Heleolaccocoris ovatus (Montandon)			x		x	x
Heleolaccocoris strabus (Montandon)			x			
Naucoris scutellaris (Stål)	x					
Nepidae						
Cercotmetus asiaticus Amyot & Serville		x	x			x
Ranatra thai Lansbury			x	x		
Ranatra sp. A			x			
Notonectidae						
Anisops nigrolineatus (Lundblad)	x			x		
Aphelonecta gavini (Lunsbury)	x			x		
Enithares ciliata (Fabricius)	x					
Nychia sappho Kirkaldy	x			x		
Gerromorpha						
Gerridae						
Amemboa armata Polhemus & Andersen	x		x	x		x
Amemboa cristata Polhemus & Andersen	x	x	x			x
Limnogonus nitidus (Mayr)	x		x	x		x
Limnometra matsudai (Miyamoto)		x	x			
Metrocoris acutus Chen & Nieser			x			
Metrocoris borneensis Polhemus		x				
Metrocoris malayensis Chen & Nieser			x			
Metrocoris nigrofasciatus Distant			x			
Metrocoris nigrofascioides (Chen & Nieser)		x	x			
Metrocoris sp. A		x	x			
Metrocoris sp. B			x			
Onychotrechus esakii Andersen			x			
Pleciogonus wongsirii Chen, Nieser & Wattanachaiyingchareon			x	x		
Ptilomera jariyae Vitheepradit & Sites	x	x	x			
Ptilomera tigrina Uhler	x	x	x	x	x	x
Rhagadotarsus kraepelini Breddin						x
Rheumatogonus intermedius Hungerford					x	x
Rheumatagonus vietnamensis Zettel & Chen					x	x

Table 2. *Cont.*

Species	Sites					
	MWF *	BK *	PWF *	MS	HSY	HPL
Ventidius modulatus Lundblad					x	x
Ventidius pulai Cheng					x	x
Hebridae						
Hebrus longisetosus Zettel	x		x			
Hyrcanus saxatilis Andersen	x					
Hydrometridae						
Hydrometra annamana Hungerford & Evans	x	x		x		
Hydrometra greeni Kirkaldy			x	x		
Hydrometra kelantan Polhemus & Polhemus		x				
Hydrometra longicapitis Torre-Bueno			x	x	x	
Hydrometra orientalis Lundblad			x			
Mesoveliidae						
Mesovelia horvathi Lundblad	x		x	x	x	x
Mesovelia vittigera (Horváth)			x	x	x	x
Veliidae						
Microvelia douglasi Scott	x	x	x	x	x	x
Microvelia genitalis Lundblad			x	x	x	x
Microvelia leveillei (Lethierry)	x					x
Microvelia sp. A		x	x	x		
Microvelia sp. B			x	x		
Neoalardus typicus (Distant)	x					
Perittopus asiaticus Zettel	x					
Rhagovelia femorata Dover	x	x	x		x	x
Rhagovelia inexpectata Zettel	x					
Rhagovelia sondaica Polhemus & Polhemus	x	x				
Rhagovelia sumatrensis Lundblad			x			
Strongylovelia setosa (Zettel & Tran)			x	x	x	
Strongylovelia sp. A	x			x		
Species of Nepomorpha	7	1	6	5	6	10
Species of Gerromorpha	17	13	27	15	12	15
Total species	24	14	33	20	18	25

Species richness was not significantly different between protected and unprotected sampling sites (Table 3). Species richness varied from season to season, appearing the highest at the beginning of the year (January 2019 and March 2020) and falling after that (November 2019 and June 2020). Species richness was nearly always lower at unprotected sampling sites (Figure 3). However, based on the results of the paired *t*-test, there was no significant difference between protected and unprotected sampling sites during each sampling event (Table 3, Figure 3).

Table 3. Paired *t*-test of species richness between three protected and three unprotected sites (n = 3 for each). Mean = average species richness of the three study sites.

Months	N	Mean ± SD		Student *t*-Test
		Protected Sampling Sites	Unprotected Sampling Sites	(*p*) *
November 2018	3	38.70 ± 4.04	28.00 ± 3.61	0.239
January 2019	3	43.30 ± 9.29	38.30 ± 4.51	0.402
April 2019	3	28.70 ± 4.16	26.70 ± 7.09	0.197
July 2019	3	24.30 ± 3.79	17.00 ± 6.56	0.19
November 2019	3	14.00 ± 6.24	14.30 ± 4.51	0.944
March 2020	3	52.30 ± 4.93	45.70 ± 8.96	0.109
June 2020	3	37.30 ± 5.69	29.39 ± 4.73	0.134

* $p > 0.05$.

The results of ordination using an nMDS analysis based on abundance data showed no clear community patterns between protected and unprotected sampling sites (Figure 4). In the protected group, the community at Mae Kra Dung La Waterfall (MWF) from June 2020 was clearly separated from other sampling sites from any other sampling event because of the positive correlation with *Enithares ciliata*. Additionally, communities at Ban Krang (BK) from every sampling event were aligned together with a positive correlation with *C. asiaticus*, *H. kelantan*, *R. inexpectata*, *Ranatra* sp. A, and *R. thai*. In the unprotected group, communities at Mae Kra Dung La Stream (MS) from April 2019 to March 2020 were separated from other communities.

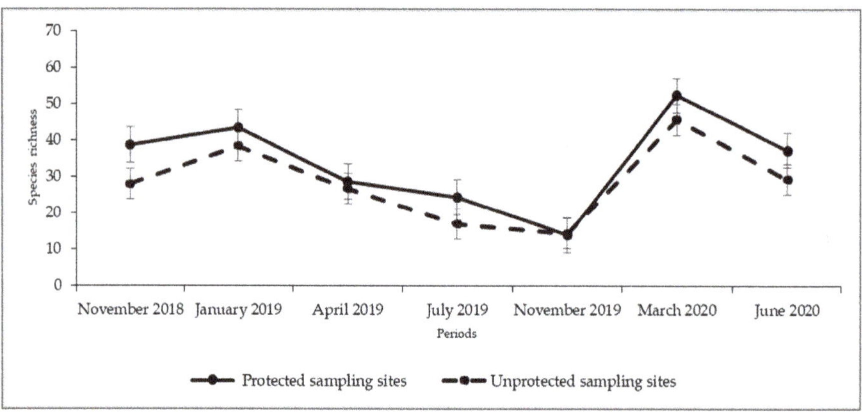

Figure 3. Species richness of protected and unprotected sampling sites during seven collecting events.

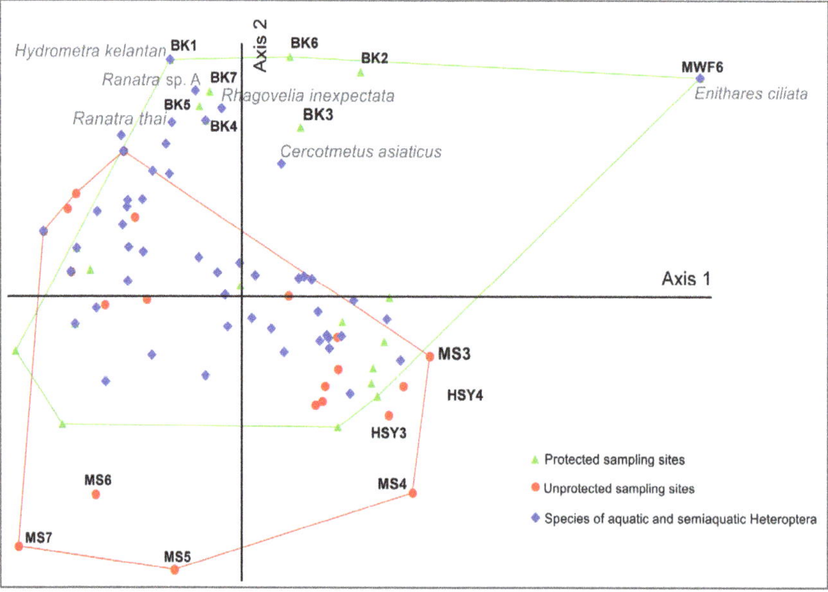

Figure 4. nMDS ordination plot samples by taxa dissimilarity (abundance data) between protected and unprotected sampling sites in the seven collecting events (stress = 14.80). Green refers to protected sampling sites, red refers to unprotected sampling sites, and blue refers to species.

4. Discussion

In general, conservation management is a vital tool to promote aquatic insect diversity, especially in preserved areas [49–51]. In protected areas, the presence of a riparian zone and marginal vegetation within aquatic ecosystems provides a wide variety of suitable microhabitats for aquatic insects [52,53]. Anthropogenic activities (e.g., agriculture, deforestation, urbanization) clearly impact the diversities of aquatic insects throughout the world [54–58]. Given the inferred large number of undescribed species and higher richness of insects, the paucity of aquatic surveys, and continuing habitat destruction in tropical areas, the aquatic insect fauna of these regions are more threatened than those of temperate regions [59]. This study focused on comparisons of species composition of aquatic and semiaquatic Heteroptera between protected and unprotected sampling sites that are located inside and outside of a national park, which reflect conservation management and the influence of human disturbance, respectively.

Season affects the insect communities of both aquatic and terrestrial ecosystems [60]. The richness and composition of aquatic insects in temperate regions fluctuates throughout the year, partially due to temperature changes among the seasons [61]. Similarly, the aquatic insect diversity in the tropical zones is strongly controlled by the seasons [61–63]. Richness and abundance of tropical aquatic insects are generally positively correlated with amount of rainfall [64,65]; however, an excessive amount of rainfall can significantly cause the decline of aquatic insects in streams [66].

The lowest richness was observed during November 2019 (Figure 3), which occurred during a drought during 2019 [67,68]. Most of the streams dried up and consisted of only stagnant pools at the sampling sites. The change in water level and flow clearly effected benthic nepomorphans, especially Aphelocheiridae and Naucoridae, since most of them are adapted to living in running waters [69,70]. Pools became the only available refuges for aquatic insects during these unfavorable periods [71–73]. These aquatic true Heteroptera have been reported to colonize new suitable habitat by short flights, and some are known to estivate during unfavorable conditions [74,75]. In this study, numerous species of semiaquatic Heteroptera were found at high densities in pools during the drought period. Gerromorpha is known to temporarily colonize these mesohabitats until lotic habitats revert to their normal stage [76–78].

Although the richness of aquatic and semiaquatic Heteroptera was not significantly different, this study was similar to previous studies that found aquatic true Heteroptera were more commonly found in altered areas [79]. Two species within each family, Aphelocheiridae and Naucoridae, were associated with gravel and sandy substrates in unprotected sampling sites (Table 2) [69,70,79], and they were not abundantly present in protected sampling sites within the national park (Table 1). Although most species of Gerridae and Veliidae are not strongly associated with specific mesohabitats [11,80], some members are found only in specific mesohabitats within aquatic systems [76]. For example, species of *Metrocoris* and *Perritopus* occur abundantly at margins and rock pools of forested streams in the highlands [25,76,81,82], whereas species of *Ventidius* are commonly found in open streams in lowlands [83]. These habitat preferences were observed in this study: species of *Metrocoris* and *Perritopus* were present only from protected sampling sites, and species of *Ventidius* were found only in unprotected sampling sites.

Based on the species abundance of aquatic and semiaquatic Heteroptera, the results of nMDS showed an unclear pattern. Nevertheless, there are several interesting arrangements of sampling sites in both the protected and unprotected group. In the protected group, the Mae Kra Dung La Waterfall in March 2020 (MWF6) was separated from other protected sampling sites because of *Enithares ciliata*. In Thailand, *E.ciliata* has only been reported from forested streams [84], which describes MWF6. *E.ciliata* was not collected at any other sampling sites. Various sampling events at Ban Krang (BK) were grouped together because of the presence of *C. asiaticus*, *H. kelantan*, *Ranatra* sp. A, and *R. thai*. Ban Krang (BK) is located deep in the national park and had a higher species richness than other sampling sites (Table 2). The cluster of Ban Krang (BK) in the nMDS is probably due to location because the diversity of aquatic insects is generally higher in remote streams in forests which are less affected by humans [85,86]. Furthermore, vegetation and root mats were abundant observed at the stream margins of Ban Krang, which provide suitable habitat for *C. asiaticus*, *H. kelantan*, *Ranatra* sp. A, and *Ranatra thai* [87–89]. In the unprotected sampling sites, several sampling events at Mae Kra Dung La Stream (MS) were separated from other sampling sites in the group. Although Mae Kra Dung La Stream (MS) is an unprotected sampling site located outside the conservation management, the physical characters of this sampling site are similar to those protected sampling sites because it is well-shaded with numerous large trees in the riparian zone and contains large emergent rocks in the stream (Table 1).

Previously, research on aquatic insect diversity and structure adjacent to different land uses (i.e., agriculture, forest, urban) indicated that both richness and abundance of streams located in forested areas are higher than those of streams associated with other land uses [90–94]. Based on this study, the use of aquatic and semiaquatic Heteroptera as bioindicators of stream habitat quality is still unclear. Additional physiochemical characteristics of water and additional physical characteristics of streams with different species richness and composition may allow for a better understanding of the relationship between aquatic and semiaquatic heteropterans and land use adjacent to stream systems [95,96]. Including additional aquatic insect orders (e.g., Ephemeroptera, Plecoptera, Trichoptera, Odonata) in assessments may assist the study as bioindicators for the quality of stream physical structure [97].

Although the aquatic and semiaquatic Heteroptera are not directly influenced by forest types or vegetative zones (they are predacious), numerous species are restricted to forests, especially gerromorphans [36,76]. The specific reason for their restricted habitat is unknown. Nonetheless, preserved forests protect habitat integrity of streams, which provide preferred or suitable habitats for aquatic and semiaquatic Heteroptera [6,30]. Therefore, conservation management is vital to protect the diversity of aquatic insects from human disturbance [98,99].

5. Summary

In total, 60 species, representing 33 genera, and 11 families of aquatic and semiaquatic Heteroptera were collected during this study. Species richness and composition did not differ significantly between protected and unprotected sampling sites. However, unprotected sampling sites tended to have lower absolute species richness than protected sampling sites. Conservation management and quality of riparian zones play a major role in shaping the composition of not only herbivorous, but also predaceous, aquatic insects. The ability to use aquatic and semiaquatic Heteroptera as indicators for habitat quality remains unclear, but this may be useful after further study.

Author Contributions: Conceptualization, A.V. and S.A.; methodology, S.A. and A.V.; software, A.V.; validation, S.A. and A.V.; investigation, S.A.; resources, A.V. and S.A.; specimen curation, A.V. and S.A.; data curation, S.A. and A.V.; writing—original draft preparation, S.A.; writing—review and editing, A.V.; funding acquisition, A.V. All authors have read and agreed to the published version of the manuscript.

Funding: This research was funded by financial support for research provided by the Kasetsart University Research and Development Institute, KURDI (Grant No. P-2.2(D) 198.61, and FF(KU) 14.64) and the Graduate School, Kasetsart University.

Institutional Review Board Statement: This research was approved by the Institutional Animal Care and Use Committee, Faculty of Agriculture, Kasetsart University, Thailand under Project number ACKU59-AGR-008.

Data Availability Statement: Data is available from the corresponding author upon request.

Acknowledgments: We are grateful to La-au Nakthong (Kasetsart University) and Prabseuk Sritipsak (Kasetsart University) for their assistance in the field. We are grateful to Boonsatien Boonsoong (Kasetsart University) and Ratchadawan Ngoen-klan for their analysis assistance. We express our gratitude to Kaeng Krachan National Park and the Department of National Park, Wildlife and Plant Conservation for permission and logistic support to conduct the fieldwork. We thank the Department of Entomology, the Faculty of Agriculture, Kasetsart University for facility support. Lastly, we thank Michael L. Ferro (Clemson University) and reviewers for their critical comments.

Conflicts of Interest: The authors declare no conflict of interest.

References

1. Department of National Park and Wildlife, DNP. Kaeng Krachan National Park. National Park, Wildlife and Plant Conservation Department. 2007. Available online: http://www.dnp.go.th/index_eng.asp (accessed on 23 December 2012).
2. UNESCO World Heritage Convention. Asia and Pacific. 2021. Available online: http://whc.unesco.org/en/list/1461/documents (accessed on 26 July 2021).
3. Dobias, R.T. *The Shell Guide to the National Parks of Thailand*; Shell Company of Thailand: Bangkok, Thailand, 1982.
4. Wildlife Conservation Society, WCS. "Kaeng Krachan". The Wildlife Conservation Society. 2012. Available online: http://www.wcsthailand.org/english/landscapekkfc-main (accessed on 23 December 2012).
5. ICEM. Thailand National Report on Protected Areas and Development. Review of Protected Areas and Development in the Lower Mekong River Region, Thailand. 2003. Available online: https://portals.iucn.org/library/sites/library/files/documents/2003-106-3.pdf (accessed on 21 May 2012).
6. Nieser, N.; Chen, P.P. The water bugs (Hemiptera: Nepomorpha and Gerromorpha) of Vanuatu. *Tijdschr. Entomol.* **2005**, *148*, 307–327. [CrossRef]
7. Chen, P.P.; Nieser, N.; Zettel, H. *The Aquatic and Semiaquatic Bugs (Heteroptera: Nepomorpha and Gerromorpha) of Malaysia*; Brill: Leiden, The Netherland; Boston, MA, USA, 2021; Volume 5, p. 546.
8. Polhemus, J.T.; Polhemus, D.A. Global diversity of true bugs (Heteroptera; Insecta) in freshwater. *Hydrobiologia* **2008**, *595*, 379–391. [CrossRef]
9. Henry, T.J. Biodiversity of Heteroptera. Insect biodiversity. *Sci. Soc.* **2017**, *1*, 279–335.
10. Stys, P.; Jansson, A. Checklist of recent family–group names of Nepomorpha (Heteroptera) of the world. *Acta Entomol. Fenn.* **1988**, *50*, 44.
11. Miyamoto, S. Gerridae of Thailand and North Borneo taken by the joint Thai–Japanese Biological Expedition 1961–62. *Environ. Nat. Resour. J.* **1967**, *5*, 217–257.
12. Nieser, N. An illustrated key to the families of Nepomorpha in Thailand. *Amemboa* **1996**, *1*, 4–9.
13. Nieser, N. A new species of *Ranatra* from Thailand. Insecta: Heteroptera: Nepidae. *Ann. Naturhist. Mus. Wien.* **1997**, *99*, 79–82.
14. Nieser, N. Introduction to the Notonectidae (Nepomorpha) of Thailand. *Amemboa* **1998**, *2*, 10–14.
15. Papáček, M.; Zettel, H. Revision of the Oriental genus *Idiotrephes* (Heteroptera: Nepomorpha: Helotrephidae). *Eur. J. Entomol.* **2000**, *97*, 201–211. [CrossRef]
16. Sites, R.W.; Polhemus, J.T. A new species of *Telmatotrephes* (Heteroptera: Nepidae) from Thailand, with distributional notes on congeners. *Aquat. Insects* **2000**, *23*, 333–340. [CrossRef]
17. Sites, R.W.; Polhemus, J.T. Distribution of Helotrephidae (Heteroptera) in Thailand. *J. N. Y. Entomol. Soc.* **2001**, *109*, 372–391. [CrossRef]
18. Sites, R.W.; Polhemus, J.T. Two new species of *Hydrometra latreille* (Heteroptera: Hydrometridae) from Thailand. *Proc. Entomol. Soc. Wash.* **2003**, *105*, 138–143.
19. Zettel, H. *Nieserius* gen. n., a new genus of the subfamily Hyrcaninae (Heteroptera: Hebridae) from Thailand, Laos, and Nepal, with the first known subaquatic species of Gerromorpha. *Aquat. Insects* **1999**, *21*, 39–52. [CrossRef]
20. Vitheepradit, A.; Sites, R.W.; Zettel, H.; Man, Y.C. Review of the Hydrometridae (Heteroptera) of Thailand, with distribution records. *Nat. Hist. Bull. Siam Soc.* **2003**, *51*, 197–223.
21. Vitheepradit, A.; Sites, S.W. A review of *Ptilomera* (Heteroptera: Gerridae) in Thailand, with descriptions of three new species. *Ann. Entomol. Soc. Am.* **2007**, *100*, 139–151. [CrossRef]
22. Vitheepradit, A. The Semiaquatic Heteroptera (Gerromorpha) of Thailand: Faunistics, Biogeography and Phylogeography. Ph.D. Dissertation, University of Missouri, Columbia, MO, USA, 2008.
23. Vitheepradit, A.; Sites, R.W. A review of *Eotrechus* Kirkaldy (Hemiptera: Heteroptera: Gerridae) of Thailand with descriptions of three new species. *Zootaxa* **2007**, *1478*, 19.
24. Sites, R.W.; Vitheepradit, A. *Heleocoris* (Heteroptera: Naucoridae: Laccocorinae) of Thailand, with description of a new species. *Zootaxa* **2011**, *2736*, 16. [CrossRef]
25. Ye, Z.; Chen, P.; Bu, W. Contribution to the knowledge on the Oriental genus *Perittopus* Fieber, 1861 (Hemiptera: Heteroptera: Veliidae) with descriptions of four new species from China and Thailand. *Zootaxa* **2013**, *3616*, 31–48. [CrossRef]
26. Vitheepradit, A. A new species of *Namtokocoris* Sites (Hemiptera: Naucoridae) from Thailand. *Zootaxa* **2017**, *4320*, 592–596. [CrossRef]

27. Cook, J.L.; Sites, R.W.; Vitheepradit, A. The Pleidae (Hemiptera, Heteroptera) of Thailand, with the descriptions of two new species and a discussion of species from Southeast Asia. *ZooKeys* **2020**, *973*, 35. [CrossRef]
28. Zettel, H.; Laciny, A. A second species of *Pleciogonus* (Hemiptera, Heteroptera, Gerridae) from Thailand. *Linz. Boil. Beitr.* **2021**, *53*, 445–449.
29. Andersen, N.M.; Weir, T.A. *Australian Water Bugs: Their Biology and Identification (Hemiptera–Heteroptera, Gerromorpha and Nepomorpha)*; Apollo Books: Melbourne, Australia, 2004; p. 341.
30. Dias-Silva, K.; Cabette, H.S.R.; Juen, L.; De Marco, P., Jr. The influence of habitat integrity and physical–chemical water variables on the structure of aquatic and semiaquatic Heteroptera. *Zool* **2010**, *27*, 918–930.
31. Odum, E.P.; Finn, J.T.; Franz, E.H. Perturbation theory and the subsidy-stress gradient. *Biosci* **1979**, *29*, 349–352. [CrossRef]
32. Buss, D.F.; Baptista, D.F.; Nessimian, J.L.; Egler, M. Substrate specificity, environmental degradation and disturbance structuring macroinvertebrate assemblages in neotropical streams. *Hydrobiologia* **2004**, *518*, 179–188. [CrossRef]
33. Merritt, R.W.; Cummins, K.W.; Berg, M.B. Trophic relationships of macroinvertebrates. In *Methods in Stream Ecology, Volume 1*; Academic Press: London, UK, 2017; pp. 413–433.
34. Vieira, T.B.; Dias-Silva, K.; Pacífico, E.D.S. Effects of riparian vegetation integrity on fish and Heteroptera communities. *Appl. Ecol. Environ. Res.* **2015**, *13*, 53–65.
35. Cunha, E.J.; de Assis Montag, L.F.; Juen, L. Oil palm crops effects on environmental integrity of Amazonian streams and Heteropteran (Hemiptera) species diversity. *Ecol. Indic.* **2015**, *52*, 422–429. [CrossRef]
36. Dias–Silva, K.; Brasil, L.S.; Juen, L.; Cabette, H.S.R.; Costa, C.C.; Freitas, P.V.; De Marco, P. Influence of local variables and landscape metrics on Gerromorpha (Insecta: Heteroptera) assemblages in Savanna streams, Brazil. *Neotrop. Entomol.* **2020**, *49*, 191–202. [CrossRef]
37. Guterres, A.P.M.; Cunha, E.J.; Juen, L. Tolerant semiaquatic bugs species (Heteroptera: Gerromorpha) are associated to pasture and conventional logging in the Eastern Amazon. *J. Insect Conserv.* **2021**, *25*, 555–567. [CrossRef]
38. Brasil, L.S.; Luiza-Andrade, A.; Calvão, L.B.; Dias-Silva, K.; Faria, A.P.J.; Shimano, Y.; Oliveira-Junior, J.M.B. Aquatic insects and their environmental predictors: A scientometric study focused on environmental monitoring in lotic environmental. *Environ. Monit. Assess.* **2020**, *192*, 10. [CrossRef]
39. Bonada, N.; Prat, N.; Resh, V.H.; Statzner, B. Developments in aquatic insect biomonitoring: A comparative analysis of recent approaches. *Annu. Rev. Entomol.* **2006**, *51*, 495–523. [CrossRef]
40. Skern, M.; Zweimuller, I.; Schiemer, K. Aquatic Heteroptera as indicators for terrestrialisation of floodplain habitats. *Limnologica* **2010**, *40*, 241–250. [CrossRef]
41. Jansson, A. Micronectinae (Heteroptera, Corixidae) as indicator of water quality in Lake Vesijarvi, Southern Finland, during the period of 1976–1986. *Biol. Res. Univ. Jyvask.* **1987**, *10*, 119–128.
42. Savage, A.A. The distribution of Corixidae in relation to the water quality of British lakes: A monitoring model. *Freshw. Forum.* **1994**, *4*, 32–661.
43. Dias-Silva, K.; Vieira, T.B.; de Matos, T.P.; Juen, L.; Simiao-Ferreira, J.; Hughes, R.M.; Junior, P.D.M. Measuring stream habitat conditions: Can remote sensing substitute for field data? *Sci. Total Environ.* **2021**, *788*, 147617. [CrossRef]
44. Spence, J.R.; Andersen, N.M. Semiaquatic bugs (Gerromorpha). In *Heteroptera of Economic Importance*; CRC Press: Boca Raton, FL, USA, 2000; pp. 601–606.
45. Nakthong, L.; Vitheepradit, A.; Sites, R.W. Key to the species of Eotrechinae (Hemiptera: Heteroptera: Gerridae) of Thailand and review of the fauna of the Phetchabun Mountain Range. *Zootaxa* **2014**, *3860*, 47–63. [CrossRef]
46. Raruanysong, S.; Vitheepradit, A.; Sites, R.W. Key to the species of Ptilomerinae (Hemiptera: Heteroptera: Gerridae) of Thailand and review of the fauna of the Tennaserim Mountain Range. *Zootaxa* **2014**, *3852*, 101–117. [CrossRef]
47. The Jamovi Project. Jamovi (Version 1.6) (Computer Software). 2021. Available online: http://www.jamovi.org (accessed on 17 May 2022).
48. McCune, B.; Mefford, M.J. *PC–ORD. Multivariate Analysis of Ecological Data, Version 6*; MjM Software: Gleneden Beach, OR, USA, 2011.
49. Niemi, G.J.; McDonald, M.E. Application of ecological indicators. *Annu. Rev. Ecol. Evol. Syst.* **2004**, *35*, 9–111. [CrossRef]
50. Rosenberg, D.M.; Resh, V.H. *Introduction to Freshwater Biomonitoring and Benthic Macroinvertebrates*; Freshwater Biomonitoring and Benthic Macroinvertebrates; Chapman and Hall: New York, NY, USA, 1993; p. 488.
51. Yoshimura, M. Effects of forest disturbances on aquatic insect assemblages. *Entomol. Sci.* **2012**, *15*, 145–154. [CrossRef]
52. Andersen, N.H.; Wallace, J.B. Habitat, life history, and behavioral adaptations of aquatic insects. In *An Introduction to the Aquatic Insects of North America*, 2nd ed.; Kendall/Hunt Publishing: Dubuque, IA, USA, 1984; pp. 38–58.
53. Erman, N.A. The use of riparian systems by aquatic insects. In *California Riparian Systems: Ecology, Conservation, and Productive Management*; University of California Press: Berkeley, CA, USA, 1984; pp. 177–1982.
54. Hellawell, J.M. Biological Indicators of Freshwater Pollution and Environmental Management. In *Pollution Monitoring Series*; Melanby, K., Ed.; Elsevier: Amsterdam, The Netherlands, 1986; p. 546.
55. Metcalfe, J.L. Biological water quality assessment of running waters based on macroinvertebrate communities. History and present status in Europe. *Environ. Pollut.* **1989**, *60*, 101–139. [CrossRef]

56. Wright, J.F.; Furse, M.T.; Armitage, P.D.; Moss, D. New procedures for identifying running-water sites subject to environmental stress and for evaluating sites for conservation, based on the macroinvertebrate fauna. *Arch. Hydrobiol.* **1993**, *127*, 319–326. [CrossRef]
57. Pinel–Alloul, B.; Méthot, G.; Lapierre, L.; Willsie, A. Macrobenthic community as a biological indicator of ecological and toxicological factors in Lake Saint-Francois (Quebec). *Environ. Pollut.* **1996**, *91*, 65–87. [CrossRef]
58. Nedeau, E.J.; Merritt, R.W.; Kaufman, M.G. The effect of an industrial effluent on an urban stream benthic community: Water quality vs. habitat quality. *Environ. Pollut.* **2003**, *123*, 13. [CrossRef]
59. Polhemus, D.A. Conservation of aquatic insects: Worldwide crisis or localized threats? *Am. Zool.* **1993**, *33*, 588–598. [CrossRef]
60. Wesner, J.S. Seasonal variation in the trophic structure of a spatial prey subsidy linking aquatic and terrestrial food webs: Adult aquatic insects. *Oikos* **2010**, *119*, 170–179. [CrossRef]
61. Hoang, D.H.; Bae, Y.J. Aquatic insect diversity in a tropical Vietnamese stream in comparison with that in a temperate Korean stream. *Limnology* **2006**, *7*, 45–55. [CrossRef]
62. Maneechan, W.; Prommi, T.O. Diversity and distribution of aquatic insects in streams of the Mae Klong Watershed, Western Thailand. *J. Entomol.* **2015**, *2015*, 912451. [CrossRef]
63. Sundar, S.; Silva, D.P.; de Oliveira Roque, F.; Simião-Ferreira, J.; Heino, J. Predicting climate effects on aquatic true bugs in a tropical biodiversity hotspot. *J. Insect Conserv.* **2021**, *25*, 229–241. [CrossRef]
64. Pramual, P.; Wongpakam, K. Seasonal variation of black fly (Diptera: Simuliidae) species diversity and community structure in tropical streams of Thailand. *Entomol. Sci.* **2010**, *13*, 17–28. [CrossRef]
65. Thamsenanupap, P.; Seetapan, K.; Prommi, T. Caddisflies (Trichoptera, Insecta) as bioindicator of water quality assessment in a small stream in northern Thailand. *Sains Malays.* **2021**, *50*, 655–665.
66. Kim, D.G.; Yoon, T.J.; Baek, M.J.; Bae, Y.J. Impact of rainfall intensity on benthic macroinvertebrate communities in a mountain stream under the East Asian monsoon climate. *J. Freshw. Ecol.* **2018**, *33*, 489–501. [CrossRef]
67. Mekong River Commission. Understanding the Mekong River's Hydrological Conditions: A Brief Commentary Note on the "Monitoring the Quantity of Water Flowing Through the Upper Mekong Basin Under Natural (Unimpeded) Conditions" study by Alan Basist and Claude Williams. Vientiane, MRC Secretariat. 2020. Available online: https://www.mrcmekong.org/resource/ajg4w9 (accessed on 29 April 2022).
68. Mekong River Commission. Situation Report on Dry Season Hydrological Conditions in the Lower Mekong River Basin: November 2020–May 2021. Vientiane: MRC Secretariat. 2021. Available online: https://www.mrcmekong.org/assets/Publications/Dry-season-situation-report-Nov-2020-to-May-2021.pdf (accessed on 29 April 2022).
69. Polhemus, D.A.; Polhemus, J.T. The Aphelocheirinae of tropical Asia (Heteroptera: Naucoridae). Department of Zoology, National University of Singapore. *Raffles. Bull. Zool.* **1988**, *36*, 167–300.
70. Sites, R.W.; Willig, M.R. Microhabitat associations of three sympatric species of Naucoridae (Insecta: Hemiptera). *Environ. Entomol.* **1991**, *20*, 127–134. [CrossRef]
71. Chester, E.T.; Robson, B.J. Drought refuges, spatial scale and recolonisation by invertebrates in non–perennial streams. *Freshw. Biol.* **2011**, *56*, 2094–2104. [CrossRef]
72. Boersma, K.S.; Bogan, M.T.; Henrichs, B.A.; Lytle, D.A. Invertebrate assemblages of pools in arid–land streams have high functional redundancy and are resistant to severe drying. *Freshw. Biol.* **2013**, *59*, 491–501. [CrossRef]
73. Herbst, D.B.; Cooper, S.D.; Medhurst, R.B.; Wiseman, S.W.; Hunsaker, C.T. Drought ecohydrology alters the structure and function of benthic invertebrate communities in mountain streams. *Freshw. Biol.* **2019**, *39*, 10401–10417. [CrossRef]
74. Lake, P.S. Ecological effects of perturbation by drought in flowing waters. *Freshw. Biol.* **2003**, *48*, 1161–1172. [CrossRef]
75. Saffarinia, P.; Anderson, K.E.; Herbst, D.B. Effects of experimental multi-season drought on abundance, richness, and beta diversity patterns in perennially flowing stream insect communities. *Hydrobiologia* **2022**, *849*, 879–897. [CrossRef]
76. Andersen, N.M. The semiaquatic bugs (Hemiptera, Gerromorpha). In *Phylogeny, Adaptations, Biogeography and Classification*; Brill: Leiden, The Netherland, 1982; p. 455.
77. Andersen, N.M. Classification, phylogeny, and zoogeography of the pond skater genus *Gerris* Fabricius (Hemiptera: Gerridae). *Can. J. Zool.* **1993**, *71*, 2473–2508. [CrossRef]
78. Taylor, S.J.; Mcpherson, J.E. Gerromorpha (Hemiptera: Heteroptera) in Southern Illinois: Species assemblages and habitats. *Gt. Lakes Entomol.* **2006**, *39*, 26.
79. Yang, C.M.; Lua, H.K.; Yeo, K.L. Semiaquatic bug (Heteroptera: Gerromorpha) fauna in the nature reserves of Singapore. *Gard Bull.* **1997**, *49*, 313–319.
80. Yule, C.M.; Sen, Y.H. *Freshwater Invertebrates of the Malaysian Region*; Academy of Sciences Malaysia: Kuala Lumpur, Malaysia, 2004; pp. 457–490.
81. Zettel, H. Five new species of *Perittopus* Fieber, 1861 (Hemiptera: Veliidae) fron Southeast Asia. *Raffles Bull. Zol.* **2001**, *49*, 109–120.
82. Chen, P.P.; Nieser, N.A. Taxonomic Revision of the Oriental Water Strider Genus Metrocoris Mayr (Hemiptera, Gerridae). Part I, II Zoological Museum. *Steenstrupia* **1993**, *19*, 1–43, 45–82.
83. Chen, P.P.; Zettel, H. A taxonomic revision of the Oriental water strider genus *Ventidius* Distant (Hemiptera, Gerromorpha, Gerridae). *Tijdschr. Entomol.* **1998**, *141*, 137–208. [CrossRef]
84. Vitheepradit, A. The Aquatic and Semiaquatic Heteroptera of the Phu Pan and Mountain Ranges of Thailand. Master's Thesis, University of Missouri, Columbia, MO, USA, 2000.

85. Dolný, A.; Ožana, S.; Burda, M.; Harabiš, F. Effects of landscape patterns and their changes to species richness, species composition, and the conservation value of Odonates (insecta). *Insects* **2021**, *12*, 478. [CrossRef]
86. Danehy, R.J.; Bilby, R.E.; Justice, T.E.; Lester, G.T.; Jones, J.E.; Haddadi, S.S.; Merritt, G.D. Biological Diversity Responses to Flood Disturbance and Forest Management in Small, Forested Watersheds. *Water* **2021**, *13*, 2793. [CrossRef]
87. Polhemus, J.T.; Polhemus, D.A. Revision of the genus Hydrometra Latereille in Indochina and the western Malay Archipelago (Heteroptera: Hydrometridae). *Bishop Museum Occasional Papers* **1995**, *43*, 9–72.
88. Lansbery, I. A review of Oriental species of *Ranatra* Fabricius (Hemiptera–Heteroptera: Nepidae). *Trans. Ent. Soc. Lond.* **1972**, *124*, 287–341. [CrossRef]
89. Lansbery, I. A review of genus *Cercotmetus* Amyot & Serville, 1843 (Hemiptera–Heteroptera: Nepidae). *Tijdschr. Entomol.* **1973**, *116*, 83–106.
90. Ometo, J.P.H.; Martinelli, L.A.; Ballester, M.V.; Gessner, A.; Krusche, A.V.; Victoria, R.L.; Williams, M. Effects of land use on water chemistry and macroinvertebrates in two streams of the Piracicaba river basin, southeast Brazil. *Freshw. Biol.* **2000**, *44*, 327–337. [CrossRef]
91. Hepp, L.U.; Santos, S. Benthic communities of streams related to different land uses in a hydrographic basin in southern Brazil. *Environ. Monit. Assess.* **2009**, *157*, 305–318. [CrossRef]
92. Hepp, L.U.; Milesi, S.V.; Biasi, C.; Restello, R.M. Effects of agricultural and urban impacts on macroinvertebrates assemblages in streams (Rio Grande do Sul, Brazil). *Zoologia* **2010**, *27*, 106–113. [CrossRef]
93. Nessimian, J.L.; Venticinque, E.M.; Zuanon, J.; De Marco, P., Jr.; Gordo, M.; Fidelis, L. Land use, habitat integrity, and aquatic insect assemblages in Central Amazonian streams. *Hydrobiologia* **2008**, *614*, 117–131. [CrossRef]
94. Batalla Salvarrey, A.V.; Kotzian, C.B.; Spies, M.R.; Braun, B. The influence of natural and anthropic environmental variables on the structure and spatial distribution along longitudinal gradient of macroinvertebrate communities in southern Brazilian streams. *J. Insect Sci.* **2014**, *14*, 23. [CrossRef] [PubMed]
95. Lock, K.; Adriaens, T.; Van De Meutter, F.; Goethals, P. Effect of water quality on waterbugs (Hemiptera: Gerromorpha & Nepomorpha) in Flanders (Belgium): Results from a large-scale field survey. *Ann. Limnol.* **2013**, *49*, 121–128.
96. Ishadi, N.A.M.; Rawi, C.S.M.; Ahmad, A.H.; Abdul, N.H. The influence of heavy metals and water parameters on the composition and abundance of water bugs (Insecta: Hemiptera) in the Kerian River Basin, Perak, Malaysia. *Trop. Life Sci. Res.* **2014**, *25*, 61.
97. Brasil, L.S.; Giehl, N.F.D.S.; Batista, J.D.; De Resende, B.O.; Cabette, H.S.R. Aquatic insects in organic and inorganic habitats in the streams on the Central Brazilian savanna. *Rev. Colomb. Entomol.* **2014**, *43*, 286–291. [CrossRef]
98. Dudgeon, D.; Arthington, A.H.; Gessner, M.O.; Kawabata, Z.I.; Knowler, D.J.; Leveque, C.; Sullivan, C.A. Freshwater biodiversity: Importance, threats, status and conservation challenges. *Biol. Rev.* **2006**, *81*, 163–182. [CrossRef]
99. Sundar, S.; Heino, J.; Roque, F.D.O.; Simaika, J.P.; Melo, A.S.; Tonkin, J.D.; Nogueira, D.G.; Silva, D.P. Conservation of freshwater macroinvertebrate biodiversity in tropical regions. *Aquat. Conserv.* **2020**, *30*, 1238–1250. [CrossRef]

Article

The Gerromorpha (Heteroptera: Gerridae, Mesoveliidae, Veliidae) of Mangroves of Central and Eastern Regions, Thailand

La-au Nakthong and Akekawat Vitheepradit *

Department of Entomology, Kasetsart University, Bangkok 10900, Thailand; laau.na@ku.th
* Correspondence: agrawv@ku.ac.th; Tel.: +66-853-703-213

Abstract: The Gerromorpha assemblages in mangroves located in the central and eastern regions of Thailand were examined, and a total of nine species belonging to six genera and three families were discovered. Four of the recorded species are new records for Thailand. *Asclepios annandalei* Distant, 1915 was the most common species and widely distributed throughout the study area. The most diverse genus was *Xenobates*, which consisted of *Xenobates argentatus* Andersen, 2000, *Xenobates mandai* Andersen, 2000, *Xenobates murphyi* Andersen, 2000, and *Xenobates singaporensis* Andersen, 2000. Three of these species are new country records. Here, we present taxonomic and ecological information of mangrove gerromorphans in the central and eastern regions of Thailand.

Keywords: species richness; marine insects; Hemiptera

Citation: Nakthong, L.-a.; Vitheepradit, A. The Gerromorpha (Heteroptera: Gerridae, Mesoveliidae, Veliidae) of Mangroves of Central and Eastern Regions, Thailand. *Diversity* **2022**, *14*, 466. https://doi.org/10.3390/d14060466

Academic Editors: Marina Vilenica, Zohar Yanai and Laurent Vuataz

Received: 30 April 2022
Accepted: 7 June 2022
Published: 10 June 2022

1. Introduction

Gerromorpha (Heteroptera) consists of eight families that occupy a wide range of habitats from rock faces of waterfalls down to open oceans [1,2]. Five families of these gerromorphans are known to inhabit marine and brackish habitats, including mangroves, intertidal zones, coastal shorelines, and open oceans [1,3]. In Southeast Asia, Gerridae from marine systems consist of approximately 23 species within three genera [4]. Ten species of *Stenobates,* two species of *Asclepios* and a single species of *Rheumatometroides* occur in mangroves, coastal marshes and seashores [4–6]. Approximately 10 species of *Halobates* occur close to shores and open oceans [7]. A few species of *Hebrus* (Hebridae) have been reported from an intertidal zone of mangroves [4]. Hermatobatidae consists of three species within genus *Hermatobates* that are restricted to intertidal zones throughout the Subregion [1,4]. Two species representing one genus of Mesoveliidae (*Nereivelia*) were found underneath logs of mangroves in Singapore and Thailand [8]. Veliidae is composed of three genera that occur in marine habitats in Southeast Asia [1,4]. Approximately thirty species of *Halovelia* and *Haloveloides* are commonly found in the intertidal zone, and twenty species of *Xenobates* inhabit the water surface in mangrove forests [4].

Mangroves in Thailand are located on both the Andaman Sea and the Gulf of Thailand. More specifically, the majority of large mangroves are distributed throughout the Andaman Sea in the southern region of Thailand, whereas mangroves in the Gulf of Thailand are fragmented and scattered from the eastern region toward the southern region [9]. The mangrove ecosystem represents a complex link between freshwater and saltwater [10]. Taxonomy of plants in mangroves is considered important research since they are the main components of the ecosystem [10]. Research on insects in mangrove forest ecosystems has focused on terrestrial insects, especially pests and beneficial insects of mangroves themselves, or attractive insects [11]. However, our knowledge about the species richness of aquatic insects in mangrove ecosystems, especially in Thailand, is still very scarce [12]. Most research on aquatic insects in Thai mangroves is the result of taxonomic studies on certain groups [6,13,14] or ecological research on faunal recovery [12]. Therefore, this is the first study on aquatic insects in Thailand, which provides a better understanding of species richness and distribution of marine insects. This valuable information can be further employed for various dimensions of research in mangroves of Thailand or Southeast Asia.

2. Materials and Methods

To determine the species richness of gerromorphans in mangroves in the central and eastern regions of Thailand, a faunistic survey was conducted from 2018 to 2020 (Figure 1, Table 1). In total, twenty-five separated mangroves distributed through the Gulf of Thailand in the central and eastern regions were sampled (Figures 1 and 2, Table 1). At each site, two mesohabitats were recognized, which were water surface and water margin of the mangrove. Eight spots of each mesohabitat were sampled for a gerromorphan fauna in each mangrove. Specimens were collected using an aquatic D-net, although the specific sampling technique differed among mesohabitats [15]. At the water margin, insects associated with emergent and submerged vegetation were collected by sweeping the net back and forth across the margin. Insects were collected from the water surface using an aquatic D-net. Collecting was continued until no recognizably new morphospecies were collected in two consecutive samples at each mesohabitat. In each sampling regime, when the D-net was up to one-third full, the organic material was transferred to a white pan, specimens were removed with soft forceps, placed into containers with 80% ethyl alcohol, and labelled.

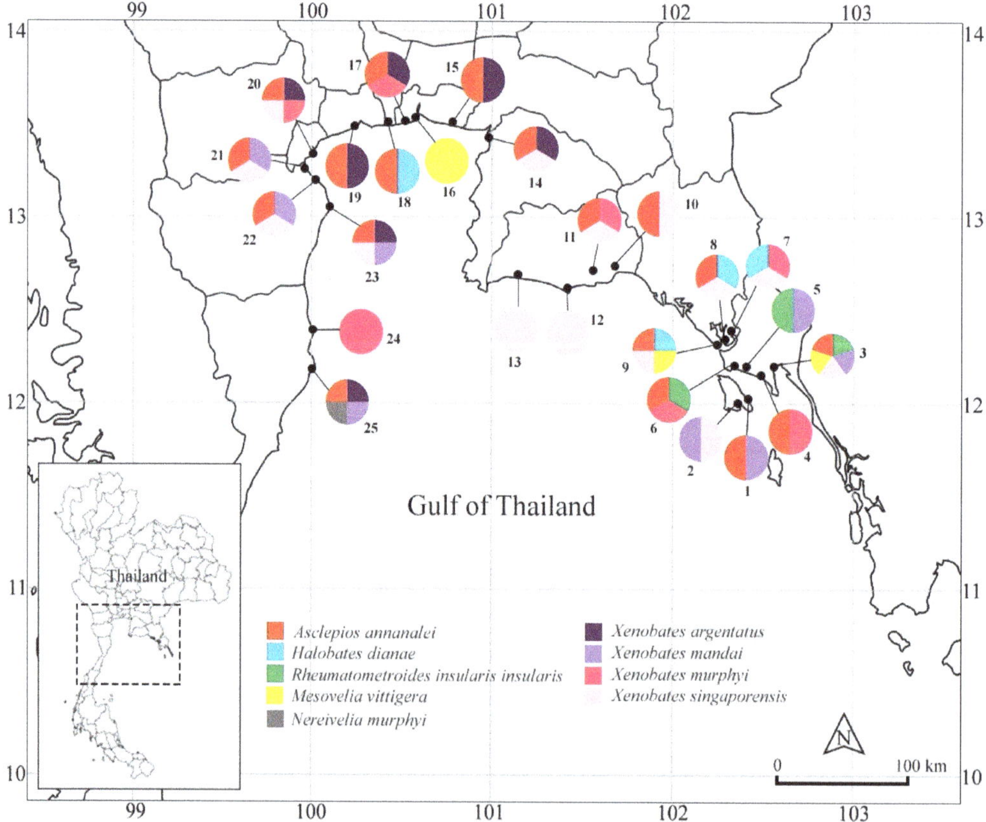

Figure 1. Map showing species richness and distribution of gerromorphans in mangroves of central and eastern regions, Thailand. Species are color-coded. Numbers refer to collecting sites (see Table 1).

Table 1. Collection data of samples in this study.

Site Number	Collecting Sites	Coordinates
1	Trat: Koh Chang Island, Ban Salak Kog	12°01.320′ N 102°23.388′ E
2	Trat: Koh Chang Island, Ban Salak Phetch	12°00.062′ N 102°22.760′ E
3	Trat: Ban Tha Rani	12°11.287′ N 102°33.410′ E
4	Trat: Ban Pret Nai	12°07.768′ N 102°30.168′ E
5	Trat: Ban Nam Chiao	12°11.086′ N 102°27.734′ E
6	Trat: Hard Sai Dum	12°10.165′ N 102°24.397′ E
7	Chantaburi: Ban Tha Sorn	12°22.097′ N 102°20.524′ E
8	Chantaburi: Leam Ward	12°25.355′ N 102°13.579′ E
9	Chantaburi: Leam Ngan	12°25.101′ N 102°14.104′ E
10	Rayong: Samea Phu	12°42.166′ N 101°43.019′ E
11	Rayong: Saphan Rak Samea	12°43.031′ N 101°39.237′ E
12	Rayong: Mangrove Forest Resource Development Station1	12°41.584′ N 101°40.500′ E
13	Rayong: Phra Chedi Klang Nam	12°40.036′ N 101°14.311′ E
14	Chonburi: Klong Tumru	13°26.370′ N 100°58.068′ E
15	Samut Prakarn: Ban Si Long	13°28.502′ N 100°51.043′ E
16	Samut Prakarn: Chulachomklao Fort	13°32.323′ N 100°34.885′ E
17	Samut Prakarn: Wat Khun Samut Chin	13°30.649′ N 100°31.957′ E
18	Bangkok: Bang Khun Thian	13°30.415′ N 100°25.409′ E
19	Samut Sakhon: Khok Kham	13°29.173′ N 100°20.060′ E
20	Samut Songkram: Khlong Khon	13°19.208′ N 099°58.812′ E
21	Phetchaburi: Bang Ta Boon	13°15.440′ N 099°56.580′ E
22	Phetchaburi: Mangrove Resource Development Station6	13°10.954′ N 100°01.182′ E
23	Phetchaburi: Laem Phak Bia	13°03.284′ N 100°05.537′ E
24	Prachaup Khiri Khan: Pranburi Forest Park	12°24.676′ N 099°59.330′ E
25	Prachaup Khiri Khan: Khlong Khao Dang	12°08.317′ N 099°57.110′ E

Morphological terminology largely follows Polhemus & Polhemus [6,16], Andersen [14], Andersen & Cheng [7], and Yang & Murphy [8]. Dried male specimens of each genus were placed under a Leica EZ4W stereomicroscope coupled with the LAS EZ program to obtain images. Images were then prepared with Photoshop CS5 (Adobe Systems Inc., San Jose, CA, USA). Specimens were deposited in the Entomology Museum, the Department of Entomology, Kasetsart University, Bangkok, Thailand.

Principle Component Analysis (PCA) was used to reveal species richness patterns of Gerromorpha in mangroves in the Central and Eastern regions based on the presence/absence data. PCA compresses richness variance into component axes to show the species richness relationships among communities. This analysis was performed using PC-ORD software 5.0 [17].

Figure 2. Photos of representative collecting sites in this study: (**A**) Ban Tha Rani (collecting site 3), (**B**) Phra Chedi Klang Nam (collecting site 13), (**C**) Chulachomklao Fort (collecting site 16), and (**D**) Pranburi Forest Park (collecting site 24).

3. Results

Nine species representing six genera and three families of Gerromorpha were collected from 25 mangroves in Central and Eastern regions (Tables 1 and 2). The most species-rich family was Veliidae with four species (Table 2). In addition, four new country records were discovered during this study. The preferred mesohabitats of each genus in this study were shown in Figure 3. *Halobates* was commonly found in the open water of estuaries. *Asclepios* and *Rheumatometroides* were collected in shallow sections of mangroves near estuaries. *Mesovelia* and *Nereivelia* were collected from margins of mangroves along the river, but *Nereivelia* occupied mangroves closer to the estuary. *Xenobates* species were found along the shorelines of mangrove streams and inside mangrove forests.

FAMILY GERRIDAE

This family consists of 75 genera worldwide that occupy a wide range of habitats [4,16,18]. More specifically, each genus has a specific preferred habitat [1], ranging from rock faces of waterfalls to open oceans [1]. Nevertheless, only five genera are considered marine gerrids: *Asclepios*, *Halobates*, *Stenobates*, *Rheumatometroides*, and *Rheumatobates* [3,4]. These genera, except *Rheumatobates*, have been reported from Southeast Asia [3,4]. In Thailand, six species representing these four genera were recorded from the Southern Region [6,7,12].

Genus *Asclepios* Distant, 1915

Members of this genus are small (3.0–4.0 mm body size) and yellowish brown with dark patterns on the head, thorax, and abdomen [7]. *Asclepios* species are commonly found in coastal areas associated with mangroves [5,7]. This genus is in the subfamily

Halobatinae and distributed in East and Southeast Asia [7]. Three species of *Asclepios* have been described, nevertheless only two of them have been reported from mangroves in Southeast Asia [7]. Specifically, *Asclepios apicalis* (Esaki, 1924) has been found in Vietnam [7]. In Thailand, *Asclepios annandalei* Distant, 1915 has been previously collected from Phuket Island [5].

Asclepios annandalei Distant, 1915: 505–506 (Figure 4A) [19].
Diagnosis: This species can be distinguished from *Halobates dianae* Zettel, 2001 by the smaller body (3.0–4.0 mm) and yellowish brown pronotum and thoracic pleura, whereas the latter species is larger (3.2–6.5 mm body size) with mainly dark pronotum and thoracic pleura. Furthermore, this species can be distinguished from *Rheumatometroides insularis insularis* (J. Polhemus & Cheng, 1982) by the middle femur distinctively longer than the middle tibia, whereas the latter species has the middle femora clearly shorter than the middle tibia [20].
Discussion: *Asclepios annandalei* was the most common gerrid in the study. This species was collected from mangroves in the Central (i.e., Phetchaburi, Samut Prakan, Samut Songkhram provinces) and Eastern (i.e., Chantaburi, Chon Buri, Rayong, Trat provinces) regions. This species was collected with *Halobates dianae* at Laem Ward, Chantaburi Province. Furthermore, this species occurred with *Rheumatometroides insularis insularis* in mangroves of the Eastern Region (Figure 1).
Material collected: Bangkok: collecting site 18; Chantaburi Province: collecting sites 8, 9; Phetchaburi Province: collecting sites 21, 22, 23; Prachaup Khiri Khan Province: collecting site 25; Rayong Province: collecting sites 10, 11, 12; Samut Prakarn Province: collecting sites 15, 17; Samut Sakhon Province: collecting site 19; Samut Songkram Province: collecting site 20; Trat Province: collecting sites 1, 3, 4, 6.

Genus *Halobates* Eschscholtz, 1822

Members of this genus are approximately 3.2–6.6 mm in length and dark in color with some small brown markings [7]. This genus is in the subfamily Halobatinae and mainly distributed in the Indo-Pacific Ocean [7], with the exceptional case of *Halobates micans* Eschscholtz, 1822, which is a cosmopolitan species [7]. There are 46 species of this genus throughout the world [7]. In Southeast Asia, eleven species were reported, whereas only five species were recorded from Thailand, from the southern region [7,12].

Halobates dianae Zettel, 2001: 1097–1102 (Figure 4B) [21].
Diagnosis: This species can be distinguished from *Asclepios annandalei* and *Rheumatometroides insularis insularis* by a larger body (2.7–6.5 mm) with mainly dark pronotum and thoracic pleura, whereas the latter species are smaller (3.0–4.0 mm body size) with yellowish brown pronotum and thoracic pleura.
Discussion: Specimens of *Halobates dianae* were collected in Chantaburi Province of the Eastern region (Figure 1). The species was collected from open sections of estuaries associated with large intact mangrove forests. Previously, it has been known only from the Philippines [18] and represents a new country record for Thailand. Within this study, *Halobates dianae* was collected syntopically with *Asclepios annandalei* and *Xenobates* species.
Material collected: Chantaburi Province: collecting sites 8, 9.

Genus *Rheumatometroides* Hungerford & Matsuda, 1958

Members of this genus are small (2.7–4.0 mm body size) and dark with dorsal yellowish markings on head, thorax, and abdomen [6]. This genus is in the subfamily Trepobatinae and distributed from Malaysia to Australia [4]. This marine genus consists of seven described species that mostly occur in shallow sections of mangroves [4,6]. In Southeast Asia, *Rheumatometroides insularis insularis* has been reported from mangroves in Malaysia and Singapore [6,20].

Rheumatometroides insularis insularis (J. Polhemus & Cheng, 1982): 225–227 (Figure 4C) [22].

Diagnosis: This species can be distinguished from *Asclepios annandalei* by the middle femora clearly shorter than the middle tibia, whereas the latter species has middle femora distinctively longer than the middle tibia [20]. Furthermore, this species can be distinguished from *Halobates dianae* by the smaller body (2.7–4.0 mm) and yellowish brown pronotum and thoracic pleura, whereas the latter species is larger (3.2–6.5 mm body size) with a dark pronotum and thoracic pleura.
Discussion: This species was collected from mangroves in Trat Province, the Eastern Region (Figure 1). Based on our field observations, it commonly occurred in large numbers in open areas of channels in mangrove forests. In addition, this species occurred syntopically with several species of *Xenobates* in the Eastern Region.
Material collected: Trat Province: collecting sites 3, 5, 6.

FAMILY MESOVELIIDAE

This small family consists of 46 species representing 12 genera and 2 subfamilies [1,23]. Interestingly, these semiaquatic bugs have been collected from a wide range of habitats, including the forest floor in tropical forests, margins of freshwater habitats, and coastal shores [23,24]. *Darwinivelia*, *Mesovelia*, *Nereivelia*, and *Speovelia* are the only genera that contain marine species [13]. In Southeast Asia, one widely distributed genus, *Mesovelia*, has been recorded [24]. Additionally, two other endemic genera, *Nereivelia* and *Cryptovelia*, are known from Indonesia, Malaysia, and Thailand [8,24].

Genus *Mesovelia* Mulsant & Rey, 1852

Members of this genus are elongated (2.0–3.5 mm body size) and yellowish with brown markings [8]. Generally, they live on aquatic vegetation and are able to walk on the water surface [1]. This genus belongs to subfamily Mesoveliinae, which contains approximately 28 species [1,8,25,26]. Furthermore, this genus is widely distributed throughout the world, including Africa, Asia, Australia, and Europe [16,27]. In Southeast Asia, *Mesovelia horvathi* Lundblad, 1933 and *Mesovelia vittigera* Horváth, 1895 are two common species recorded from various aquatic habitats [8,15,16,27]. They were collected from streams, ponds, rice paddies, peatswamps, and blacklight traps in Thailand [12,15]. To distinguish the species of Thai *Mesovelia*, male specimens are needed for diagnostic genitalic features [25].

Mesovelia vittigera Horváth, 1895: 160 (Figure 4D) [28].
Diagnosis: This species can be distinguished from *Nereivelia murphyi* J. Polhemus & D. Polhemus, 1989 by a larger and more slender body (2.0–3.5 mm), whereas the latter has a smaller and stouter body (1.7–2.2 mm) [8,13].
Discussion: In this study, this species was collected from a small fragmented mangrove in Samut Prakarn Province in the Central Region, and in the large intact mangroves in Chantaburi and Trat provinces in the Eastern Region. This species has been reported from Africa, Asia, Australia, and Europe [27]. Although this species commonly inhabits freshwater habitats, it was also collected from brackish waters in Southeast Asia [8,12]
Material collected: Chantaburi Province: collecting site 9; Samut Prakarn Province: collecting site 16; Trat Province: collecting site 3.

Genus *Nereivelia* J. Polhemus & D. Polhemus, 1989

This genus is robust (1.7–2.2 mm body size) and yellowish brown without brown patterns [13]. Two species of this genus are reported from Southeast Asia [8]. Decayed logs in the intertidal zone of mangroves represent their preferred microhabitat [8].

Nereivelia murphyi J. Polhemus & D. Polhemus, 1989: 75–77 (Figure 4E) [13].
Diagnosis: This species can be distinguished from *Mesovelia vittigera* by a smaller and stouter body (1.7–2.2 mm body size), whereas the latter has a larger and slender body (2.0–3.5 mm) [8,13].
Discussion: A single male of *Nereivelia murphyi* was collected in this study. This rare species has only been reported from Singapore and Thailand [8,13]. The holotype was collected in

a log in a river associated with mangroves on the Andaman Seaside in Ranong Province, Southern Thailand [8].
Material collected: Prachaup Khiri Khan Province: collecting site 25.

FAMILY VELIIDAE

This family is the most diverse family of Gerromorpha with 57 genera distributed throughout the world [1,16]. Most veliids inhabit the margins of fresh water habitats (i.e., ponds, streams) and are associated with vegetation [29]. Five genera extend their habitats to marine systems: *Trochopus*, *Husseyella*, *Xenobates*, *Halovelia*, and *Haloveloides* [3]. Three of these genera (i.e., *Xenobates*, *Halovelia*, and *Haloveloides*) have been reported from Southeast Asia [4]. In Thailand, species of *Halovelia* and *Haloveloides* were reported only from intertidal zones, whereas species of *Xenobates* are only known from mangroves [12,30,31].

Genus *Xenobates* Esaki, 1927

Members of this genus are very small (1.45–1.80 mm body size) and dark with some brown markings [14]. This genus belongs to the subfamily Haloveliinae and contains twenty-one species distributed from the Oriental to the Australian regions [14]. They are commonly found at tidal channels of mangrove forests [14]. In Southeast Asia, six species were reported from Singapore, Malaysia, and Thailand [14]. In Thailand, *Xenobates argentatus* Andersen, 2000 is the only species recorded from the Southern Region [12,14].

Xenobates argentatus Andersen, 2000: 280–281 [14].
Diagnosis: This species can be distinguished from other species of Thai congeners by the middle femora with short hairs, whereas the other Thai congeners have the middle femora with a row of long hairs.
Discussion: This species has previously been collected from mangroves in Southern Thailand [12,14]. In this study, it is the most wide-spread species of *Xenobates*, which were collected from Chon Buri, Prachuap Khiri Khan, Phetchaburi, Samut Prakan, Samut Sakhon, and Samut Songkhram provinces (Figure 1). This species co-occurred with *Xenobates murphyi* Andersen, 2000 and *Xenobates singaporensis* Andersen, 2000 at different collecting sites in Central and Eastern regions.
Material collected: Chon Buri Province: collecting site 14; Prachuap Khiri Khan Province: collecting site 25; Phetchaburi Province: collecting site 23; Samut Prakan Province: collecting sites 15, 17; Samut Sakhon Province: collecting site 19; Samut Songkram Province: collecting site 20.

Xenobates mandai Andersen, 2000: 278–280 (Figure 4F) [14].
Diagnosis: This species can be distinguished from *Xenobates argentatus* by the middle femora with a row of long hairs, whereas the latter species has the middle femora with short hairs. It can also be distinguished from *Xenobates singaporensis* by males with an unmodified sternum VII, whereas males of the latter species have a deep depression on sternum VII. This species can be distinguished from *Xenobates murphyi* by antennal segments II and III with long hairs anteriorly, whereas the latter species has antennal segments II and III with short hairs anteriorly.
Discussion: This species has been previously recorded from Singapore [14]. Our results represent a new country record for Thailand. In this study, this species was collected from Bangkok, Phetchaburi, and Trat provinces (Figure 1). Specifically, it was the most common species on Kho Chang Island, Trat Province. This species was collected with *Xenobates singaporensis* at several mangrove sites.
Material collected: Bangkok: collecting site 18; Phetchaburi Province: collecting sites 21, 22, 23; Trat Province: collecting sites 1, 2, 3, 5.

Xenobates murphyi Andersen, 2000: 277–278 [14].
Diagnosis: This species can be distinguished from *Xenobates argentatus* by the middle femora with a row of long hairs, whereas the latter species has the middle femora with short

hairs. It can be distinguished from *Xenobates singaporensis* by males with an unmodified sternum VII, whereas males of the latter species have a deep depression on sternum VII. This species can be distinguished from *Xenobates mandai* by antennal segments II and III with short hairs anteriorly, whereas the latter species has antennal segment II and III with long hairs anteriorly.

Discussion: This species has been previously recorded from Indonesia, Malaysia, Philippines and Singapore [14]. Our findings represent a new country record of this species in Thailand. In this study, it was widely distributed throughout the study area (Figure 1). It occurred syntopically with *Xenobates argentatus* at several mangrove forests.

Material collected: Chantaburi Province: collecting site 7; Prachuap Khiri Khan Province: collecting sites 24, 25; Rayong Province: collecting site 11; Samut Prakan Province: collecting site 17; Samut Songkram Province: collecting site 20; Trat Province: collecting sites 4, 6.

Xenobates singaporensis Andersen, 2000: 275–277 [14].

Diagnosis: This species can be distinguished from *Xenobates argentatus* by the middle femora with a row of long hairs, whereas the latter species has the middle femora with short hairs. This species can be distinguished from *Xenobates mandai* and *Xenobates murphyi* by males with a deep depression on sternum VII, whereas males of the latter species have an unmodified sternum VII.

Discussion: *Xenobates singaporensis* has been previously recorded from Singapore [14], while our findings represent a new country record of this species in Thailand. In this study, it was widely distributed in investigated mangroves (Figure 1). This species occurred syntopically with *Xenobates argentatus* and *Xenobates murphyi*.

Material collected: Chantaburi Province: collecting sites 7, 8, 9; Chon Buri Province: collecting site 14; Phetchaburi Province: collecting sites 21, 22, 23; Rayong Province: collecting sites 10, 11, 12, 13; Samut Songkram Province: collecting site 20; Trat Province: collecting sites 2, 3.

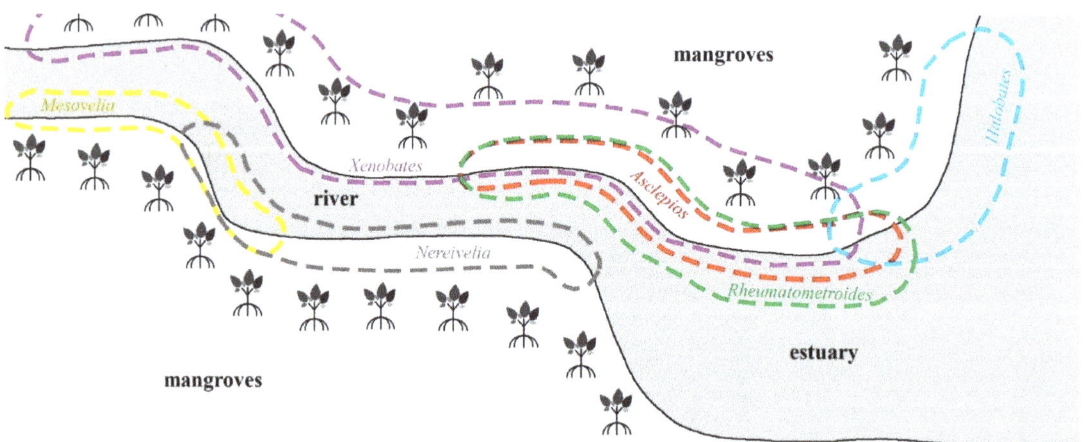

Figure 3. Preferred mesohabitats of marine Gerromorpha in mangroves of central and eastern regions.

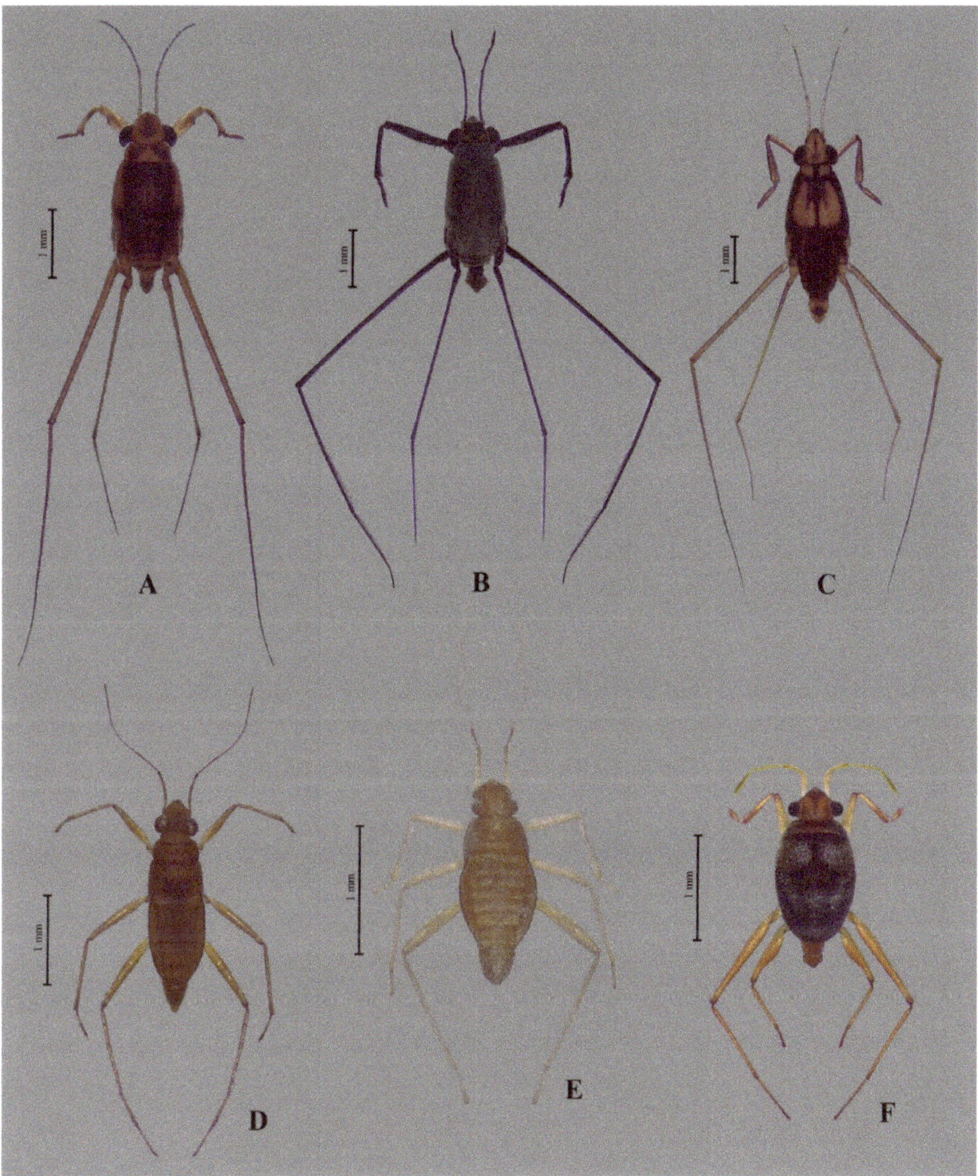

Figure 4. Habitus of a male of each genus collected for the 25 mangroves in Thailand: (**A**) *Asclepios annandalei*, (**B**) *Halobates dianae*, (**C**) *Rheumatometroides insularis insularis*, (**D**) *Mesovelia vittigera*, (**E**) *Nereivelia murphyi*, and (**F**) *Xenobates mandai*.

Principle Component Analysis (PCA) showed little distinct grouping consistent with geographic expectations based on the presence/absence data of species richness of Gerromorpha from 25 mangroves in the eastern and central regions (Figure 5). Most of the collecting sites of both regions were aligned together in the middle of axes 1 and 2. Nevertheless, two collecting sites from Chantaburi Province (e.g., collecting sites 8, 9) and three collecting sites from Trat Province (e.g., collecting sites 3, 5, 6) were arranged in the parameter of the eastern region group. Meanwhile, collecting site 20 in Phetchaburi Province and a collecting site 25 in Prachaup Khiri Khan Province were separated from other collecting sites in the central region group.

Table 2. Checklist of Gerromorpha taxa found in this study.

Family	Species	Central Region	Eastern Region
Gerridae	*Asclepios annandalei* Distant, 1915	X	X
	Halobates dianae Zettel, 2001 *		X
	Rheumatometroides insularis insularis (J. Polhemus & Cheng, 1982)		X
Mesoveliidae	*Mesovelia vittigera* Horváth, 1895	X	X
	Nereivelia murphyi J. Polhemus & D. Polhemus, 1989	X	
Veliidae	*Xenobates argentatus* Andersen, 2000	X	X
	Xenobates mandai Andersen, 2000 *	X	X
	Xenobates murphyi Andersen, 2000 *	X	X
	Xenobates singaporensis Andersen, 2000 *	X	X

* = new country record.

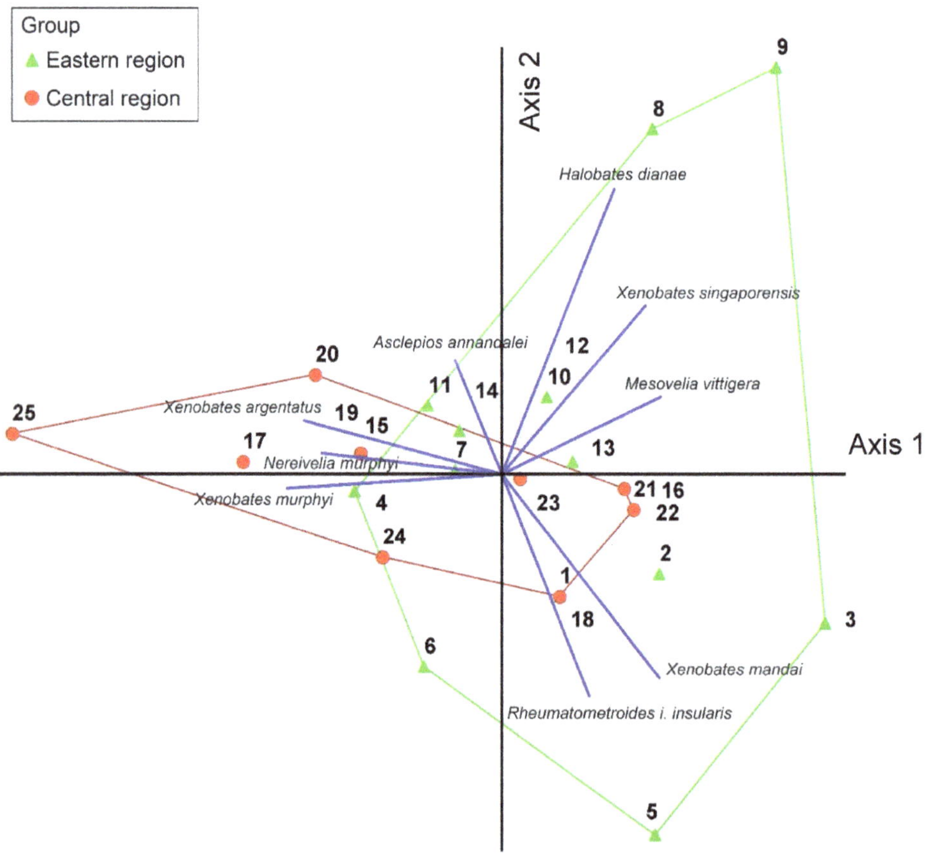

Figure 5. Principle component analysis (PCA) on species richness of Gerromorpha of each collecting site. The first and second PC axes explain 24.370% (eigenvalue: 2.193) and 16.909% (eigenvalue: 1.522) of the variation in the data set, respectively. Numbers refer to collecting sites (see Table 1).

4. Discussion

Species of *Asclepios* are commonly found in coastal areas associated with mangroves [7]. In Malaysia and Singapore, *Asclepios annandalei* is relatively rare and only collected in mangroves [20]. In this study, this species was common and distributed throughout the study area. Based on field observations, members of this species are normally found in a pair, where a male is riding a female without genital contact (a mate guarding behavior) in shaded areas. A large number of individuals was found skating against currents in tidal streams of mangroves and an irrigation canal associated with a mangrove plantation at high tide in several collecting sites located in the Central Region. A large population of this species was observed in a small fragmented mangrove at Bangtaboon, Phetchaburi Province. Thorough examination of specimens revealed the two morphological forms of this species. The first form represents majority of specimens collected and perfectly matches the description of *Asclepios annandalei*. The second form was found only at three collecting sites in the Central Region. These two forms can be distinguished in the following manner. The first form has a solid large dark pattern on the meso- and metathorax, whereas the second form has a thin dark stripe on the meso- and metathorax. Additionally, males of the first form have a distinct large tooth in the middle of the profemur, whereas males of the second form have a small tooth in the middle of the profemur. Male genital structures of the two forms are similar to each other. This phenomenon was observed in a previous project in populations in the mangroves in the Andaman Sea [12]. Representatives of both forms will be sent for identification confirmation by the experts. Additionally, specimens of these two forms were kept frozen for further molecular analyses.

Members of *Halobates* were commonly found near shores of marine habitats, nevertheless some species of this genus occur in the open oceans [7]. In this study, *Halobates dianae* was collected from the open water of estuaries associated with large intact mangrove forests in the eastern region (Figure 3). This species was previously collected from the open water of a sea and a river associated with mangroves in the Philippines [21]. A large population of adults and nymphs was observed swiftly moving in a zigzag pattern on the water surface of the open section of a mangrove lined river in this study. Species of *Rheumatometroides* mostly occur in shallow sections of mangroves [4,6] (Figure 3). In this study, individuals commonly occurred in large numbers in shaded areas of channels in mangrove forests, similar to the mesohabitat used by *Asclepios*. Specimens of *Mesovelia vittigera* have been commonly collected from streams, ponds, rice paddies, peat swamps, and blacklight traps in Thailand [12,15]. Although this species commonly inhabits freshwater habitats, it has also been collected from brackish waters in Southeast Asia [8,12,32]. In this study, the species was collected from water margins of mangroves. Species of *Nereivelia* generally hide in crevices of logs filled with air bubbles at the margin of mangrove streams during high tide and come out searching for food during low tide [8,13]. Due to the cryptic behavior and preferred habitat, *Nereivelia murphyi* is rarely collected in Thailand [8,13]. In this study, a single male was captured on a mud flat next to a river in mangrove in Khlong Khao Daeng, Prachaup Khiri Khan Province, Central Region (Figure 3). This species was previously collected at Ranong Province on the Andaman Sea [13]. Therefore, this cryptic species predictably occurs in mangroves of both sides of the peninsula of Thailand. Members of *Xenobates* are commonly found at tidal channels of mangrove forests [14]. In this study, adults and nymphs were commonly found around mangrove trees and boardwalk poles in mangrove forest during high tide and in mud puddles during low tide (Figure 3). *Xenobates argentatus* was distributed throughout the study area, while the other congeners were restricted to a certain region or even a single province.

Although gerromorphans are adapted to live on water surfaces or at margins of aquatic systems, they display a wide variety of habitat preferences [3]. In general, members of marine Gerridae and Veliidae occur on the water surface [3], whereas marine Hydrometridae are found at the banks of mangrove rivers [33]. In this study, each group of Gerromorpha clearly showed the mesohabitat preference pattern throughout the study area (Figure 3). Members of *Halobates* (Gerridae) occurred in the open water of mangrove streams or the shorelines where strong waves and turbulence are present, whereas members of *Asclepios* (Gerridae) and *Rheumatometroides* (Gerridae) were found only in shaded areas on the side of mangrove streams characterized by presence of low wave and turbulence (Figure 3). Species of *Xenobates* (Veliidae) were found near mangrove trees or boardwalk poles away from the shorelines, where they are protected from water movement (Figure 3). Members of Mesoveliidae were collected only from the marginal sections of mangroves (Figure 3). Although there is no clear explanation for the habitat preference of aquatic insects in mangroves, salinity, waves, temperature, predation capability and food availability were suggested as key factors that influence the distribution of semiaquatic gerromorphans in mangroves [34]. The effect of anthropogenic disturbance on insect diversity has received significant attention during the past decade [35–37]. For example, mangroves associated with a higher level of human activities, such as agriculture, urbanization, and tourism, have lower Gerromorpha species richness (e.g., collecting sites 12, 13, 16, 24) in this study. On the other hand, the remote mangroves, with low accessibility were characterized by a higher species richness among the studied insects (e.g., collecting sites 3, 9, 23, 25) (Figure 2A–D).

The biogeographic patterns of Gerromorpha from mangroves in the Central and Eastern regions were not clearly displayed based on the PCA results. This could have been influenced by the distribution patterns of gerromorphans recorded in this study. Most of Gerromorpha species are widely distributed in the study area, such as members of *Xenobates*, the most speciose genus in this study [14], whereas *Nereivelia murphyi* and *Halobates dianae* are two endemic species that have been reported from specific areas [8,21]. The restricted distribution of *Nereivelia murphyi* and *Halobates dianae* could have resulted in clear separation of collecting sites 8, 9, and 25 from other collecting sites of the same region (Figure 5). More specifically, *Nereivelia murphyi* was collected only from collecting site 25 in Prachaup Khiri Khan Province, Central Region, whereas *Halobates dianae* was found at collecting sites 8 and 9 in Chantaburi Province, Eastern Region. Therefore, the clear biogeographic patterns in mangroves may not be displayed based on Gerromorpha in this study due to their wide distribution.

Although the Thai government has been strongly protecting the mangrove areas from deforestation and habitat degradation, the rate of habitat destruction still occurs to an excessive degree [10]. Furthermore, land alteration for agricultural purposes in mangrove areas in Thailand is still taking place at a rapid speed [38]. Human disturbances (e.g., oil spills, sewage, garbage dumps, and pesticide drainage) have been reported to reduce the diversity of animals in Thai mangroves [39]. Thailand is located in Southeast Asia, where the fauna of marine insects is highly diverse [14,33]. Nevertheless, the taxonomic knowledge on aquatic and semiaquatic insects in mangroves of Thailand is still in its beginnings. The results of this study provide fundamental knowledge of the Thai mangrove ecosystem, which is currently undergoing extensive anthropogenic disturbance. Species in this fragile environment may be at risk of exaptation or extinction, and studies such as this are needed to record and monitor species presence and composition for the future.

5. Conclusions

The species of Gerromorpha in mangroves in the central and eastern regions of Thailand were documented for the first time. Although this study focused on mangrove aquatic insects in the small study area in Thailand, nine species of gerromorphans were discovered with four new country records. However, the larger part of remaining Thai mangrove fauna is still unexplored. We suggest including physiochemical water properties and anthropogenic pressures in future research, which might provide a better understanding of the relationship between aquatic insects and land use adjacent to mangroves.

Author Contributions: Conceptualization, A.V. and L.-a.N.; methodology, A.V. and L.-a.N.; software, A.V.; validation, L.-a.N. and A.V.; investigation, L.-a.N.; resources, A.V. and L.-a.N.; specimen curation, A.V. and L.-a.N.; data curation, L.-a.N. and A.V.; writing—original draft preparation, L.-a.N. and A.V.; writing—review and editing, A.V.; funding acquisition, A.V. All authors have read and agreed to the published version of the manuscript.

Funding: This research was funded by the Kasetsart University Research and Development Institute, KURDI (Grant No. K-S (D) 40.58, P-Y (D) 140.60, and FF (KU) 14.64) and the Thailand Science Research and Innovation (RSA6280047).

Institutional Review Board Statement: This research was approved by the Institutional Animal Care and Use Committee, Faculty of Agriculture, Kasetsart University, Thailand, under Project number ACKU59-AGR-009.

Data Availability Statement: Data are available from the corresponding author upon request.

Acknowledgments: We thank Sajeemat Attawanno (Kasetsart University) and Prabseuk Sritipsak (Kasetsart University) for their assistance in the field. We are grateful to Boonsatien Boonsoong (Kasetsart University) for his analysis assistance. We extend our gratitude to the Department of National Parks, Wildlife and Plant Conservation for permission and logistic support to conduct the fieldwork. We would also like to thank the Department of Entomology, Faculty of Agriculture, Kasetsart University, for the facility support. Lastly, we thank Michael L. Ferro (Clemson University), the editors, and the reviewers for their critical comments.

Conflicts of Interest: The authors declare no conflict of interest.

References

1. Andersen, N.M. *The Semiaquatic Bugs (Hemiptera, Gerromorpha)*; Entomonograph; Brill: Klampenborg, Denmark, 1982; Volume 3, p. 445.
2. Vitheepradit, A.; Sites, R.W. A review of *Eotrechus* Kirkaldy (Hemiptera: Heteroptera: Gerridae) of Thailand with descriptions of three new species. *Zootaxa* **2007**, *1478*, 1–19.
3. Andersen, N.M.; Polhemus, J.T. Water-strider (Hemiptera, Gerridae, Veliidae). In *Marine Insects*; Cheng, L., Ed.; North-Holland Publishing Company: Amsterdam, The Netherlands, 1976; pp. 187–224.
4. Chen, P.P.; Nieser, N.; Zettel, H. *The Aquatic and Semi-Aquatic Bugs (Heteroptera: Nepomorpha & Gerromorpha) of Malesia*; Brill: Lei-den-Boston, The Netherlands, 2005; Volume 5, p. 546.
5. Andersen, N.M.; Foster, W.A. Sea skaters of India, Sri Lanka and the Maldives, with a new species and a revised key to Indian Ocean species of *Halobates* and *Asclepios* (Hemiptera, Gerridae). *J. Nat. Hist.* **1992**, *26*, 533–553. [CrossRef]
6. Polhemus, J.T.; Polhemus, D.A. The Trepobatinae (Hemiptera, Gerridae) of New Guinea and surrounding regions, with a review of the World fauna. Part 4. The marine Tribe Stenobatini. *Entomol. Scand.* **1996**, *27*, 279–346. [CrossRef]
7. Andersen, N.M.; Cheng, L. The marine insect *Halobates* (Heteroptera: Gerridae): Biology, adaptations, distribution, and phylogeny. Oceanography and marine biology. *Annu. Rev.* **2004**, *42*, 119–180.
8. Yang, C.M.; Murphy, D. Guide to the aquatic Heteroptera of Singapore and Peninsular Malaysia. 6. Mesoveliidae, with description of a new *Nereivelia* species from Singapore. *Raffles Bull. Zool.* **2011**, *59*, 53–60.
9. Pumijumnong, N. Mangrove forests in Thailand. In *Mangrove Ecosystems of Asia*; Springer: New York, NY, USA, 2014; pp. 61–79.
10. Faridah-Hunum, I.; Latiff, A.; Hakeem, K.R.; Ozturk, M. *Mangrove Ecosystems of Asia: Status, Challenges and Management Strategies*; Springer: New York, NY, USA, 2014; pp. 81–92.
11. Veenakumari, K.; Prashanth, M. A note on the entomofauna of mangrove associates in the Andaman Island (Indian Ocean: India). *J. Nat. Hist.* **2009**, *43*, 807–823. [CrossRef]
12. Sites, R.W.; Vitheepradit, A. Recovery of the freshwater lentic insect fauna in Thailand following the tsunami of 2004. *Raffles Bull. Zool.* **2010**, *58*, 329–348.
13. Polhemus, J.T.; Polhemus, D.A. The new mesoveliid genus and two new species of *Hebrus* (Heteroptera: Mseveliidae, Hebridae) from intertidal habitats in South-east Asia mangrove swamps. *Raffles Bull. Zool.* **1989**, *37*, 73–82.
14. Andersen, N.M. The marine Haloveliinae (Hemiptera: Veliidae) of Singapore, Malaysia and Thailand, with six new species of *Xenobates* Esaki. *Raffles Bull. Zool.* **2000**, *48*, 273–292.
15. Vitheepradit, A. The Semiaquatic Heteroptera (Gerromorpha) of Thailand: Faunistics, Biogeography, and Phylogeography. Ph.D. Thesis, University of Missouri, Columbia, MO, USA, 2008.
16. Andersen, N.M.; Weir, T.A. Mesoveliidae, Hebridae, and Hydrometridae of Australia (Hemiptera: Heteroptera: Gerromorpha), with a reanalysis of the phylogeny of semi-aquatic bugs. *Invertebr. Syst.* **2004**, *18*, 467–522. [CrossRef]
17. McCune, B.; Mefford, M.J. *PC-ORD Multivariate Analysis of Ecological Data, Version 6*; MjM Software Design: Gleneden Beach, OR, USA, 2011.
18. Chen, P.P.; Zettel, H. Key to the genera and subgenera of Gerridae (Gerromorpha) of Thailand and adjacent countries, with a check-list of species known from Thailand. *Amemboa* **1998**, *2*, 24–41.
19. Distant, W.L. A few undescribed Rhynchota. *Ann. Mag. Nat.* **1915**, *15*, 503–507. [CrossRef]
20. Cheng, L.; Yang, C.M.; Andersen, N.M. Guide to the aquatic Heteroptera of Singapore and Peninsular Malaysia. I. Gerridae and Hermatobatidae. *Raffles Bull. Zool.* **2001**, *49*, 129–148.
21. Zettel, H. *Halobates dianae* (Heteroptera: Gerridae), a new sea skater from the Philippines. *Linz. Boil. Beitr.* **2001**, *33*, 1097–1102.
22. Polhemus, J.T.; Cheng, L. Notes on marine water-striders with descriptions of new species. *Pac. Insects* **1982**, *24*, 219–227.

23. Damgaard, J.; Moreira, F.F.F.; Hayashi, M.; Weir, T.A.; Zettel, H. Molecular phylogeny of the pond treaders (Insecta: Hemiptera: Heteroptera: Mesoveliidae), discussion of the fossil record and a checklist of species assigned to the family. *Insects Syst. Evol.* **2012**, *43*, 175–212. [CrossRef]
24. Andersen, N.M.; Polhemus, J.T. Four new genera of Mesoveliidae (Hemiptera, Gerromorpha) and the phylogeny and classification of the family. *Insects Syst. Evol.* **1980**, *11*, 369–392. [CrossRef]
25. Andersen, N.M.; Polhemus, D.A. A new genus of terrestrial Mesoveliidae from the Seychelles (Hemiptera, Gerromorpha). *J. N. Y. Entomol. Soc.* **2003**, *111*, 12–21. [CrossRef]
26. Floriano, C.F.B.; Moreira, F.F.F.; Bispo, P.C.; Morales, L.; Molano-Rendon, F. A new species of *Mesovelia* (Heteroptera: Gerromorpha: Mesoveliidae) from South America, with identification key and note on Colombian species. *Zootaxa* **2016**, *4175*, 345–352. [CrossRef]
27. Polhemus, J.T.; Polhemus, D.A. The genus *Mesovelia* Mulsant & Rey in New Guinea (Heteroptera: Mesoveliidae). *J. N. Y. Entomolo. Soc.* **2000**, *108*, 205–230.
28. Horváth, G. Hemipteres nouveaux de l'Europe et des pays limitrophes. *Revue Entomol.* **1895**, *14*, 152–165.
29. Hecher, C. Key to the genera of Veliidae (Gerromorpha) of Thailand and adjacent countries, with a check-list of genera and species known from Thailand. *Amemboa* **1998**, *2*, 3–7.
30. Andersen, N.M. The coral bugs, genus *Halovelia* Bergroth (Hemiptera, Veliidae). I. History, classification, and taxonomy of species except the *H. Malaya*-group. *Entomol. Scand.* **1989**, *20*, 75–120. [CrossRef]
31. Andersen, N.M. Cladistic Biogeography of Marine Waterstrider (Insecta, Hemiptera) in the Indo-Pacific. *Aust. Syst. Bot.* **1991**, *4*, 151–163.
32. Zettel, H.; Tran, A.D. First inventory of the water bugs (Heteroptera: Nepomorpha: Gerromorpha) of Langkawi Island, Kedah, Malaysia. *Raffles Bull. Zool.* **2009**, *57*, 279–295.
33. Andersen, N.M. The evolution of marine insects: Phylogenetic, ecological and geographical aspects of species diversity in marine water striders. *Ecography* **1999**, *22*, 98–111. [CrossRef]
34. Conjard, S.; Garrouste, R.; Gustave, S.D.; Gros, O. Mangrove semiaquatic bugs (Hemiptera: Gerridea) from Guadeloupe in Lesser Antilles: First records and new data on species distribution. *Aquat. Insects* **2021**, *42*, 239–246. [CrossRef]
35. Ramírez, A.; Engman, A.; Rosas, K.G.; Perez-Reyes, O.; Martinó-Cardona, D.M. Urban impacts on tropical island streams: Some key aspects influencing ecosystem response. *Urban Ecosyst.* **2012**, *15*, 315–325. [CrossRef]
36. Dolný, A.; Harabiš, F.; Bárta, D.; Lhota, S.; Drozd, P. Aquatic insects indicate terrestrial habitat degradation: Changes in taxonomical structure and functional diversity of dragonflies in tropical rainforest of East Kalimantan. *Trop. Zool.* **2012**, *25*, 141–157. [CrossRef]
37. Hepp, L.U.; Restello, R.M.; Milesi, S.V.; Biasi, C.; Molozzi, J. Distribution of aquatic insects in urban headwater streams. *Acta Limnol. Bras.* **2013**, *25*, 1–9. [CrossRef]
38. Khemnark, C. Ecology and Management of Mangrove Restoration and Regeneration in East and Southeast Asia. In Proceedings of the IV Ecotone, Surat Thani, Thailand, 18–22 January 1995; Amarin Co. Ltd.: Bangkok, Thailand, 1995; p. 339.
39. Aksornkoae, S.; Paphavasit, N.; Wattayakorn, G. Mangroves of Thailand: Present status of conservation, use and management. In *The Economic and Environmental Values of Mangrove Forests and Their Present State of Conservation in the South-East Asia/Pacific Regions*; Clough, B.F., Ed.; International Society for Mangrove Ecosystems: Okinawa, Japan, 1993; pp. 83–133.

 diversity

MDPI

Article

Water Quality Analysis in a Subtropical River with an Adapted Biomonitoring Working Party (BMWP) Index

Guillermo Magallón Ortega [1], Carlos Escalera Gallardo [1], Eugenia López-López [2], Jacinto Elías Sedeño-Díaz [3], Martín López Hernández [4], Miriam Arroyo-Damián [5] and Rodrigo Moncayo-Estrada [6,*]

1 Instituto Politécnico Nacional, Centro Interdisciplinario de Investigación para el Desarrollo Integral Regional, Unidad Michoacán, Justo Sierra 28, Jiquilpan 59510, Mexico; gmagallono1700@alumno.ipn.mx (G.M.O.); cescalera@ipn.mx (C.E.G.)
2 Instituto Politécnico Nacional, Escuela Nacional de Ciencias Biológicas, Prolongación de Carpio y Plan de Ayala, Col. Sto. Tomás, Ciudad de México 11340, Mexico; eulopez@ipn.mx
3 Instituto Politécnico Nacional, Coordinación Politécnica para la Sustentabilidad, Av. Wilfrido Massieu esq. IPN, Col. Zacatenco, Ciudad de México 07738, Mexico; jsedeno@ipn.mx
4 Universidad Nacional Autónoma de México, Instituto de Ciencias del Mar y Limnología, Circuito Exterior S/N, Ciudad Universitaria, Ciudad de México 04510, Mexico; martinl@cmarl.unam.mx
5 Universidad de La Ciénega del Estado de Michoacán de Ocampo, Avenida Universidad 3000, Col. Lomas de la Universidad, Sahuayo 59103, Mexico; marroyo@ucienegam.edu.mx
6 Instituto Politécnico Nacional, Centro Interdisciplinario de Ciencias Marinas, Avenida Instituto Politécnico Nacional S/N, Col. Playa de Santa Rita, La Paz 23096, Mexico
* Correspondence: rmoncayo@ipn.mx; Tel.: +52-6121234658

Citation: Magallón Ortega, G.; Escalera Gallardo, C.; López-López, E.; Sedeño-Díaz, J.E.; López Hernández, M.; Arroyo-Damián, M.; Moncayo-Estrada, R. Water Quality Analysis in a Subtropical River with an Adapted Biomonitoring Working Party (BMWP) Index. *Diversity* **2021**, *13*, 606. https://doi.org/10.3390/d13110606

Academic Editors: Marina Vilenica, Zohar Yanai, Laurent Vuataz and Michael Wink

Received: 14 September 2021
Accepted: 12 November 2021
Published: 22 November 2021

Publisher's Note: MDPI stays neutral with regard to jurisdictional claims in published maps and institutional affiliations.

Abstract: Subtropical rivers in developing countries often lack adequate monitoring, which makes it difficult to comprehensively determine their water quality when faced with different anthropic impacts. There are no proper protocols in the regulations to incorporate indicators and adapt them to different biogeographic regions, limiting the potential success of conservation and restoration of river ecosystems. This study proposes implementing macroinvertebrates as bioindicators of water quality in river ecosystems, and modifying the calibration of the widely used Biomonitoring Working Party (BMWP) index for its adaptation in a subtropical river. The Duero River, Mexico, was used as an example in this study. Data were explored with multivariate statistics, and the water quality and habitat values were averaged to obtain the families' bioindication values and the index categories. The BMWP adequately described a deterioration gradient from the origin to the river mouth (from fair to extremely polluted), with some intermediate recovery points related to the presence of springs. Its performance was compared with other biological indices and exhibited a positive relationship with all of them. In addition, how BMWP changed over time was analyzed by examining previous samples, and highlighted increased river deterioration over time. A calibrated BMWP will allow for long-term monitoring at a low cost.

Keywords: developing country; multivariate statistics; bioindication value; index scores; WQI; HQI; EPT

1. Introduction

One of the most commonly used biological indices to monitor the quality of lotic aquatic ecosystems is the biological monitoring working party (BMWP) [1–5]. The BMWP is a biotic approach because it includes taxonomic groups, considering their sensitivity or tolerance to pollution, and both aspects are incorporated into an index [3]. The index describes and analyzes the macroinvertebrate community at the taxonomic family level [6]. The characteristics that make macroinvertebrates good bioindicators are their wide distribution, limited mobility, numerical abundance, sensitivity and response to distinct types of environmental conditions and stressors, and ease of finding, quantifying and standardizing them [7]. The application of the BMWP and other indices that use bioindicator organisms provides complementary information on biotic and abiotic conditions of lotic ecosystems in addition to traditional monitoring techniques (physical, chemical, and bacteriological variables). The development of such indices, arises from the need to systematically reflect changes in riparian and fluvial ecosystems and to express the environmental and habitat factors in an integral way, with the expectation of long-term management [8].

The BMWP assigns a bioindication value for each family and is the minimum perceived value of the tolerance of macroinvertebrates to organic contamination. It ranges from 1 (very tolerant) to 10 (very sensitive). Consequently, the total index score for a sample is defined as the sum of the minimum perceived value of the tolerance of all the families present [9]. In general, if the sum is approximate or greater than 100, the rivers and streams are deemed healthy, but if the sum is less than 10 they are considered highly polluted [6]. The BMWP has been used in various regions worldwide, and has been adapted, mainly due to the absence of some taxa used in the original version and the presence of others not initially included [10,11]. In addition, the modifications include the combination of families due to taxonomic difficulties and changes in some families' values, which is sometimes related to their frequency of occurrence and the degree of saprobity [1,12]. A more recent calibration and validation of the index has been proposed to relate it more closely to a river's specific environmental characteristics [13].

Despite these advantages, there are also some disadvantages. The extrapolation of the BMWP in large regions or countries results in poor generalizations because the sensitivity of some organisms changes over space and time [14]. There is subjectivity in the allocating tolerance intervals because fine taxonomic levels can bias the index, since the macroinvertebrate genera and species within the families have different sensitivities to distinct types of environmental degradation [9]. Therefore, a calibration process has been implemented, which was also suggested in the original proposal to establish suitable bioindication values [4,15]. Accordingly, once an adequate monitoring scheme is established and the protocol for obtaining and reviewing samples in the whole system is developed (e.g., Cornejo et al. [16]), with an adequately adapted index proposal (e.g., Ruiz-Picos et al. [17]), then the results will better reflect water quality condition of the aquatic ecosystem. Particularly, it is important to implement frequent monitoring and measure the BMWP regularly to satisfactorily describe the ecosystem conditions.

In this context, different countries, mainly in North America, Europe, Asia, and Australasia, have adopted the use of macroinvertebrates in their environmental regulations and have long-term monitoring programs [18–21]. More recently, some developing countries have also assumed this approach [16,22,23], but several nations still lack official biomonitoring programs. This deficiency is not necessarily related to the absence of biological community studies, but to the inadequate standardization and sporadic monitoring depending on the river typology, habitat type, and time of year [24].

Although a protocol has been developed to sample macroinvertebrates in Mexico and includes the application of the BMWP to be adopted in the official normative about ecological flow (NMX-AA-159-SCFI-2012, [25]), it concludes that the bioindication values and the water quality ranges should be adapted according to the sampled system. This adaptation should consider the different biogeographic, physiographic, environmental, and climatic characteristics as well as the anthropogenic impacts that affect aquatic ecosystems in the country. Some isolated studies in tropical systems of Mexico have already implemented the BMWP using the values and ranges from other places [26,27] or implemented the calibration and validation process according to environmental characteristics [17].

We focused on the subtropical Duero River because it is located within a different climate region than the previous studies and belongs to the so-called Mexican Transition Zone, which joins the biogeographical Nearctic and Neotropical regions [28]. This biotic crossroad is characterized by the presence of several hydric resources with high biodiversity and endemicity and is also related to ancient prehispanic human intervention. Additionally, previous studies on the Duero River have analyzed water quality and entomofauna during different time periods since the 1980s [29–31]. In this river, environmental conditions have shown contrasting results, with some upstream localities and sites related to springs showing good water quality (e.g., control sites); in contrast, other localities experience different impacts with high levels of organic and bacteriological contamination, and are associated with urban development and agricultural and livestock production [32]. Moreover, another study included the addition of other bioindicators, such as fish, to describe the system's biotic integrity in specific decades [33]. In this context, the current study's main aim was to develop a calibrated BMWP according to the river environmental conditions and compare the results in different periods and with other biological indices. We hypothesized that (1) the ecosystem presents a general spatial gradient from least polluted at the origin to severely polluted at the mouth, related to human impacts, and contains intermediate recoveries by occasional tributaries from the springs, (2) the category of the BMWP scores does not necessarily coincide with the category of the index of water quality and the index of habitat quality because some habitat modifications (not included in the index of water quality) and pollution (not included in the index of habitat quality) affect the macroinvertebrates distribution and abundance directly, and (3) the river has experienced a degradation in water quality over time.

2. Materials and Methods

2.1. Study Area

The Duero River is located on the Central Plateau of Mexico and it is an affluent of the Lerma River, which finally enters Chapala Lake, the largest natural lake in the country. The river has a catchment area of 2531 km^2, a length of 85 km, and a total annual flow of 457.8 hm^3 [34]. The source of the river is at 1860 m.a.s.l. and has an altitude change of 340 m [35]. The catchment of the Duero River is divided into four sub basins (Figure 1). The upper sub basin (Cañada) consists of volcanic rocks (andesites, basalts), and it is the primary recharge zone, while in the middle and lower zones there are three sub basins containing valleys (Guadalupe, Zamora, and Ciénega) and where vertisol soils predominate, supporting agricultural land use (143,333 ha) [31]. Although there are 52 springs, and theoretically there is enough water available to meet the needs of different users, this is not the case due to inadequate management and distribution; consequently, the Duero River has a deficit of 44.7 hm^3 s^{-1} [34].

2.2. Sampling and Data Analysis

We selected 14 sites, some of which have good water and habitat conditions, whereas others were affected by anthropogenic impacts associated with principal human activities and urban development (Figure 1). Most of the sites are located in the main river course, but three sites are near springs in tributaries, La Toma (site 3), the national park Camécuaro Lake (site 6), and Verduzco (site 9). Water and macroinvertebrates were sampled under different flow conditions: low velocity in the dry season (May), high in rainy season, and intermediate in transition season (November) in 2019.

In the three seasons, we analyzed water parameters at each site related to the organic load (dissolved oxygen, five-day biochemical oxygen demand [BOD_5], and chemical oxygen demand [COD]), nutrients (nitrate [NO_3], ammonium-nitrogen [NH_4]), ionic composition (salinity, conductivity, total dissolved solids, pH, and chloride), physical aspects (temperaturé, turbidity, river discharge, depth, and transparency), and bacterial load (total and fecal coliforms, and *Escherichia coli*). The parameters were analyzed according to the methods of the American Public Health Association [36] and the Mexican Norms (NMX-AA-042-SCFI-2015; NMX-AA-113-SCFI-2012).

We used the multihabitat protocol proposed by Cornejo et al. [16] to sample macroinvertebrates. In a 100 m stretch of the river or tributary, we sampled the main aquatic habitats (e.g., bed substrate, aquatic and riparian vegetation, areas of different water characteristics). The number of samples per habitat depended on the percentage coverture. Samples covered an area of approximately 1 m^2, and the organisms were captured with two nets; in fast-flowing riffles and pools, we used a kick net and for floating, submerged, and riverine vegetation a type-D net (both with a mesh size of 500 μm). We preserved the macroinvertebrates collected with 70% alcohol. Organisms were identified to the family level according to different taxonomical keys [37–39].

Figure 1. Duero River watershed with the location of the sub basins and sampling sites: (1) Kunio, (2) Tacuro, (3) La Toma, (4) El Rastro (5) Etúcuaro, (6) Camécuaro, (7) Antes Central de Abastos, (8) Después Central de Abastos, (9) Verduzco, (10) Estanzuela, (11) San Cristóbal, (12) Capulín, (13) Cumuato, and (14) Ibarra.

Some of the water parameters were used to determine the water quality index developed by the National Sanitation Foundation (NSFWQI) [40]. The habitat quality index (HQI) related to the physical structure of the river and its surroundings was to describe each sampling event following the criteria proposed by Barbour et al. [41]. The biological quality of the river was evaluated with the use of macroinvertebrates according to three indices: the percentage of Ephemeroptera (minus the Baetidae family), Plecoptera, Trichoptera individuals index (EPT-B) [42], the Hill's numbers that describe biological diversity [43], and the BMWP adapted to Mexican rivers [13]. The EPT index was modified because Masese and Raburu [42] proposed the subtraction of the Baetidae, Caenidae, and Hydropsychidae families; however, we only subtracted Baetidae because this family was the only one that showed low bioindication values in different tropical areas [1,12,16,17]. Hill numbers describe the effective number of species based on a sampling framework with rarefaction and extrapolation methods and derives species richness (q = 0), Shannon's entropy index exponential (q = 1), and the inverse of Simpson's concentration index (q = 2) [43].

2.3. Index Calibration and Comparison

Although the BMWP has been used in different rivers in Mexico, Ruiz-Picos et al. [13] emphasize the necessity to adapt the index, not only to the macroinvertebrate families present in the region, but also to the specific environmental characteristics of the ecosystem studied. They propose a complete protocol to calibrate the BMWP, which includes four main steps. First, the physicochemical quality index (Pcq) is calculated. From the average physicochemical data set (seasons), the variables that better describe the sites are chosen using statistical methods and according to the parameter's range and the way it describes impacts. The water parameters selected are standardized and combined on a scale of 0 to 10, from poor to better water quality, and sites are arranged within this scale. Second, the bioindication value is obtained. The macroinvertebrate abundance is structured into five classes from zero (no individuals) to 5 (>100 individuals). Every family abundance class is assigned to the different Pcq intervals according to their presence or absence in the sites. The bioindication value of every family is then derived from the 5th percentile of each abundance class. Third, these bioindication values are replaced at every site and in every study period for each family found, and they are summed to evaluate the BMWP index. Fourth, the general scale of the index in the whole system is established from the median and tenth percentile values of the sites with better water quality. Above the median is excellent quality, in the tenth percentile is good quality, and below this value, the amount is divided by four to get the rest of the categories: regular, polluted, very polluted, and extremely polluted (Figure 2).

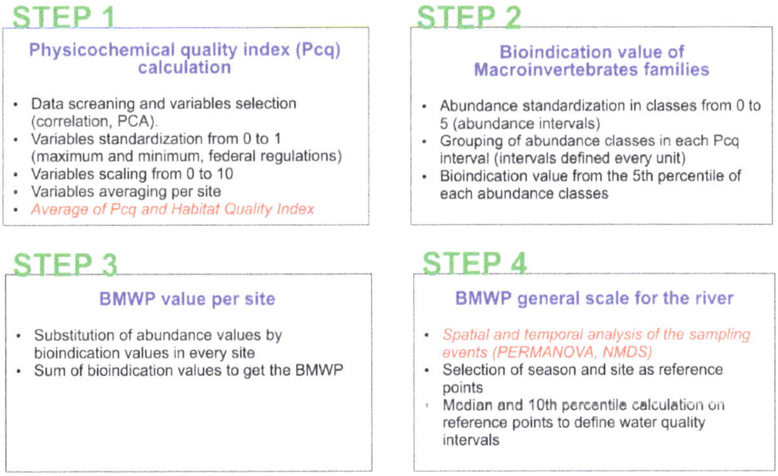

STEP 1

Physicochemical quality index (Pcq) calculation

- Data screaning and variables selection (correlation, PCA).
- Variables standardization from 0 to 1 (maximum and minimum, federal regulations)
- Variables scaling from 0 to 10
- Variables averaging per site
- *Average of Pcq and Habitat Quality Index*

STEP 2

Bioindication value of Macroinvertebrates families

- Abundance standardization in classes from 0 to 5 (abundance intervals)
- Grouping of abundance classes in each Pcq interval (intervals defined every unit)
- Bioindication value from the 5th percentile of each abundance classes

STEP 3

BMWP value per site

- Substitution of abundance values by bioindication values in every site
- Sum of bioindication values to get the BMWP

STEP 4

BMWP general scale for the river

- *Spatial and temporal analysis of the sampling events (PERMANOVA, NMDS)*
- Selection of season and site as reference points
- Median and 10th percentile calculation on reference points to define water quality intervals

Figure 2. BMWP calibration protocol adapted from Ruiz-Picos et al. [13]. In red and italics, the modification implemented in the present study.

We followed this protocol to obtain BMWP-calibrated values for the Duero River, but we also implemented two modifications. In the first step, we average the Pcq calculated with the HQI values to combine the information from both indices. The HQI was rescaled from 0 to 10 to have compatible values with the Pcq because the original scale was from 0 to 200 [41]. We assumed this initiative because the physical and chemical parameters of the water do not always entirely reflect the characteristics of the sites (i.e., good water quality but bad habitat conditions). Consequently, this index reflected not only the organic pollution but also the habitat quality in every site. In the fourth step, from the multivariate analysis of the differences and trends in the environmental and community variables, the season and sites were selected to establish the general BMWP scale for the river. We used the results from the calibration to compare different sites at distinct time periods (1984 and 1986 [29]; and 2013 [30]). Moreover, we related the BMWP with the physical and chemical parameters to identify which of them better explain the index variations, and with other biotic indices, such as the EPT-B, and Hill's numbers [43], to identify the similar or dissimilar response to pollution [6,44].

2.4. Statistical Analyses

From the total samples (N = 42), two methods were used to select the environmental variables that best described the sites and to obtain the Pcq index. First, Spearman's rho correlation was used to identify redundancy and collinearity among the variables ($\geq$0.9 and $p < 0.0001$) [45]. The selection of one of the variables with similar contributions was made, on the one hand, according to their magnitudes and, on the other, to reduce the influence of specific water characteristics in the subsequent analyses (several variables measuring similar aspects, like ion composition). Second, we performed a principal component analysis (PCA) to reduce the number of variables and to identify those variables that have a higher contribution to the variance in the dataset (a threshold to the chosen principal components > 75% and a variable loading > 0.7). From 15 variables, 5 were excluded: salinity, total dissolved solids, chloride, depth, and percentage saturation of DO.

We used a distance-based permutational multivariate analysis of variance (PERMANOVA) [46] in a two-way factorial design to test differences among sub basins and seasons. The information was log-transformed to reduce the effect of distinct units and magnitudes, and then we used the Euclidian dissimilarity coefficient in the environmental matrix; in the community dataset we used the Bray–Curtis dissimilarity due to the high frequency of zeros [47]. Prior to the PERMANOVA and because the data had an unbalanced design (different number of sampled sites per sub basin), we ran a method to test the homogeneity of multivariate dispersions (PERMDISP), since heterogeneity could heavily influence the PERMANOVA results [48]. In both datasets, the PERMDISP showed no significant difference: environment (sub basins $F = 1.11$ and $p = 0.35$; Seasons $F = 1.57$ and $p = 0.21$) and community (sub basins $F = 2.58$ and $p = 0.07$; Seasons $F = 0.02$ and $p = 0.99$).

To find the predominant trends in the dataset and to identify the spatial ordination of the sites in different seasons in relation to the macroinvertebrate families, we used nonparametric multidimensional scaling (NMDS) ordination with Bray–Curtis dissimilarity. We calculated three-dimensional solutions from 250 random starts of real data, with up to 1000 iterations, to evaluate stability, and obtained a final stress of 0.14. In the plot, we incorporated the environmental variables; although they were not part of the NMDS analysis, they were integrated as vectors, scaled by their correlation with the axes, and their significance was assessed using permutations [49].

The BMWP, the physicochemical parameters, and the biological indices (EPT-B, Hill's numbers) were correlated using Spearman's test ($p < 0.05$). The Kruskal–Wallis test was used to compare historical BMWP values. The Spearman correlations were computed using the 'Hmisc' package (v. 4.5-0, [50]); PCA, PERMANOVA, PERMDISP, and NMDS were computed using the vegan package (v. 2.5-7, [49]), and the Kruskal–Wallis test was performed with the function 'kruskal.test', all in the R language [51].

3. Results

We found 62 families at the different sites and in different seasons (Table A1). The most frequent families were Baetidae, Chironomidae, Gammaridae, and Lumbriculidae. The most abundant families were Baetidae, Chironomidae, Gammaridae, and Lymocytheridae. The PERMANOVA test for the physicochemical variables showed differences in the sub basins ($F = 5.24$, $p = 0.0005$) and seasons ($F = 6.46$, $p = 0.0006$), but not the interaction of both factors ($F = 1.92$, $p = 0.065$). The Ciénega sub basin in the lower portion of the basin differed spatially from all the others (with Cañada: $F = 15.84$, $p = 0.006$; with Guadalupe: $F = 8.19$, $p = 0.006$; and with Zamora: $F = 7.05$, $p = 0.012$). Only September and November showed a temporally significant difference ($F = 8.49$, $p = 0.003$). In the case of the macroinvertebrate families' abundances, we found differences among the sub basins ($F = 2.56$, $p = 0.00001$) but not among the seasons ($F = 1.25$, $p = 0.18$) or the interaction of both ($F = 0.86$, $p = 0.8$).

Zamora differed from the other zones but Cañada, which is located in the upper portion of the basin (with Guadalupe: $F = 3.49$, $p = 0.006$; with Ciénega: $F = 3.75$, $p = 0.006$) and Cañada only with Ciénega ($F = 2.5$, $p = 0.018$; Figure 1).

The NMDS ordination showed an indirect gradient, and the first axis separated in the left to right direction, with sites reflecting better (higher oxygen and transparency, and the presence of more sensitive families) to worse conditions (higher Turbidity, BOD_5, nitrogen nutrients, and mostly tolerant families; Figure 3). In this context, according to the amount of information, we duplicated the plot in the same figure, emphasizing this distinction. To the left, we mainly found that most sites in the different seasons related to the origin of the river in the Cañada sub basin (Kunio, Tacuro, La Toma, but El Rastro and Etúcuaro) or to the presence of springs (Verduzco and Camécuaro). On the right side of the plot, the sites were mainly located near the river mouth in the Ciénega sub basin (Estanzuela, Capulín, San Cristóbal, Cumuato, but Ibarra). The environmental variables which most correlated to the axes and that showed significance using permutations were temperature ($r^2 = 0.17$, $p = 0.04$), conductivity ($r^2 = 0.19$, $p = 0.001$), dissolved oxygen ($r^2 = 0.34$, $p = 0.001$), ammonium ($r^2 = 0.34$, $p = 0.001$), nitrate ($r^2 = 0.14$, $p = 0.001$), turbidity ($r^2 = 0.47$, $p = 0.001$), and BOD_5 ($r^2 = 0.18$, $p = 0.02$).

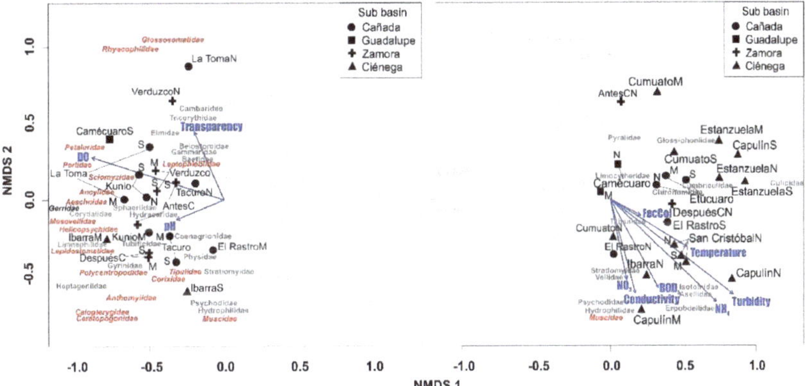

Figure 3. NMDS ordination axes 1 and 2 of the Duero River, including sampling sites within sub-basins, macroinvertebrates families, and environmental variables (stress = 0.14). The same plot is divided into two parts to describe a trend from good (**left**) to poor (**right**) conditions. After the site's name, the capital letter represents the season (M = May, S = September, and N = November) AntesC = Antes Central de Abastos and DespuésC = Después Central de Abastos. DO = Dissolved Oxygen, FecCol = Fecal Coliforms. Family names in red have high bioindication values (6 to 9), and those in gray have low values (3 to 5).

The water parameters showed a general increasing trend in turbidity, BOD_5, nitrogen nutrients, conductivity, and temperature, but dissolved oxygen and transparency, which were also related to the indirect gradient in the first axis. In addition, fluctuations in the physicochemical parameters along the river are common because of the presence of springs in the different sub basins, aspect that is evident by low values of turbidity, BOD_5, and nitrogen nutrients (Table 1). Of particular note are the high values of fecal coliform bacteria present at most sites (average 9488–7,334,062 MPN/100 mL), which indicates the discharge of wastewater without treatment, and even very high values were found in one of the springs (Camécuaro from 19,000 to 5,333,333 MPN/100 mL).

Table 1. Three-month average of chemical and bacteriological parameters recorded in the fourteen sites of the Duero river. The Grey line indicates sites related to springs. ColFec = fecal coliforms bacteria, MPN = most probable number, and AntesC = Antes Central de Abastos and DespuésC = Después Central de Abastos.

Sites	Temperature °C	Conductivity µS cm^{-1} 20 °C	pH	Dissolved Oxygen mg L^{-1} O_2	NH^+_4-N mg L^{-1}	NO_3-N mg L^{-1}	Turbidity UTM	Transparency Proportion	BOD_5 mg L^{-1} O_2	Discharge m s^{-1}	ColFec MPN 100 mL^{-1}
Kuinio	16.8	180.4	6.6	9.3	1.2	1.4	0.2	1	0.4	5.4	277,817.0
Tacuro	17.3	184.6	7.3	9.4	1.0	2.0	5.9	1	1.6	5.8	700,851.1
Toma	17.7	163.9	6.7	9.0	2.2	1.0	2.5	1	0.6	1.1	11,122.6
Rastro	18.4	197.9	6.9	8.4	3.1	5.8	23.0	1	25.6	3.3	57,037.2
Etúcuaro	19.0	199.4	6.9	8.3	1.8	1.9	19.8	1	15.4	4.2	80,601.1
Camécuaro	20.8	196.0	7.1	8.0	1.6	0.8	5.0	1	1.2	9.9	2,061,888.9
AntesC	20.6	149.9	7.4	8.3	1.8	1.2	29.2	0.9	17.2	2.3	9488.9
DespuesC	20.5	250.7	7.5	8.9	2.4	2.4	68.3	0.7	27.1	2.3	220,144.4
Verduzco	21.4	175.9	7.7	9.7	1.4	1.2	14.7	1	1.4	6.6	57,466.7
Estanzuela	21.9	330.5	6.8	1.3	6.1	1.6	63.2	0.2	20.8	22.1	7,334,062.2
San Cristóbal	23.0	353.7	6.9	2.7	7.3	1.8	1033.9	0.2	18.1	29.4	264,494.4
Capulín	23.5	371.1	6.9	1.2	13.0	1.3	2003.7	0.3	20.2	0	17,533.3
Cumuato	25.6	376.5	7.4	2.5	4.4	0.8	44.4	0.3	22.6	0	107,555.6
Ibarra	26.5	599.5	7.8	4.7	10.0	1.5	55.2	0.1	27.5	0	100,866.7

To calibrate the BMWP for the Duero River, each family's bioindication scores were obtained, representing the minimum tolerance value to organic pollution (Table 2) [17]. Despite the scale ranging from 1 to 10, no families were classified as a '1' because no site was highly polluted with few macroinvertebrate families, or as a '10' due to a lack of sites without apparent alteration. Additionally, the general quality categories for the index were defined. Temporally, we decided to establish this general BMWP scale of the river according to the measurements from the dry season (May) because it does not differ from the other seasons in the physicochemical and macroinvertebrate variables according to the PERMANOVA results, and it had a continuous moderate flow and stable conditions, unlike the rainy or intermediate seasons. Spatially, we selected four sites with better water quality, from which the median and tenth percentile were obtained. These sites were located at different parts of the river: two represent small creeks from springs (La Toma and Verduzco), and the other two were located in the main river with little contamination (Kunio and Antes Central de Abastos). Above the median, the water quality was deemed excellent, and between the tenth percentile and the median, the water quality was deemed good. Then, the tenth percentile is divided by four to obtain the rest of the water quality categories: regular, polluted, very polluted, and extremely polluted (Table 3) [52].

Table 2. Bioindication values were modified and adapted to the different families found in the Duero river as part of the BMWP calibration.

Families	Score
Calamoceratidae, Corydalidae, Leptoceridae, Naucoridae, Nepidae, and Petaluridae	9
Anthomyiidae, Leptophlebidae, Perlidae, Polycentropodidae, Rhyacophilidae, and Sciomyzidae	8
Aeschnidae, Caenidae, Gerridae, Glossosomatidae, Lepidostomatidae, Mesoveliidae, and Syrphidae	7
Ancylidae, Athericidae, Calopterygidae, Corixidae, Helicopsychidae, Muscidae, and Tipulidae	6
Hydropsychidae, Tabanidae, and Tricorythidae	5
Agriidae, Baetidae, Cordulegastridae, Culicidae, Elmidae, Empididae, Glossiphoniidae, Pyralidae, Simuliidae, and Stratiomyidae	4
Asellidae, Belostomidae, Cambaridae, Ceratopogonidae, Chironomidae, Coenagrionidae, Dryopidae, Dytiscidae, Erpobdellidae, Gammaridae, Gomphidae, Gyrinidae, Heleidae, Heptageniidae, Herpobdellidae, Hygrobatidae, Hydrophilidae, Hydroptilidae, Isotomidae, Lestidae, Libellulidae, Limnephilidae, Limnocytheridae, Lumbriculidae, Macroveliidae, Margaritiferidae, Nematodos, Notonectidae, Physidae, Planariidae, Planorbidae, Pseudothelphusidae, Psychodidae, Sphaeriidae, Tubificidae, and Veliidae	3

Table 3. Water quality categories of the Duero river from the BMWP.

Statistic	Range	Quality
Median	≥126	Excellent
10th Percentile	97–125	Good
3/4 10th Percentile	73–96	Regular
1/2 10th Percentile	49–72	Polluted
1/4 10th Percentile	24–48	Very polluted
	<24	Extremely polluted

The water, habitat, and biological quality described by the different indices during the dry season are presented in Table 4. No sites showed good conditions, and most were regular, but two were extremely polluted. Regarding the water quality variables, the BMWP had a positive correlation with dissolved oxygen (Spearman = 0.87, $p = 0.00001$) and a negative correlation with temperature (Spearman = −0.6, $p = 0.02$), conductivity (Spearman = −0.57, $p = 0.03$), and ammonium (Spearman = −0.64, $p = 0.01$). In the correlation of the BMWP with the biological indices, we found positive correlation with the EPT (Spearman = 0.67, $p = 0.008$) and the Hill numbers of Shannon's entropy index (Spearman = 0.88, $p = 0.00001$) and the inverse of Simpson's index (Spearman = 0.88, $p = 0.00001$). When we compared the historical BMWP values to try to identify changes over time, the analysis did not show a statistically significant difference among the years ($\chi^2 = 7.54$, $p = 0.11$). However, when plotting the values at the different sites, sites at the origin of the river tended to decrease their water quality in recent years (Figure 4).

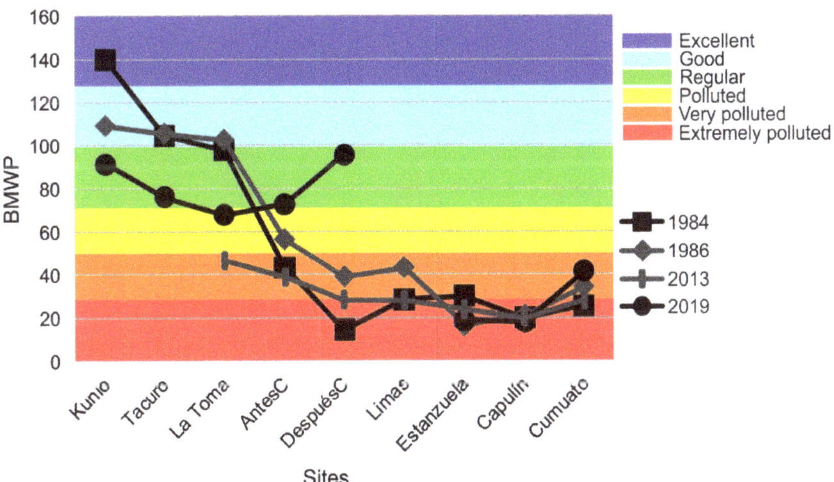

Figure 4. Comparison of the BMWP in different years in the Duero river. Sites are presented in order from upstream to downstream. AntesC = Antes Central de Abastos and DespuésC = Después Central de Abastos. BMWP scores shown in color.

Table 4. Average values of the Habitat Quality Index (HQI), the Water Quality Index (NSFWQI), the Ephemeroptera, Plecoptera, Trichoptera index minus Baetidae (EPT-B), and the BMWP, including their interpretation. AntesC = Antes Central de Abastos y DespuésC = Después Central de Abastos. The Grey line indicates sites with contrasting interpretation of the indices.

Sites	HQI	Significance	NSFWQI	Significance	EPT-B	BMWP	Significance
Kuinio	16	Optimal	84	Good	7	91	Regular
Tacuro	15	Suboptimal	77	Good	3	76	Regular
La Toma	17	Optimal	82	Good	4	68	Polluted
El Rastro	11	Suboptimal	63	Medium	0.1	63	Polluted

Table 4. *Cont.*

Sites	HQI	Significance	NSFWQI	Significance	EPT-B	BMWP	Significance
Etúcuaro	8	Marginal	64	Medium	2	33	Very polluted
Camécuaro	17	Optimal	73	Good	1	37	Very polluted
AntesC	12	Suboptimal	68	Medium	14	73	Regular
DespuésC	12	Suboptimal	62	Medium	2	93	Regular
Verduzco	16	Optimal	74	Good	35	74	Regular
Estanzuela	7	Marginal	44	Bad	0.1	19	Extremely polluted
San Cristóbal	6	Marginal	49	Bad	0.2	31	Very polluted
Capulín	5	Poor	41	Bad	1	18	Extremely polluted
Cumuato	5	Poor	43	Bad	1	41	Very polluted
Ibarra	4	Poor	48	Bad	6	58	Polluted

4. Discussion

The calibrated BMWP was defined according to the community structure (abundance intervals of macroinvertebrates) associated with the river's environmental conditions. We characterized the general physicochemistry and nutrient contents related to organic loads as defined in other studies and protocols [53,54]. However, the Pcq index showed higher values at sites with an evidently transformed habitat (e.g., river rectification, human-made ponds, and small dams in the riverbed), such as Etúcuaro and Antes de la Central (sites 5 and 7, respectively), which placed them in a higher quality interval than expected. In contrast, we obtained small values of Pcq at sites with good habitat quality, such as the national park Camécuaro (site 6). Consequently, the combined Pcq and HQI value, better reflected the condition of the river at the different sites.

Adapting the biological index from the original version is a regular procedure because it establishes a better description of the conditions found in the ecosystem analyzed [55,56]. However, the way through which BMWP is calibrated is an important process, reducing subjectivity and taking into account the way in which macroinvertebrate families respond to environmental conditions in every system [17]. In this context, we used and modified the calibration procedure by Ruiz-Picos et al. [13] to propose a quantitative protocol. We based the analysis spatially on a low but representative number of sites, describing each sub basin's hydrology, water quality, habitat structure, and human impacts. In each site, we implemented a rigorous multi-habitat sampling to characterize all habitats in different stretches of the river. Temporally, the study covered the most contrasting climatological conditions present in the Duero river to describe better the water and habitat quality and the macroinvertebrates community. In addition, this represents the adapted BMWP in an important biogeographical zone (transition zone) within a highly productive and crowded basin in the central part of Mexico (Lerma River basin).

As expected from our first hypothesis, the ecosystem showed a general decrease down the stream in the water quality gradient, with some intermediate recoveries in the environmental variables and the macroinvertebrate community (Figure 3). The BMWP clearly revealed this pattern, since in more transparent, oxygen-rich, and fast current sites, we found higher scores and index categories (regular; Table 4). In these sites we found macroinvertebrate families with higher bioindication values such as Ephemeroptera (Leptophlebiidae, Tricorythidae), Plecoptera (Perlidae), and Trichoptera (Glossosomatidae, Lepidostomatidae). These families are considered the most sensitive to pollution, and most have also been described in other subtropical and tropical rivers with equivalent bioindication values [13,16,57]. In contrast, the water quality decreased mainly in the final section of the river, where we found sites with high turbidity, pollution (high nutrients, DBO_5, and fecal coliforms), low flow, and with some of the lowest BMWP categories (very polluted and extremely polluted; Table 4). In this area, although some of the macroinverte-brates found were present in all sites along the river, their dominance was higher in this section of the river, showing low bioindication values in the Duero River and other rivers, such as in the case for Diptera (Chironomidae), Oligochaeta (Lumbriculidae), and Isopoda (Asellidae) [54,58]. Other organisms also had low bioindication values at the river mouth, but they are from families with low frequency and abundance values, such as Hirudinea (Erpobdellidae, Glossiphonidae) and Diptera (Culicidae).

Two aspects are important to mention. First, some macroinvertebrates showed a different bioindication value compared with other studies [13,16,57]. For instance, the Ephemeroptera families, Heptageniidae and Leptophlebiidae, had lower and higher values in the Duero River (three and eight, respectively). However, the latter family value was similar to another study where they calculated the bioindication score according to the degrees of saprobity [12]. These results are part of the adaptation process of the BMWP and because of the calibration. For instance, Heptageniidae, which had low frequency and abundance was absent in most sites with better water quality but site 8 (Después de la Central). However, it had higher abundances at site 14 (Ibarra), the site with the lowest Pcq-HQI combined index. Second, regarding the intermediate recoveries, the springs improved the water quality at adjacent sites. For instance, the national park Camécuaro (site 6) has the highest discharge value in the watershed ($Q = 1.75\ m^3\ s^{-1}$, [34]), and it is related to the better water quality in Antes and Después Central de Abastos (sites 7 and 8, respectively).

For our second hypothesis, the results showed mixed trends among indices. For instance, Camécuaro showed contrasting results between the water and habitat quality indices (high values) and the EPT and BMWP (low value; Table 4). Low values could be related to habitat modifications for recreational use, such as the removal of submerged macrophytes to promote swimming and the introduction of non-native fish species to promote sport fishing; these aspects are also linked to the disappearance of other organisms at the site, such as the native fish *Skiffia multipunctata* [59]. Additionally, natural aspects in different zones, such as the higher water flow and the presence of roots and shade by ahuehuetes (cypress), restricts the growth of submerged vegetation and diminishes habitat diversity. Another site with different values was Ibarra (site 14) at the mouth of the river. This area of the river has been modified to be a small reservoir and the flow is controlled by a gate adjacent to the union with the Lerma River. The higher BMWP value at this site, in contrast to the other indices, could be related to continuous flow (the gates were open in all seasons) and the presence of water hyacinths that contribute to overall biotic diversity because they offer different habitats for macroinvertebrates [60].

In the third hypothesis, the lack of statistically significant seasonal or annual differences in water quality may result from a temporally stable gradient of water quality from mild pollution in the upper basin to severe pollution in the river mouth. However, on one hand, historically the Duero River had a macroinvertebrate composition of 72 families and only 62 are found currently. On the other hand, when comparing each site, there was a marked difference in the first sub basin related to habitat modification. Some springs were channelized to provide more water to nearby towns, or isolated with infrastructure to avoid degradation because several of these places are used as recreation areas, or are used to provide water for agriculture. According to the BMWP categories obtained in the present study, the main difference is that Kuinio, Tacuro, and La Toma (sites 1 to 3) showed that the average water quality was excellent and good in 1984 and 1986, respectively, and only regular in 2019. The higher values in the middle of the river, related to Antes and Después de la Central (sites 7 and 8), could be affected more by the sampling plan than by the conspicuous restoration of the habitat.

Unfortunately, this river has been used since pre-Hispanic times, and there are no major initiatives for its restoration, so there are no very high-quality sites. However, those that were used of better quality as controls allowed to make the calibration and the BMWP adequately reflect the Duero River conditions. In future studies, it is important to validate the BMWP by including additional sites and samples collected from multiple years to prove that the model fits well in the Duero River. A critical aspect of Duero River conservation is the maintenance of water flow to promote the river's biotic integrity because, in some years, different parts of the river dry up and lose continuity [33].

Supplementary Materials: The following is available online at https://www.mdpi.com/article/10.3390/d13110606/s1, Supplementary material: Abundance of the macroinvertebrate families and environmental variables measured in the Duero river, including the three sampling events.

Author Contributions: Conceptualization, G.M.O., C.E.G. and R.M.-E.; methodology, C.E.G., E.L.-L., J.E.S.-D., M.L.H. and M.A.-D.; software, G.M.O., J.E.S.-D. and R.M.-E.; validation, E.L.-L., J.E.S.-D. and M.A.-D.; formal analysis, G.M.O., C.E.G., J.E.S.-D., M.L.H. and R.M.-E.; investigation, G.M.O., C.E.G. and R.M.-E.; resources, E.L.-L., M.L.H., M.A.-D. and R.M.-E.; data curation, G.M.O., C.E.G. and M.L.H.; writing—original draft preparation, G.M.O., C.E.G. and R.M.-E.; writing—review and editing, C.E.G., E.L.-L., J.E.S.-D., M.L.H., M.A.-D. and R.M.-E.; visualization, G.M.O. and M.A.-D.; supervision, C.E.G., E.L.-L. and R.M.-E.; project administration, R.M.-E.; funding acquisition, E.L.-L. and R.M.-E. All authors have read and agreed to the published version of the manuscript.

Funding: This research was funded by Secretaría de Investigación y Posgrado, Instituto Politécnico Nacional, grant number SIP2045, SIP-20196116, SIP 20200576 and SIP20210131.

Institutional Review Board Statement: Not applicable.

Data Availability Statement: Data is contained within the article and the Supplementary Material.

Acknowledgments: G.M.O. was supported by Consejo Nacional de Ciencia y Tecnología (CONACYT) and Beca de Estímulo Institucional de Formación de Investigadores (BEIFI). R.M.-E., C.E.G., E.L.-L. and J.E.S.-D. are supported by the Comisión de Operación y Fomento de Actividades Académicas (COFAA) and Estímulo al Desempeño Académico (EDI).

Conflicts of Interest: The authors declare no conflict of interest.

Appendix A

Table A1. Taxonomic list of the macroinvertebrate found in the Duero river. Asterisks indicate the families not found in the year 2019.

Class	Order	Family	Genera
Turbellaria	Tricladida	Planariidae	*Dugesia*
Gastropoda	Basommatophora	Planorbidae	*Helisoma*
		Ancylidae	*Ferrisia*
			Hebetancylus
		Physidae	*Physa*
Pelecypoda	Eulamellibranchia	Margaritiferidae	*Margaritifera*
	Heterodonta	Sphaeriidae	*Eupera*
			Pisidium
			Sphaerium
Oligochaeta	Lumbriculida	Lumbriculidae	*Lumbriculus*
	Haplotaxida	Tubificidae	*Branchiura*
			Tubifex
Hirudinea	Pharyngobdellida	Erpobdellidae	*Moreobdella*
	Rhynchobdellida	Glossiphoniidae	*Hellobdella*
Crustacea	Podocopa	Limocytheridae	*Limnocythere*
	Isopoda	Asellidae	*Asellus*
	Amphipoda	Gammaridae	*Gammarus*
	Decapoda	Cambaridae	*Cambarellus*
			Procambarus
		Pseudothelphusidae	*Pseudothelphusa*
Insecta	Collembola	Isotomidae	*Isotomorus*
	Ephemeroptera	Baetidae	*Baetis*
			Baetodes
			Callibaetis
			Dactylobaetis
		Heptageniidae	*Epeorus*
			Heptagenia
		Leptophlebiidae	*Leptophlebia*
			Paraleptophlebia
			Traulodes

Table A1. *Cont.*

Class	Order	Family	Genera
		Tricorythidae	*Leptohyphes*
			Tricorhythodes
		Caenidae *	*Caenis*
	Odonata	Coenagrionidae	*Argia*
			Coenagrion
			Enallagma
		Lestidae *	*Archilestes*
		Agriidae	*Agrion*
		Calopterygidae	*Hetaerina*
		Petaluridae	*Tanypteryx*
		Cordulegastridae *	*Cordulegaster*
		Gomphidae	*Erpetogomphus*
		Aeschnidae	*Aeschna*
		Libellulidae *	*Libellula*
	Plecoptera	Perlidae	*Acroneuria*
	Hemiptera	Naucoridae *	*Ambrysus*
		Belostomatidae	*Belostoma*
		Corixidae	*Graptocorixa*
			Trichocorixa
		Notonectidae	*Buenoa*
		Mesovellidae	*Mesovelia*
		Macrovellidae	*Macrovelia*
		Gerridae	*Gerris*
		Vellidae	*Microvelia*
			Rhagovelia
		Nepidae *	*Ranatra*
	Megaloptera	Corydalidae	*Corydalus*
	Trichoptera	Polycentropodidae	*Polycentropus*
		Hydropsychidae	*Hydropsyche*
			Leptonema
		Rhyacophilidae	*Atopsyche*
			Rhyacophila
		Glossosomatidae	*Protoptila*
		Hydroptilidae	*Hydroptila*
			Leucotrichia
		Limnephilidae	*Limnephilus*
		Lepidostomatidae	*Lepidostoma*
		Helicopsychidae	*Helicopsyche*
		Calamoceratidae *	*Phylloicus*
		Leptoceridae *	*Oecetis*
	Lepidoptera	Pyralidae	*Paragyractis*
	Coleoptera	Dytiscidae	*Dytiscus*
			Laccophilus
		Gyrinidae	*Dineutus*
			Gyrinus
		Hydroptilidae	*Hydrobius*
			Hydrophilus
		Dryopidae	*Helichus*
		Elmidae	*Cylloepus*
			Heterelmis
			Microcylloepus
			Phanocerus
	Diptera	Tipulidae	*Tipula*
		Culicidae	*Culex*
		Ceratopogonidae	*Dasyhelea*
		Chironomidae	*Chironomus*
			Pentaneura
		Simullidae	*Simulium*

Table A1. *Cont.*

Class	Order	Family	Genera
		Stratiomyidae	*Odontomyia*
		Tabanidae	*Tabanus*
		Rhagionidae *	*Atheris*
		Empididae	*Hemerodromia*
		Syrphidae *	*Eristalis*
		Muscidae	*Limnophora*
		Anthomyiidae	
		Psychodidae	
		Sciomyzidae	
Arachnida	Acarina	Hydracaridae	

References

1. Mustow, S.E. Biological monitoring of rivers in Thailand: Use and adaptation of the BMWP score. *Hydrobiologia* **2002**, *479*, 191–229. [CrossRef]
2. Roche, K.F.; Queiroz, E.P.; Righi, K.O.; De Souza, G.M. Use of the BMWP and ASPT indexes for monitoring environmental quality in a neotropical stream. *Acta Limnol. Bras.* **2010**, *22*, 105–108. [CrossRef]
3. Wyżga, B.; Oglęki, P.; Hajdukiewicz, H.; Zawiejska, J.; Radecki-Pawlik, A.; Skalski, T.; Mikuś, P. Interpretation of the invertebrate-based BMWP-PL index in a gravel-bed river: Insight from the Polish Carpathians. *Hydrobiologia* **2013**, *712*, 71–88. [CrossRef]
4. Paisley, M.F.; Trigg, D.J.; Walley, W.J. Revision of the Biological Monitoring Working Party (BMWP) score system: Derivation of present-only and abundance-related scores from field data. *River Res. Appl.* **2014**, *30*, 887–904. [CrossRef]
5. Mao, F.; Zhao, X.; Ma, P.; Chi, S.; Richards, K.; Hannah, D.M.; Krause, S. Revision of biological indices for aquatic systems: A ridge-regression solution. *Ecol. Ind.* **2019**, *106*, 105478. [CrossRef]
6. Armitage, P.D.; Moss, D.; Wright, J.F.; Furse, M.T. The performance of a new biological water quality score system based on macroinvertebrates over a wide range of unpolluted running-water sites. *Water Res.* **1983**, *17*, 333–347. [CrossRef]
7. Hilty, J.; Merenlender, A. Faunal indicator taxa selection for monitoring ecosystem health. *Biol. Conserv.* **2000**, *92*, 185–197. [CrossRef]
8. Akyildiz, G.K.; Duran, M. Evaluation of the impact of heterogeneous environmental pollutants on benthic macroinvertebrates and water quality by long-term monitoring of the buyuk menderes river basin. *Environ. Monit. Assess.* **2021**, *193*, 280. [CrossRef]
9. Hawkes, H.A. Origin and development of the biological monitoring working party score system. *Water Res.* **1997**, *32*, 964–968. [CrossRef]
10. Alba-Tercedor, J.; Sánchez-Ortega, A. Un método rápido y simple para evaluar la calidad biológica de las aguas corrientes basado en el de Hellawell (1978). *Limnética* **1988**, *4*, 51–56.
11. Roldán-Pérez, G. *Bioindicación de la Calidad del Agua en Colombia: Uso del Método BMWP/Col*, 1st ed.; Editorial Universidad de Antioquia: Medellín, Colombia, 2003; pp. 1–170.
12. Junqueira, V.M.; Campos, S.C.M. Adaptation of the "BMWP" method for wáter quality evaluation to Rio Das Velhas watershed (Minas Gerais, Brazil. *Acta Limnol. Bras.* **1998**, *10*, 125–135.
13. Ruiz-Picos, R.A.; Sedeño-Díaz, J.E.; López-López, E. Calibrating and validating the Biomonitoring Working Party (BMWP) index for the Bioassessment of water quality in neotropical streams. In *Water Quality*, 1st ed.; Tutu, H., Ed.; IntechOpen: London, UK, 2017; pp. 39–58.
14. Monaghan, K.A. Four reasons to question the accuracy of a biotic index; the risk of metric bias and the scope to improve accuracy. *PLoS ONE* **2016**, *11*, e0158383. [CrossRef]
15. Walley, W.J.; Hawkes, H.A. A computer-based reappraisal of the Biological Monitoring Working Party scores using data from the 1990 river quality survey of England and Wales. *Water Res.* **1996**, *30*, 2086–2094. [CrossRef]
16. Cornejo, A.; López-López, E.; Sedeño-Díaz, J.E.; Ruiz-Picos, R.A.; Macchi, P.; Kohlmann, B.; Correa-Araneda, F.; Boyero, L.; Bernal-Vega, J.; Ríos, T.; et al. *Protocolo de Biomonitoreo Para la Vigilancia de la Calidad del Agua en Afluentes Superficiales de Panamá*, 1st ed.; Instituto Conmemorativo Gorgas de Estudios de la Salud: Panamá, Panama, 2019; pp. 1–81.
17. Ruiz-Picos, R.A.; Kohlmann, B.; Sedeño-Díaz, J.E.; López-López, E. Assessing ecological impairments in Neotropical rivers of Mexico: Calibration and validation of the Biomonitoring Working Party Index. *Int. J. Environ. Sci. Technol.* **2017**, *14*, 1835–1852. [CrossRef]
18. Lee, S.; Hwang, S.; Lee, J.; Jung, D.; Park, Y.; Kim, J. Overview and application of the National Aquatic Ecological Monitoring Program (NAEMP) in Korea. *Ann Limnol. Int. J. Lim.* **2011**, *47*, S3–S14. [CrossRef]
19. Water: Monitoring & Assessment. Chapter 4 Macroinvertebrates and Habitat. United States Environmental Protection Agency. Available online: https://archive.epa.gov/water/archive/web/html/vms40.html (accessed on 20 August 2021).
20. Surface Water Quality Monitoring. 4.2 Biological Assessment of River Quality. European Environment Agency. Available online: https://www.eea.europa.eu/publications/92-9167-001-4/page021.html (accessed on 20 August 2021).

21. Framework for Assessment of River and Wetland Health (FARWH), WetlandInfo Website. Department of Environment and Science, Australia. Available online: https://wetlandinfo.des.qld.gov.au/wetlands/resources/tools/assessment-search-tool/framework-for-assessment-of-river-and-wetland-health-farwh/ (accessed on 20 August 2021).
22. MINAE-S (Ministerio del Ambiente y Energía—Ministerio de Salud). *Reglamento Para la Evaluación y la Clasificación de la Calidad de Cuerpos de Agua Superficiales. Decreto Ejecutivo 33903*, 1st ed.; Diario Oficial La Gaceta 178: San José, Costa Rica, 2007; pp. 1–16.
23. River Eco-Status Monitoring Programme. Department of Water and Sanitation Republic of South Africa. Available online: https://www.dws.gov.za/iwqs/rhp/default.aspx (accessed on 20 August 2021).
24. Eriksen, T.E.; Brittain, J.E.; Søli, G.; Jacobsen, D.; Goethals, P.; Friberg, N. A global perspective on the application of riverine macroinvertebrates as biological indicators in Africa, South-Central America, Mexico and Southern Asia. *Ecol. Indic.* **2021**, *126*, 107609. [CrossRef]
25. Pineda, L.R.; Pérez, M.R.M.; Mathuriau, C.; Villalobos, H.J.L.; Barba, A.R. *Protocolo de Muestreo de Macroinvertebrados en Aguas Continentales Para la Aplicación de la Norma de Caudal Ecológico (NMX-AA-159-SCFI-2012)*, 1st ed.; Programa Nacional de Reservas de Agua: Mexico City, Mexico, 2014; pp. 1–47.
26. Rosas-Acevedo, J.L.; Ávila-Pérez, H.; Sánchez-Infante, A.; Rosas-Acevedo, A.Y.; García-Ibañez, S.; Sampedro-Rosas, L.; Granados-Ramírez, J.G.; Juárez-López, A.L. Índice BMWP, FBI y EPT para determinar la calidad del agua en la laguna de Coyuca de Benítez, Guerrero, México. *Rev. Iberoam. Cienc.* **2014**, *1*, 81–88.
27. Rosas-Acevedo, J.L.; Rosas-Acevedo, A.Y.; Sánchez-Infante, A.; Sampedro-Rosas, L.; Juárez-López, A.L. Evaluación del medio físico y calidad del agua por medio de insectos bioindicadores, en el brazo derecho del cauce Aguas Blancas, Acapulco, Gro., México. *Entomol. Mex.* **2015**, *2*, 689–694.
28. Morrone, J.J. Regionalización biogeográfica y evolución biótica de México: Encrucijada de la biodiversidad del Nuevo Mundo. *Rev. Mex. Biod.* **2019**, *90*, e902980. [CrossRef]
29. López-Hernández, M. Caracterización Limnológica del Río Duero. Ph.D. Thesis, Universidad Nacional Autónoma de México, Mexico City, Mexico, 1997.
30. Rivera-Rivas, M.V. Relación de Parámetros Fisicoquímicos y Presencia de Macroinvertebrados Para Determinar la Calidad del Agua del río Duero, Michoacán. Master's Thesis, Instituto Politécnico Nacional, Mexico City, Mexico, 2016.
31. Moncayo-Estrada, R.; Zarazúa, J.A. *Saneamiento Integral de una Cuenca Hidrográfica. Entorno Ambiental y Socioeconómico, Problemática y Soluciones*, 1st ed.; Instituto Politécnico Nacional: Mexico City, Mexico, 2016; pp. 1–446.
32. Silva, J.T.; Moncayo, R.; Ochoa, S.; Cruz-Cárdenas, G.; Escalera, C.; Villalpando, F.; Nava, J.; Ramos, A.; López, M. Calidad química del agua subterránea y superficial en la cuenca del río Duero, Michoacán. *Tecnol. Cienc. Agua* **2013**, *4*, 127–146.
33. Moncayo-Estrada, R.; Lyons, J.; Ramírez-Herrejón, J.P.; Escalera-Gallardo, C.; Campos-Campos, O. Status and trends in biotic integrity in a sub-tropical river drainage: Analysis of the fish assemblage over a three decade period. *River Res. Appl.* **2015**, *31*, 808–824. [CrossRef]
34. Silva, J.T.; Ochoa, S.; Cruz-Cárdenas, G.; Nava, J.; Villalpando, F. Manantiales de la cuenca del río Duero Michoacán: Operación, calidad y cantidad. *Rev. Int. Contam. Ambient.* **2016**, *32*, 55–68.
35. Méndez-Toribio, M.; Zermeño-Hernández, I.; Ibarra-Manríquez, G. Effect of land use on the structure and diversity of riparian vegetation in the Duero river watershed in Michoacán, Mexico. *Plant Ecol.* **2014**, *215*, 285–296. [CrossRef]
36. APHA. *Standard Methods for the Examination of Water and Wastewater*, 23rd ed.; American Public Health Association/American Water Works Association/Water Environment Federation: Washington, DC, USA, 2017; p. 9260. [CrossRef]
37. Pennak, R.W. *Fresh-Water Invertebrates of the United States*, 2nd ed.; John Wiley & Sons: Hoboken, NJ, USA, 1978; pp. 1–803.
38. Merritt, R.W.; Cummins, K.W.; Berg, M.B. *An Introduction to the Aquatic Insects of North America*, 4th ed.; America Kendall Publishing: Dubuque, IA, USA, 2008; pp. 1–1158.
39. Thorp, J.; Covich, A. *Ecology and Classification of North American Freshwater Invertebrates*, 3rd ed.; Academic Press: Cambridge, MA, USA, 2009; pp. 1–1021.
40. Brown, R.M.; McClelland, N.I.; Deininger, R.A.; Landwehr, J.M. Validating the WQI. In *National Meeting of American Society of Civil Engineers on Water Resources Engineering*; American Society of Civil Engineers: Washington, DC, USA, 1973.
41. Barbour, M.T.; Gerritsen, J.; Snyder, B.D.; Stribling, J.B. *Rapid Bioassessment Protocols for Use in Streams and Wadeable Rivers: Periphyton, Benthic Macroinvertebrates and Fish*, 2nd ed.; EPA 841-B-99-002; U.S. Environmental Protection Agency, Office of Water: Washington, DC, USA, 1999; pp. 1–337.
42. Masese, F.O.; Raburu, P.O. Improving the performance of the EPT Index to accommodate multiple stressors in Afrotropical streams. *Afr. J. Aquat. Sci.* **2017**, *42*, 219–233. [CrossRef]
43. Chao, A., Gotelli, N.J.; Hsieh, T.C.; Sander, E.L.; Ma, K.H.; Colwell, R.K.; Ellison, A.M. Rarefaction and extrapolation with Hill numbers: A framework for sampling and estimation in species diversity studies. *Ecol. Monogr.* **2014**, *22*, 45–67. [CrossRef]
44. Zamora-Muñoz, C.; Sáinz-Cantero, C.E.; Sánchez-Ortega, A.; Alba-Tercedor, J. Are biological indices BMWP' and ASPT' and their significance regarding water quality seasonally dependent? Factors explaining their variations. *Water Res.* **1995**, *29*, 285–290. [CrossRef]
45. Kennen, J.G.; Sullivan, D.J.; May, J.T.; Bell, A.H.; Beaulieu, K.M.; Rice, D.E. Temporal changes in aquatic-invertebrate and fish assemblages in streams of the north-central and northeastern US. *Ecol. Indic.* **2012**, *18*, 312–329. [CrossRef]
46. Anderson, M.J. A new method for non-parametric multivariate analysis of variance. *Austral. Ecol.* **2001**, *26*, 32–46. [CrossRef]

47. McCune, B.; Grace, J.B.; Urban, D.L. *Analysis of Ecological Communities*, 1st ed.; MjM Software Design, Wild Blueberry Media LLC: Corvallis, OR, USA, 2000; pp. 1–284.
48. Anderson, M.J. Permutational Multivariate Analysis of Variance (PERMANOVA). In *Wiley StatsRef: Statistics Reference Online*, 1st ed.; Balakrishnan, N., Colton, T., Everitt, B., Piegorsch, W., Ruggeri, F., Teugels, J.L., Eds.; John Wiley & Sons, Ltd.: Hoboken, NJ, USA, 2017; pp. 1–15.
49. Oksanen, J.; Blanchet, F.G.; Kindt, R.; Legendre, P.; O'Hara, R.B.; Simpson, G.L.; Solymos, P.; Stevens, M.H.H.; Wagner, H. Vegan: Community Ecology Package. R Package Version 2.5-7, 2020. Available online: https://cran.r-project.org/web/packages/vegan/index.html (accessed on 15 October 2020).
50. Harrell, F.E., Jr. Hmisc: Harrell Miscellaneous. R Package Version 4.5-0, 2021. Available online: https://cran.r-project.org/web/packages/Hmisc/Hmisc.pdf (accessed on 15 October 2020).
51. R Core Team. *R: A Language and Environment for Statistical Computing*; R Foundation for Statistical Computing: Vienna, Austria, 2021; Available online: https://www.R-project.or.rg/ (accessed on 16 August 2021).
52. Pond, G.J.; McMurray, S.E. *A Macroinvertebrate Bioassesment Index for Headwater Streams of the Eastern Coalfield Region, Kentucky*, 1st ed.; Kentucky Department for Environmental Protection: Frankfort, KY, USA, 2002; pp. 1–48.
53. Torrisi, M.; Scuri, S.; Dell'Uomo, A.; Cocchioni, M. Comparative monitoring by means of diatoms, macroinvertebrates and chemical parameters of an Apennine watercourse of central Italy: The river Tenna. *Ecol. Indic.* **2010**, *10*, 910–913. [CrossRef]
54. Pardo, I.; Costas, N.; Méndez-Fernández, L.; Martínez-Madrid, M.; Rodríguez, P. Changes in invertebrate community composition allow for consistent interpretation of biodiversity loss in ecological status assessment. *Sci. Total Environ.* **2020**, *715*, 136995. [CrossRef] [PubMed]
55. Von der Ohe, P.C.; Prüß, A.; Schäfer, R.B.; Liess, M.; de Deckere, E.; Bracka, W. Water quality indices across Europe—A comparison of the good ecological status of five river basins. *J. Environ. Monit.* **2007**, *9*, 970–978. [CrossRef]
56. Roldán-Pérez, G. Los macroinvertebrados como bioindicadores de la calidad del agua: Cuatro décadas de desarrollo en Colombia y Latinoamerica. *Rev. Acad. Colomb. Cienc. Exactas Físicas Nat.* **2016**, *40*, 254–274. [CrossRef]
57. Andrade, I.C.P.; Krolow, T.K.; Boldrini, R. Diversity of EPT (Ephemeroptera, Plecoptera, Trichoptera) along streams fragmented by waterfalls in the Brazilian Savanna. *Neotrop. Entomol.* **2020**, *49*, 203–212. [CrossRef]
58. Ochien, H.; Odong, R.; Okot-Okumu, J. Comparison of temperate and tropical versions of Biological Monitoring Working Party (BMWP) index for assessing water quality of River Aturukuku in Eastern Uganda. *Glob. Ecol. Conserv.* **2020**, *23*, e01183. [CrossRef]
59. Lyons, J.; Piller, K.R.; Artigas-Azas, J.M.; Dominguez-Dominguez, O.; Gesundheit, P.; Köck, M.; Medina-Nava, M.; Mercado-Silva, N.; García, A.R.; Findley, K.M. Distribution and current conservation status of the Mexican Goodeidae (Actinopterygii, Cyprinodontiformes). *ZooKeys* **2019**, *885*, 115–158. [CrossRef] [PubMed]
60. Poi, A.S.G.; Neiff, J.J.; Casco, S.L.; Gallardo, L.I. Macroinvertebrates of Eichhornia crassipes (Pontederiaceae) roots in the alluvial floodplain of large tropical rivers (Argentina). *Int. J. Trop. Biol.* **2020**, *68*, 104–115. [CrossRef]

diversity

MDPI

Article

Characterization of the Multidimensional Functional Space of the Aquatic Macroinvertebrate Assemblages in a Biosphere Reserve (Central México)

Alexis Joseph Rodríguez-Romero [1], Axel Eduardo Rico-Sánchez [2], Jacinto Elías Sedeño-Díaz [3] and Eugenia López-López [1,*]

[1] Laboratorio de Evaluación de la Salud de los Ecosistemas Acuáticos, Escuela Nacional de Ciencias Biológicas, Instituto Politécnico Nacional, Prolongación de Carpio y Plan de Ayala s/n, Col. Santo Tomás, Miguel Hidalgo C.P., Ciudad de México 11340, Mexico; ajrodriguezr@ipn.mx

[2] Programa de Hidrociencias, Colegio de Postgraduados, Km 35.5 Carretera México-Texcoco, Texcoco 56230, Mexico; rico.axel@colpos.mx

[3] Coordinación Politécnica para la Sustentabilidad, Instituto Politécnico Nacional, Av. Instituto Politécnico Nacional s/n, Esq. Wilfrido Massieu, Col. San Pedro Zacatenco, Gustavo A. Madero C.P., Ciudad de México 07738, Mexico; jsedeno@ipn.mx

* Correspondence: eulopez@ipn.mx

Abstract: The analysis of functional diversity has shown to be more sensitive to the effects of natural and anthropogenic disturbances on the assemblages of aquatic macroinvertebrates than the classical analyses of structural ecology. However, this ecological analysis perspective has not been fully explored in tropical environments of America. Protected Natural Areas (PNAs) such as biosphere reserves can be a benchmark regarding structural and functional distribution patterns worldwide, so the characterization of the functional space of biological assemblages in these sites is necessary to promote biodiversity conservation efforts. Our work characterized the multidimensional functional space of the macroinvertebrate assemblages from an ecosystemic approach by main currents, involving a total of 15 study sites encompassing different impact and human influence scenarios, which were monitored in two contrasting seasons. We calculated functional diversity indices (dispersion, richness, divergence, evenness, specialization, and originality) from biological and ecological traits of the macroinvertebrate assemblages and related these indices to the physicochemical characteristics of water and four environmental indices (Water Quality Index, habitat quality, Normalized Difference Vegetation Index, and vegetation cover and land use). Our results show that the indices of functional richness, evenness, and functional specialization were sensitive to disturbance caused by salinization, concentration of nutrients and organic matter, and even to the occurrence of a forest fire in the reserve during one of the sampling seasons. These findings support the conclusion that the changes and relationships between the functional diversity indices and the physicochemical parameters and environmental indices considered were suitable for evaluating the ecological conditions within the reserve.

Keywords: water quality; functional richness; functional specialization; functional evenness; impact of mining and forest fire

Citation: Rodríguez-Romero, A.J.; Rico-Sánchez, A.E.; Sedeño-Díaz, J.E.; López-López, E. Characterization of the Multidimensional Functional Space of the Aquatic Macroinvertebrate Assemblages in a Biosphere Reserve (Central México). *Diversity* **2021**, *13*, 546. https://doi.org/10.3390/d13110546

Academic Editors: Marina Vilenica, Zohar Yanai and Laurent Vuataz

Received: 30 September 2021
Accepted: 22 October 2021
Published: 29 October 2021

1. Introduction

Freshwater ecosystems are considered the most threatened natural systems globally since water is extracted from them to meet human needs [1]. These ecosystems have diverse natural, economic, cultural, aesthetic, and scientific resource values, among others [2]; they are considered biological diversity hotspots because they are home to approximately 10% of the known species worldwide [3]. However, freshwater ecosystems are affected by water extraction, flow regulation, wastewater discharges, overfishing, invasion of exotic species, and climate change, all of which degrade freshwater bodies and threat biodiversity [4]. In addition, biodiversity in the New World is far from being extensively known from a taxonomical standpoint. Nowadays, biodiversity loss is on the rise due to severe disturbances at regional and global scales [5]. This will likely lead to massive extinction rates, particularly in Protected Natural Areas (PNAs) such as the Sierra Gorda Reserve Biosphere (SGRB), where multiple species may disappear over a short period of time. This may be related to the impact of tourism, local mining extraction and the pollution associated with it, and the presence of invasive species [6].

The maintenance of aquatic ecosystems depends on physical, chemical, and biological processes sustained by different groups of organisms [7]. Aquatic macroinvertebrates play key functions in these ecosystems, participating in processes associated with energy flow across food webs in their roles as herbivores, predators, and filter-feeders. In addition, they participate in the decomposition of detritus and the mineralization of nutrients [8], and are consumed as food by other trophic levels [9]. The loss of species and biological resources, including macroinvertebrates, impairs the functioning and services supplied by aquatic ecosystems [10].

The analysis of functional diversity has shown to be more sensitive to the effects of natural and anthropogenic disturbances on aquatic macroinvertebrate assemblages than the classical analyses of structural ecology [8]. Little is known about how the functional diversity of macroinvertebrates changes with the characteristics of aquatic systems, particularly in the intertropical regions of America [11]. In these environments, some studies have investigated the composition and taxonomic diversity of macroinvertebrate assemblages [12], as well as the environmental conditions, in various rivers [13]. However, diversity measures such as the number of species do not contribute to understanding the functional traits or functional diversity of these assemblages. In the context of functional diversity, it is relevant to know how the functional traits of macroinvertebrates depend on the characteristics of rivers, especially if the aim is to maintain the functionality of these systems and, consequently, the ecosystem services they provide. Additionally, macroinvertebrates have been used as bioindicators of water quality due to their diverse responses when facing different types of impacts. It is widely recognized that the structure of macroinvertebrate assemblages reflects their ecological condition, habitat heterogeneity, and water quality [14–16]. Several studies have described the trophic functional groups and their relationship with the physical and chemical characteristics of river ecosystems in intertropical regions of America [17–19]. However, as far as we know, few studies have addressed the effect of environmental variables on the functional traits and functional diversity of aquatic macroinvertebrate assemblages in tropical rivers of America [20,21].

Functional traits are defined as physiological, morphological, or phenological characteristics related to how organisms interact with their environment [22]. For this reason, the study of functional traits allows understanding how biological diversity and ecosystem functioning are governed by environmental conditions, and how functional diversity is affected by human activities. Furthermore, functional diversity brings information about how niche space is shared and partitioned by species within an assemblage [23]. Thus, functional composition and diversity are useful approximations to exploring ecosystem imbalances.

As one of the main megadiverse countries, Mexico has developed a strategy to conserve its multiple natural ecosystems based on the establishment of Protected Natural Areas (PNAs), including the Sierra Gorda Biosphere Reserve (SGBR). This PNA is located in the Central Plateau of Mexico, an area influenced by the two biogeographical regions converging in the Mexican territory—the Nearctic and the Neotropical regions—and is part of the so-called Transition Zone [24]. Despite its high biological diversity, this PNA is affected by anthropic activities like agriculture and mining, as well as human settlements [25]. Consequently, the SGBR comprises zones influenced by anthropogenic activities and areas with low disturbance levels, thus being an ideal region for analyzing functional diversity.

This study involved two approaches. The first explores the functional composition of aquatic macroinvertebrate assemblages, i.e., the assessment of river ecosystems in the SGBR based on multifunctional features of the components of macroinvertebrate assemblages.

The other approach includes analyses of several indices that quantify the distribution of functional traits of macroinvertebrate assemblages. In both cases, the relationships with environmental (physical and chemical) parameters of the river systems in the SGBR were investigated.

2. Materials and Methods

2.1. Study Area

The SGBR stretches across 3834 km^2 in the Mexican Central Plateau. The Tamuín river runs through this PNA, including the Concá, Ayutla, Santa María, and Jalpan main streams (Figure 1). Moreover, a section of the Moctezuma river and the Extoraz river flow into the southern area of the reserve; both tributaries converge downstream into the Panuco river, which flows into the Gulf of Mexico. The altitude in the SGBR ranges between 300 and 3100 m a.s.l., hosting grasslands and mountain forests [26]. Since 1997, the Mexican authorities have established 11 core zones aiming to maximize the conservation of the natural conditions, which jointly represent about 7% of the total surface of the SGBR [27]. Other areas are considered buffer zones where agriculture and forestry exploitation are substantially reduced [28]. Several towns are located within the SGBR, with a total population that does not exceed 95,000 inhabitants [29] dedicated primarily to mining and agriculture. Approximately 116 mines are located in the SGBR [25].

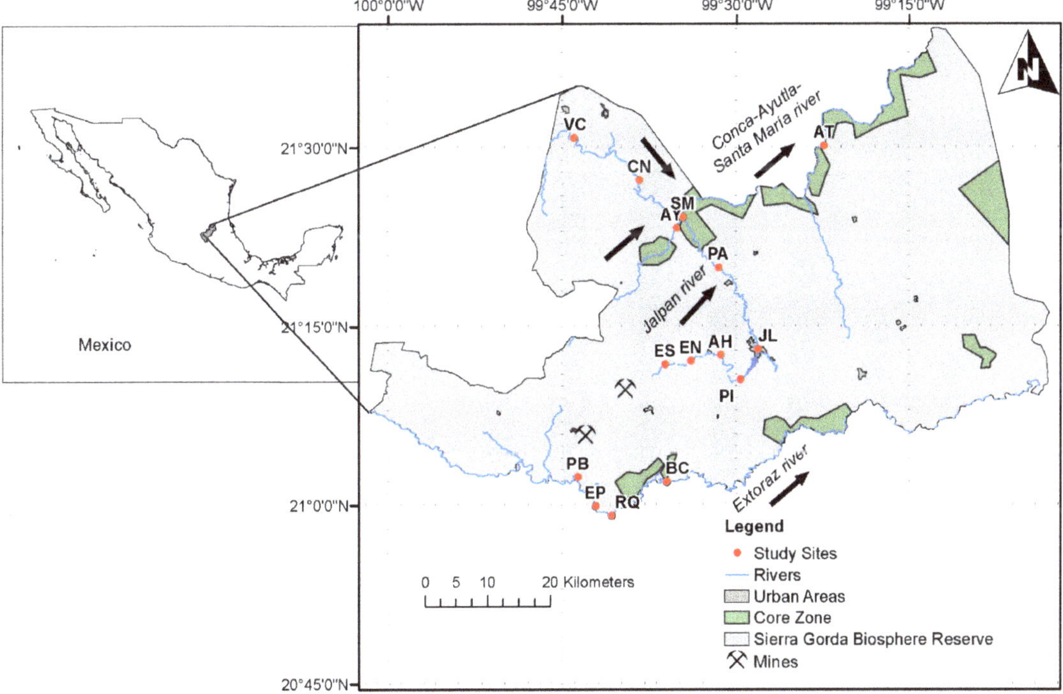

Figure 1. Study area in the Sierra Gorda Biosphere Reserve. Fifteen sampling sites were selected along three streams (Extoraz, Jalpan, and Concá-Ayutla-Santa María rivers): four sites in core zones (BC: Bucareli, AY: Ayutla, SM: Santa María, AT: Autopista 190) and 11 in buffer zones (PB: Peña Blanca, EP: El Paraíso, RQ: Rancho Quemado, ES: Escanela, EN: Escanelilla, AH: Ahuacatlán, PI: Pizquintla, JL: Jalpan, PA: Purísima de Arista, VC: Vegas Cuatas, CN: Concá). The Extoraz river includes PB, EP, RQ and BC; the Jalpan river includes ES, EN, AH, PI, JL, and PA; the Concá-Ayutla-Santa María river includes VC, CN, AY, SM, and AT.

2.2. Field Sampling and Environmental Variables

Fifteen sampling sites were selected in three main streams: Extoraz (four study sites), Jalpan (six study sites), and Concá-Ayutla-Santa María (five study sites) (Figure 1). In all the study sites, sampling and environmental monitoring were carried out in February 2017 and July 2017. In addition, the six sites along the Jalpan river, which concentrates the largest urban localities of the SGBR, were monitored in June 2019. The months monitored correspond to contrasting climatic seasons, i.e., cold dry (February 2017) and warm rainy (July 2017 and June 2019) seasons (Figure 2). In each study site, physicochemical variables were recorded in situ, such as dissolved oxygen (mg/L), oxygen saturation (%), pH, conductivity (ms/cm), salinity (UPS), suspended solids (mg/L), and temperature (°C), using a Quanta (Hydrolab)® (Sheffield, UK) multiparametric probe. Water samples were collected in 500 mL flasks in duplicate, plus a 100 mL sample placed in a Whirlpack® (Madison, USA) bag, to measure physicochemical parameters and run microbiological testing in the laboratory. Samples were transported refrigerated and protected from direct sunlight. In the laboratory, water samples were processed to determine total nitrogen (TN, mg/L), nitrites (NO_2, mg/L), nitrates (NO_3, mg/L), ammonia nitrogen (NH_3, mg/L), sulfates (SO_4, mg/L), orthophosphates (PO_4, mg/L), total phosphorus (PT, mg/L), color (Pt-Co units), and total suspended solids (TSS, mg/L) using a HACH® (Sheffield, UK) DR3900 spectrophotometer (HACH, 2001), and hardness ($CaCO_3$, mg/L) by titration. In addition, alkalinity ($CaCO_3$, mg/L), chlorides (Cl, mg/L), biochemical oxygen demand over 5 days (BOD_5, mg/L), and total and fecal coliforms (MPN/100 mL) were determined following APHA techniques [30] (Table 1).

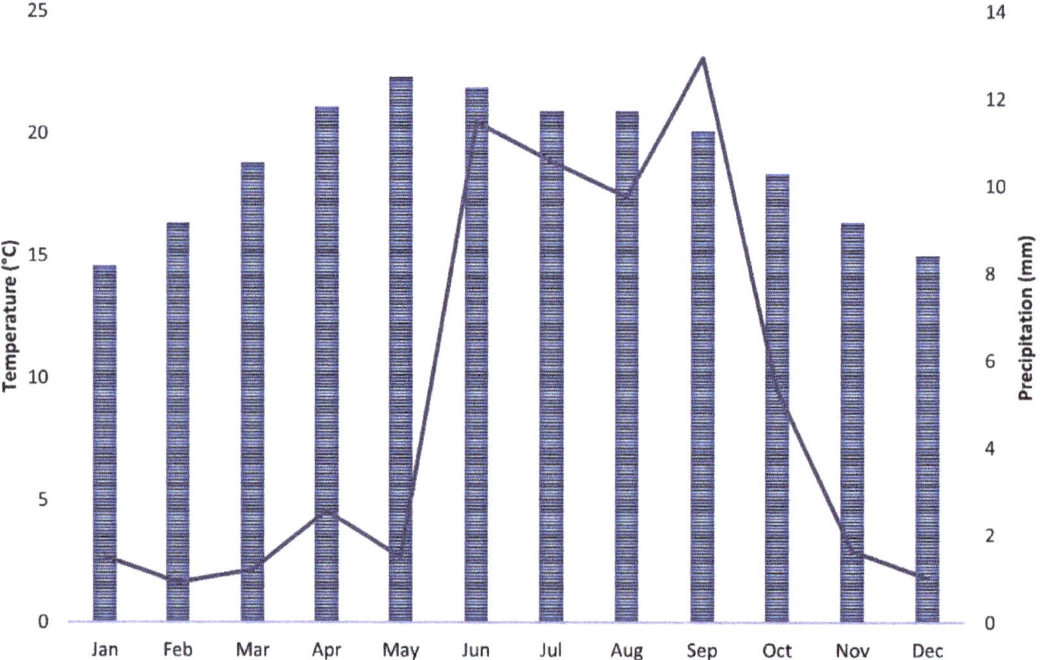

Figure 2. Average monthly temperature (line) and precipitation (bar) values in the Sierra Gorda Biosphere Reserve for the period 2017–2019.

Table 1. Mean values and SE (±) of the physicochemical environmental variables recorded in the three main streams of SGRB.

Environmental Variables/Mainstream	Extoraz_ Feb_17	Extoraz_ Jul_17	Jalpan_ Feb_17	Jalpan_ Jul_17	Jalpan_ Jun_19	Concá_ Ayutla_StMaría_ Feb_17	Concá_ Ayutla_StMaría_ Jul_17
Temperature (°C)	22.65 ± 1.01	25.31 ± 1.03	17.57 ± 0.67	22.00 ± 1.35	24.52 ± 1.85	23.22 ± 0.81	24.94 ± 0.47
Conductivity (ms/cm)	0.70 ± 0.22	0.89 ± 0.17	0.33 ± 0.01	0.37 ± 0.03	0.38 ± 0.02	0.53 ± 0.03	0.28 ± 0.02
Disolved oxygen (mg/L)	9.60 ± 0.55	7.30 ± 0.33	9.41 ± 0.30	7.63 ± 0.32	7.67 ± 1.04	9.68 ± 0.32	8.10 ± 0.20
Oxygen saturation (%)	110.00 ± 6.66	92.21 ± 1.65	103.00 ± 3.05	89.94 ± 3.72	94.62 ± 11.52	106.46 ± 3.98	93.74 ± 1.84
pH	8.06 ± 0.07	7.77 ± 0.12	8.07 ± 0.06	8.01 ± 0.15	8.48 ± 0.08	7.84 ± 0.06	7.83 ± 0.06
Turbidity (NTU)	17.07 ± 10.91	251.57 ± 130.43	15.87 ± 6.65	41.16 ± 21.40	7.69 ± 2.75	14.78 ± 6.30	1002.42 ± 403.33
Salinity (UPS)	0.32 ± 0.08	0.43 ± 0.08	0.16 ± 0.00	0.18 ± 0.01	0.18 ± 0.01	0.25 ± 0.01	0.15 ± 0.01
NO_2 (mg/L)	0.01 ± 0.00	0.03 ± 0.01	0.12 ± 0.09	0.06 ± 0.06	0.07 ± 0.07	0.01 ± 0.00	0.03 ± 0.01
NO_3 (mg/L)	1.28 ± 0.29	3.55 ± 2.64	1.63 ± 0.17	0.43 ± 0.15	0.97 ± 0.20	1.57 ± 0.11	1.16 ± 0.32
NH_3 (mg/L)	0.20 ± 0.04	1.32 ± 0.35	0.56 ± 0.27	0.42 ± 0.11	0.22 ± 0.17	0.79 ± 0.65	3.89 ± 2.73
Total Nitrogen (mg/L)	3.03 ± 0.97	8.07 ± 1.63	2.65 ± 0.34	6.63 ± 0.56	2.99 ± 0.24	1.95 ± 0.38	13.45 ± 3.27
PO_4 (mg/L)	0.16 ± 0.04	0.47 ± 0.28	0.26 ± 0.05	0.35 ± 0.09	0.52 ± 0.27	0.18 ± 0.08	0.70 ± 0.25
Total Phosphorous (mg/L)	1.51 ± 0.98	1.35 ± 0.41	0.34 ± 0.04	1.40 ± 0.82	1.13 ± 0.35	0.37 ± 0.08	1.34 ± 0.52
SO_4 (mg/L)	81.12 ± 14.06	92.37 ± 20.83	12.70 ± 0.51	15.50 ± 0.85	16.00 ± 0.85	72.30 ± 18.84	21.80 ± 7.70
Chlorides (mg/L)	20.36 ± 7.24	21.24 ± 6.08	8.99 ± 0.40	7.28 ± 0.76	0.99 ± 0.38	10.29 ± 1.31	8.69 ± 1.77
Alkalinity (mg/L)	193.12 ± 8.40	233.00 ± 13.77	195.80 ± 11.49	224.50 ± 18.62	183.63 ± 6.66	192.40 ± 6.76	109.40 ± 30.28
Hardness (mg/L)	126.75 ± 43.33	244.50 ± 81.43	59.40 ± 8.66	179.66 ± 13.18	159.40 ± 3.18	99.60 ± 33.17	99.80 ± 17.61
Suspended solids (mg/L)	14.25 ± 11.60	281.25 ± 123.76	1.24 ± 0.51	30.66 ± 17.09	13.40 ± 5.29	4.62 ± 3.36	733.00 ± 289.88
Color (Pt/Co U.)	2.75 ± 1.18	20.75 ± 4.17	1.00 ± 0.01	7.83 ± 3.45	9.20 ± 3.15	2.00 ± 0.63	40.60 ± 16.15
Fecal coliforms (MPN/100 mL)	24.00 ± 6.64	645.75 ± 265.32	301.00 ± 186.00	658.83 ± 205.06	243.42 ± 214.80	111.40 ± 88.13	133.60 ± 82.32
BOD_5 (mg/L)	3.15 ± 0.91	3.03 ± 0.32	5.14 ± 0.78	3.18 ± 0.47	0.30 ± 0.13	2.57 ± 0.46	2.62 ± 0.32

2.3. Characterization of Sites with Environmental Indices

The protocol for characterizing habitat quality was applied in each monitoring station [31,32]. The percentage of each land use (natural vegetation, grassland, secondary vegetation, induced grassland, agriculture, and human settlements-urban areas) influencing each site was estimated inside a buffer area of 2 km upstream and 0.5 km to the sides of each study site, following the criteria of [33]. Buffer sites were set using the available information from a map of land use and vegetation at a 1:250,000 scale provided by the National Institute of Statistics and Geography of Mexico (INEGI, 2021) [34] and using the software QGIS version 3.20.3 (Open-Source Geospatial Foundation, Chicago, IL, USA). Additionally, the Normalized Difference Vegetation Index (NDVI) was calculated based on Landsat 8 OLI TIRS images from the USGS viewer [35], using the following equation [36]:

$$NDVI = \frac{(B5 - B4)}{(B5 + B4)}$$

where $B4$ and $B5$ correspond to the bands of the Landsat 8 OLI TIRS satellite image.

In addition, the Water Quality Index proposed by [37] was calculated with the following equation:

$$WQI = \prod_{i=1}^{n} I_i^{Wi}$$

where WQI = Water Quality Index (0 to 100); Ii = subindex of the i_{th} parameter (0 to 100); Wi = weighting value of the i_{th} parameter (0 to 1); n = number of parameters.

2.4. Macroinvertebrate Monitoring

Aquatic macroinvertebrates were collected using two types of sampling gear, namely, a scoop-type net (for riparian vegetation and ponds) and a kicking net (for riffles and zones with laminar and turbulent flow), both with a 500 µm mesh. Sampling was carried out according to the multi-habitat monitoring proposal [31,32], considering all the potential habitats where these organisms thrive; four replicate samples (the area in each sample was 2.5 m^2) were obtained, two for each collection method, with a sampling effort of 10–20 min per study site. The macroinvertebrates sampled were preserved in 70% ethanol, and the identification and quantification of each taxon were carried out at family level (refer to the Table S1 of Supplementary Material). The functional diversity analysis was performed at family level following [38], which found that functional attributes based on biological and ecological traits, such as type of feeding, reproductive strategy, and trophic status, were strongly correlated with the composition of the assemblages at family level (ρ = 0.64–0.85). These attributes indicate that taxonomic sufficiency was universally applicable within taxonomic groups for different habitats within a biogeographical region, and that aggregation to family or order was adequate to quantify biodiversity and environmental gradients. The identification was based on specialized taxonomic keys [39–41] and using a Nikon® (Tokio, Japan) SMZ 745T stereo microscope.

2.5. Characterization of the Multifunctional Space

Functional diversity was calculated from the combination of two matrices. The first included the abundances of taxa throughout study sites, streams, and sampling seasons; the second considered four ecological traits (food availability, cross-sectional distribution, habitat preference, and tolerance) and six biological traits (life cycle, life stage, respiratory mode, nutritional status, functional group, and body size) obtained from databases and published works [42–45] (refer to the Table S2 of Supplementary Material). Traits were coded using a 'fuzzy' approach, in which a value given to each trait category indicates whether the taxon has no (0), weak (1), moderate (2), or strong (3) affinity for the trait. Affinities were determined based on observations (taxon-specific information from the literature) [42–45]. Fuzzy coding can incorporate intra-taxon variability when trait profiles differ between genera within a family, early and late stages of a species, or individuals of a species living in different environments [46]. Six functional diversity indices were calculated from the multidimensional space of the features, considering the relative abundance of each taxon: Functional dispersion (FDis), Functional richness (FRic), Functional divergence (FDiv), Functional evenness (FEve), Functional specialization (FEsp), and Functional originality (FOri). An increase in FDis, FRic, and FDiv values indicates a greater amplitude of the niche space occupied by the taxa and a broader divergence in the distribution of abundances across the niche space [23]. The multidimensional space of traits was constructed, and functional diversity indices were calculated from the R script proposed by [5,47], available at: http://villeger.sebastien.free.fr/Rscripts.html (accessed on 6 September 2021).

2.6. Statistical Analysis

The average value and standard error of each functional diversity index were calculated. First, we computed the value for each study site and season, and then the mean values for each main stream per study season. Significant differences between average values of the functional diversity indices calculated in each main stream and study season were analyzed using the Kruskal-Wallis test with a significance value of $p < 0.05$ and the Mann-Whitney U test for multiple comparisons. A database was created for environmental variables, with the indices of functional diversity and physicochemical parameters as active variables (i.e., those that are subject to manipulation or experimentation) and the environmental indices as supplementary variables, to run a Principal Component Analysis (PCA). Groups were defined *a priori*, each corresponding to the sites located in the three main streams (Extoraz, Jalpan, and Concá-Ayutla-Santa María) and monitoring seasons (February 2017, June 2017, and July 2019). Those environmental variables with a significance value greater than 0.5 in a previous Factor Analysis were maintained. All data were previously processed from ln $(x + 1)$, and the XLStat (2020) package was used for all statistical analyses.

3. Results

The multidimensional space of the functional traits that was characterized in the first place corresponds to the total number of SGRB study sites (Figure 3a), considering all sites in the three main streams and all monitoring events. This procedure allowed us to identify the broadest spectrum of functional diversity within the reserve (multidimensional functional space). The value of FRic was 1, which is expected since all functional traits were present; however, FDiv and FDis were not necessarily equal to 1, although they were greater than 0.5, which indicates the broad spectrum of the functional niche occupied by aquatic macroinvertebrate assemblages in the entire reserve (Figure 3b–d). FEve for the total reserve was low (Figure 3e) since some functions (body size 0.25–0.5 cm; collectors and very tolerant taxa mainly distributed in riparian zones) were more abundant than other macroinvertebrates, associated with high abundances of some taxa (Baetidae, Chironomidae, Elmidae, and Leptophlebidae), which are concentrated at the lower left quadrant of the functional space (Figure 3a,e). Finally, FSpec and FOri were also greater than 0.5, indicating the importance of specific functions within the assemblages (Figure 3f,g).

Average values of functional diversity indices for each main stream throughout the monitoring seasons are shown in Figure 4. FDis (functional dispersion) is a multivariate measure of the dispersion of assemblages' members across the trait space, estimated as the mean distance of all species to the weighted centroid of the assemblages in the trait space, equivalent to the multivariate dispersion. FDis values (Figure 4a) were above 0.5, with significant differences ($p < 0.05$) between monitoring seasons in the Extoraz and Concá-Ayutla-Santa María rivers in July 2017.

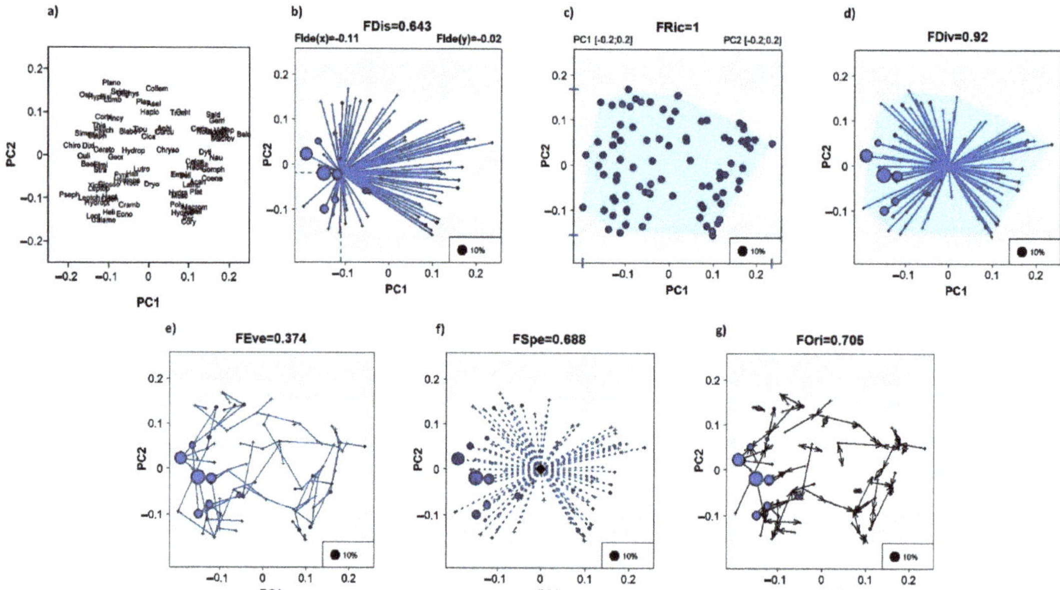

Figure 3. Multidimensional functional space of the aquatic macroinvertebrate assemblages in the SGRB and functional diversity indices for the entire SGRB throughout the monitoring seasons. (**a**) The box in the upper left corner includes the location of taxa (families) in the functional space of the entire SGRB. The meaning of abbreviations is found in the supplementary material. The diameter of the blue dots indicates the abundance of the respective taxon; (**b**) FDis; (**c**) FRic; (**d**) FDiv; (**e**) FEve; (**f**) FSpe; and (**g**) FOri indices for the entire SGRB.

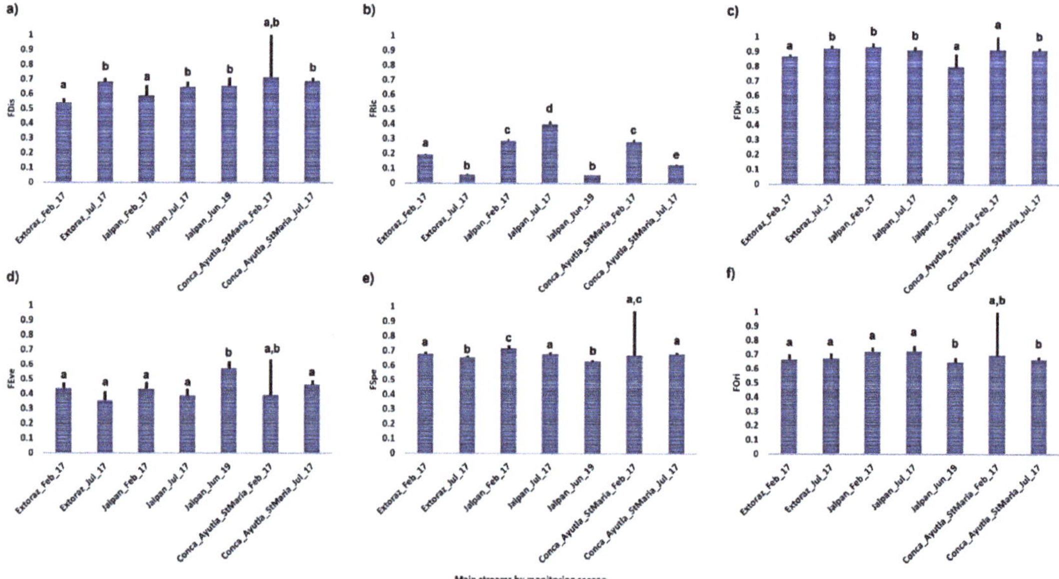

Figure 4. Average values of functional diversity indices for the main streams and monitoring seasons studied. (**a**) FDis, (**b**) FRic, (**c**) FDiv, (**d**) FEve, (**e**) FSpe and (**f**) FOri. Letters above the dispersion values of all indices (**a–e**) indicate statistically significant differences ($p < 0.05$).

FRic (functional richness) represents the range of the functional space occupied by the assemblages, estimated as the number of combinations of functional traits in the assemblage. FRic values (Figure 4b) in the Jalpan river ranged from 0.058 ± 0.0007 in June 2019 to 0.4 ± 0.0175 in July 2017, which are significantly different ($p < 0.05$) between each other and also compared to the Extoraz river during February 2017 and the Concá-Ayutla-Santa María river in the two seasons. FRic values (Figure 4b) lower than 0.5 indicate that the range of functions is unique to a given stream relative to the multifunctional space of the entire SGRB. However, the Extoraz river in July 2017 and the Jalpan river in June 2019 showed very low values that contained less than 10% of the spectrum of functions of the entire reserve. FDiv (functional divergence) represents the proportion of the total abundance supported by taxa with the most extreme trait values within the assemblage. FDiv (Figure 4c) showed very high values (>0.799) in all streams and monitoring seasons.

FEve (functional evenness) represents the uniformity of the distribution and relative abundance of taxa in the functional space of a given assemblage. Higher FEve values indicate a more uniformly occupied niche space. FEve values (Figure 4d) fluctuated from 0.354 ± 0.039 in Extoraz in July 2017 to 0.576 ± 0.030 in Jalpan in June 2019. Values close to 0.5 indicate that the distribution of trait abundances are relatively evenly distributed in the functional space. This indicates that, overall, there are no dominant groups of macroinvertebrates performing similar functions or showing similar attributes. Finally, FSpe and FOri are defined as the mean distance of a taxon and the level of isolation of a taxon, respectively, relative to the functional space occupied by a certain assemblage. FSpe and FOri (Figure 4e,f) showed a similar behavior because these indices indicate the level of specialization of the functions, reaching values above 0.5 that peaked in the Jalpan river in February 2017 (0.718 ± 0.019) and July 2017 (0.728 ± 0.036).

The PCA of the variables and environmental indices that defined the ranking of streams and seasons studied are shown in Figure 5. The first two PCA components accounted for 57.29% of the variance and showed a main environmental gradient on the horizontal axis that clusters the monitoring points into two large groups: the first, on the left side of the biplot (Extoraz_Feb_2017, Jalpan_Feb_17, Conca_Ayutla_StMaria_2017, and Jalpan_Jul_17) is characterized by high oxygen levels (percent saturation), related to the highest FRic, FSpe, and FOri values. The monitoring points on the right side of the biplot (Extoraz_Jul_17, Conca_Ayutla_StaMaria_Jul_17, and Jalpan_Jun_19) are characterized by the highest values of color (9.2–40.6 Pt/Co U.), suspended solids, and turbidity, related to the highest FDis. The main environmental gradient along the horizontal axis denotes the physicochemical properties associated with well-oxygenated waters, in contrast with the study sites with higher contents of solid materials and organic matter. These results are closely related to the monitoring season because the streams positioned to the left were monitored in February 2017 (dry season), except for the Jalpan river in July 2017, while streams positioned to the right were monitored in July 2017 and June 2019 in the Jalpan river (rainy season).

A second environmental gradient is represented on the vertical axis, showed on the upper quadrants of the biplot (Jalpan_Feb_17, Jalpan_Jun_19, and Conca_Ayutla_StaMaria_Jul_17) with the highest values of pH, NDVI, secondary vegetation, natural vegetation, water quality, and habitat quality, related to the highest FEve values. The Extoraz_Feb_17, Extoraz_Jul_17, Jalpan_Jul_17, and Concá_Ayutla_StMaría_Feb_17 monitoring points are located at the lower portion of the biplot, characterized by the highest concentrations of chlorides, conductivity, salinity, sulfates, fecal coliforms, hardness, nitrite nitrogen, nitrates, total phosphorus, hardness, nitrite nitrogen, nitrates, total phosphorus, and nitrogen, related to human settlements, urban areas, and agriculture. Consequently, the second gradient refers to properties related to environmental quality, ranging from better water quality, habitat quality, and well-preserved vegetation cover (in the upper portion of the biplot) to higher contents of minerals, organic matter, and nutrients derived from human activities (at the bottom of the biplot).

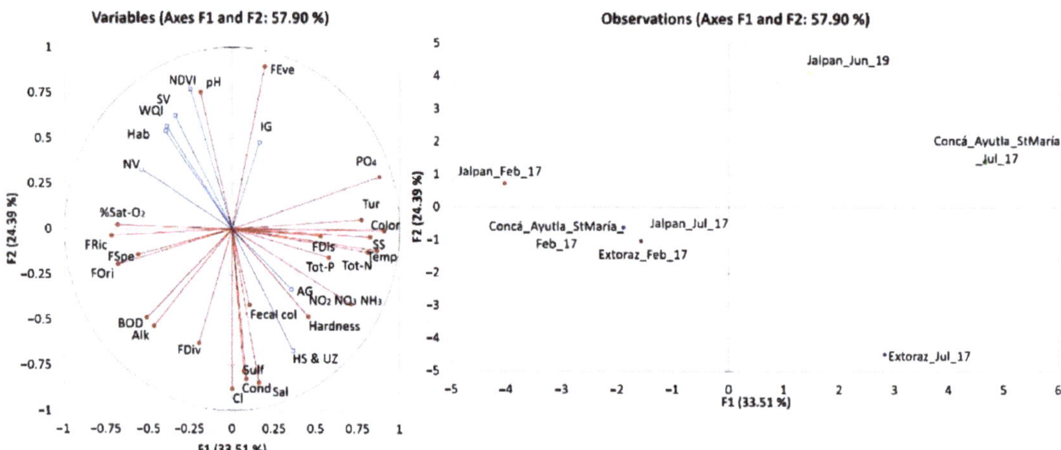

Figure 5. Biplot of the PCA of study sites and seasons (observations) and environmental variables measured in situ and in the laboratory, as well as functional diversity and environmental indices. Upper left quadrant. pH = pH values, NDVI = Normalized Difference Vegetation Index, SV = Secondary vegetation, WQI = Water Quality Index, Hab = Habitat quality, NV = Natural vegetation, %Sat-O_2 = Oxygen saturation (%); Lower left quadrant. FRic = Functional richness, FSpe = Functional specialization, FOri = Functional originality, BOD = Biochemical oxygen demand, Alk = Alkalinity, FDiv = Functional divergence; Upper right quadrant. FEve = Functional evenness, IG = Induced grasslands, PO_4 = Orthophosphates, Tur = Turbidity, Color = Color; Lower right quadrant. SS = Suspended solids, FDis = Functional dispersion, Temp = Water temperature, Tot-P = Total phosphorus, Tot-N = Total nitrogen, AG = Agriculture, NO_2, NO_3, NH_3 = Nitrites, nitrates, and ammonia nitrogen, Fecal col = Fecal coliforms, Hardness = Hardness, HS & UZ = Human Settlements and Urban Zones, Sulf = Sulfates, Cond = Conductivity, Sal = Salinity, Cl = Chlorides.

4. Discussion

The past decade has witnessed an increase in the number of studies focused on changes in functional diversity using aquatic macroinvertebrate assemblages [48] since these show noticeable changes when facing impacts from human activities [49,50]. Our study proposes the use of functional diversity indices calculated from the characterization of the multifunctional space of macroinvertebrate assemblages in a biosphere reserve located in a tropical latitude, and its comparison within the reserve according to the streams and seasons studied. The perspective for the analysis of functional diversity used in our study, based on [5,47,49], is applied for the first time in Mexico, as far as we known [11].

The macroinvertebrate assemblages sampled in streams running across the SGRB comprised a total of 88 families (refer to the Table S3 of Supplementary Material), evidencing the high taxonomic diversity of aquatic macroinvertebrates in the SGRB. On the other hand [51], who studied some of the streams in the SGBR, reported a similar taxonomic richness, with 86 families identified. Based on taxonomic diversity, our work used a database of 52 functional traits divided into four categories of ecological attributes and six of biological attributes [41–45]. Compared to other reports [52,53], this study used a markedly lower number of traits, which highlights the scarcity of autoecology studies addressing macroinvertebrate groups in tropical areas of America [8,54]. Similarly to other authors [46,55,56], we used fuzzy coding to score the affinity of a given trait to each of our taxa; it has been shown that biological functions or attributes related to functional processes in ecosystems are not binomial in nature, but commonly result from multiple responses by a given assemblage [57].

In general, the reserve showed very high values for almost all the functional diversity indices, except for FEve, which is explained by the occurrence of dominant functions throughout the reserve. In this case, these dominant functions were related to nutrient-enrichment processes since collector organisms were present in all streams and monitoring seasons. Besides, the taxa to which these functions are associated showed high overall abundances (>50 individuals). In our study areas, as well as in other Neotropical rivers, high nutrient levels are mainly due to the incorporation of fine particulate organic matter, which is consistent with [58].

The functional multidimensional space of aquatic macroinvertebrate assemblages was evaluated for each stream within the SGRB in different monitoring seasons to identify variations in the functional diversity indices and explore how these changes are related to the functions within the reserve and the environmental variables and indices measured in the streams and monitoring seasons. It was observed that FDis values (Figure 4a) tend to be higher in Concá-Ayutla-Santa María in both seasons, likely because this is located in the mid-terminal portion of the stream. Here, the macroinvertebrate assemblages show generalist trophic habits and adaptations to avoid extreme hydraulic conditions related to their life cycle such as small body sizes that facilitate searching for shelters to avoid being dragged by strong currents [59], (extreme conditions were detected during the rainy season, with high values of suspended solids and turbidity, due to the incorporation of materials from the upper tributaries of this river, in contrast with those seasonal variations in Extoraz river where there are a lower number of tributaries; see Table 1). FRic (Figure 4b) may be considered one of the most important functional diversity indices because it indicates the variation of the functional space [23] in the streams and seasons monitored. This index showed the lowest values in Extoraz in July 2017 and Jalpan in June 2019. In both cases, this may be an effect of the rainy season as described by [60], who demonstrated that high-flow events caused by rains significantly reduced the richness of the macroinvertebrate assemblages. In the Extoraz river, lower FRic values may also reflect the effect of mining pressure (note the proximity of mining activities to the Extoraz river in Figure 1), mainly from mercury extraction in this area [61,62]. The Extoraz river showed significant differences in FRic values between February and July 2017; however, the low values recorded may be related to the local climate and type of vegetation in the basin. The Extoraz river is a stream located in an area with semi-arid climate and surrounded by xeric shrubland. According to [63], currents flowing across semi-arid environments show spatial and temporal changes that modify the vegetation in the riverbanks and riparian zones. Hence, these currents do not offer enough shelter for macroinvertebrates. Given the scarce habitat availability, the effect of the surrounding mining operations may have been intensified during February 2017, probably leading to marked reductions in the number of functions in this stream. In addition, increased conductivity (see the lower right quadrant of the PCA in Figure 5) affects the taxonomic and functional structure of the macroinvertebrate assemblage, as reported by [64]. In the Jalpan river in June 2019, the forest fires that occurred in that year [65] had a significant adverse effect ($p < 0.05$) when compared to this same stream in July 2017; noteworthily, the latter date reached the highest FRic value for the rainy season.

The effects of fires on macroinvertebrate assemblages have been rarely addressed. The reports by [66] showed that fire adversely affects FRic, as observed in the Jalpan river in June 2019. However, FDiv values (Figure 4c) remained relatively unchanged, similar to the findings reported by [67], i.e., this index did not decrease despite environmental and anthropogenic stressors. Other authors suggest that FDiv shows less variations in the presence of urban or agricultural land uses [68], as observed mainly in the Jalpan river. Moreover, high FDiv values (>0.799) indicate that the range of functions may be unique to each stream, with no niche overlap [23]. The highest FEve values (Figure 3d) were recorded in the Jalpan river in June 2019, when a disturbance event caused by forest fires occurred. According to [68], disturbance effects tend to increase functional evenness due to the concentration of the combinations of the most similar traits that result from the presence of tolerant and dominant species over the rest of the assemblage [69]. Functional specialization and originality (Figure 4e,f) have been little addressed in macroinvertebrate assemblages [8]; our results showed high values of functional specialization and originality along the streams. According to [70], functional specialization is an indicator that is sensitive to environmental disturbance. The values observed in this study suggest that these functional diversity indices are seemingly not compromised within the SGRB; the exception is the Jalpan river in June 2019, when the lowest values for these indices were observed.

Finally, the ordination analysis (PCA) (Figure 5) showed two environmental gradients along which the aquatic macroinvertebrate assemblages responds regarding its characterization of the multidimensional functional space and functional diversity indices. On the one hand, the gradient marked by good oxygenation levels associated with the highest FRic, FSpe, and FOri in the main streams is similar to the one reported by [71]. The second gradient, represented by the best vegetation conditions in terms of NDVI, habitat quality, and water quality, was found to be related to high functional diversity indices. In contrast, the Extoraz stream in July 2017 showed the most severe disturbance impacts. Besides, this stream showed high conductivity values (due to the calcareous nature of the basin), which is consistent with [72]. Although the Jalpan river (in June 2019) was located to the upper left quadrant of the biplot associated with the highest FEve, we propose that the lowest FRic recorded was mainly due to the effect of fires in the reserve that year. According to [73], the effects of urbanization on macroinvertebrate assemblages are still poorly understood. The Jalpan river runs across the most urbanized area of the reserve, where structural and functional diversity are subject to multiple stressors, including the adverse effects of forest fires on the aquatic macroinvertebrate assemblages, as reported by [74].

5. Conclusions

Our results represent the first approximation to characterize the multidimensional functional space of the aquatic macroinvertebrate assemblages in a Neotropical biosphere reserve. In general, the functional space of this assemblages within the SGBR is characterized by high values of functional diversity indices. However, some indices, such as functional richness, evenness, and specialization, were sensitive to disturbances in the Extoraz river in February 2017 and the Jalpan river in June 2019. Both findings add to the few published reports about the adverse effects of salinization from mining activities on the structure and function of the aquatic macroinvertebrate assemblages, as well as the impact of forest fires. The approach in this study integrated the responses of functional diversity indices across environmental gradients, which allowed us to identify the major drivers of functional diversity within the SGBR. The highest values of the functional richness, specialization, and originality indices were associated with the best water quality (well-oxygenated waters and low values of PO_4, turbidity, suspended solids, and color) and the best habitat quality, NDVI, and natural vegetation cover. The responses of functional evenness and dispersion were correlated with the streams and seasons that showed impacts from mineralization (Extoraz river in February 2017) or forest fires (Jalpan river, June 2019). Finally, our results revealed that the relationships between the functional diversity indices and the different physicochemical parameters and environmental indices are suitable indicators to evaluate the conditions within the reserve.

Supplementary Materials: The following are available online at https://www.mdpi.com/article/10.3390/d13110546/s1, Table S1: List of macroinvertebrate family and abbreviations, Table S2: Functional traits, Table S3: Macroinvertebrate family per mainstream and study period.

Author Contributions: Conceptualization, E.L.-L., A.J.R.-R. and J.E.S.-D.; data curation, A.E.R.-S. and A.J.R.-R.; formal analysis, A.J.R.-R. and A.E.R.-S.; funding acquisition, E.L.-L. and J.E.S.-D.; investigation, A.J.R.-R. and E.L.-L.; methodology, A.J.R.-R. and A.E.R.-S.; supervision, E.L.-L. and J.E.S.-D.; writing—original draft, A.J.R.-R., A.E.R.-S. and E.L.-L.; writing—review and editing, E.L.-L. All authors have read and agreed to the published version of the manuscript.

Funding: This research was funded by the Secretariat of Research and Post-grade Studies at Instituto Politécnico Nacional SIP 20210132 and SIP 20210133.

Institutional Review Board Statement: Not applicable.

Informed Consent Statement: Not applicable.

Data Availability Statement: Not applicable.

Acknowledgments: We thank Maria Elena Sanchez-Salazar, who edited the English manuscript. We also thank CONACyT for granting a Ph.D. scholarship to one of the authors (A.J.R.-R.).

Conflicts of Interest: The authors declare no conflict of interest. "The funders had no role in the design of the study; in the collection, analyses, or interpretation of data; in the writing of the manuscript, or in the decision to publish the results".

References

1. Dudgeon, D. Multiple threats imperil freshwater biodiversity in the Anthropocene. *Curr. Biol.* **2019**, *29*, R960–R967. [CrossRef]
2. Dudgeon, D.; Arthington, A.H.; Gessner, M.O.; Kawabata, Z.-I.; Knowler, D.J.; Lévêque, C.; Naiman, R.J.; Prieur-Richard, A.-H.; Soto, D.; Stiassny, M.L.J.; et al. Freshwater biodiversity: Importance, threats, status and conservation challenges. *Biol. Rev.* **2006**, *81*, 163–182. [CrossRef]
3. Strayer, D.; Dudgeon, D. Freshwater biodiversity conservation: Recent progress and future challenges. *J. North Am. Benthol. Soc.* **2010**, *29*, 344–358. [CrossRef]
4. Vörösmarty, C.J.; McIntyre, P.B.; Gessner, M.O.; Dudgeon, D.; Prusevich, A.; Green, P.; Glidden, S.; Bunn, S.E.; Sullivan, C.A.; Liermann, C.R.; et al. Global threats to human water security and river biodiversity. *Nature* **2010**, *467*, 555–561. [CrossRef] [PubMed]
5. Mouillot, D.; Graham, N.; Villéger, S.; Mason, N.W.; Bellwood, D.R. A functional approach reveals community responses to disturbances. *Trends Ecol. Evol.* **2013**, *28*, 167–177. [CrossRef] [PubMed]
6. Rico-Sánchez, A.E.; Sundermann, A.; López-López, E.; Torres-Olvera, M.J.; Mueller, S.A.; Haubrock, P.J. Biological diversity in protected areas: Not yet known but already threatened. *Glob. Ecol. Conserv.* **2020**, *22*, e01006. [CrossRef]
7. Almeida, B.D.A.; Green, A.J.; Sebastián-González, E.; Dos Anjos, L. Comparing species richness, functional diversity and functional composition of waterbird communities along environmental gradients in the neotropics. *PLoS ONE* **2018**, *13*, e0200959. [CrossRef]
8. Schmera, D.; Heino, J.; Podani, J.; Erős, T.; Dolédec, S. Functional diversity: A review of methodology and current knowledge in freshwater macroinvertebrate research. *Hydrobiol.* **2017**, *787*, 27–44. [CrossRef]
9. Allan, J.D.; Castillo, M.M.; Capps, K.A. *Stream Ecology: Structure and Function of Running Waters*; Springer Nature: Berlin/Heidelberg, Germany, 2020. [CrossRef]
10. Cardinale, B.J.; Duffy, J.E.; Gonzalez, A.; Hooper, D.U.; Perrings, C.; Venail, P.; Narwani, A.; Mace, G.M.; Tilman, D.; Wardle, D.A.; et al. Biodiversity loss and its impact on humanity. *Nature* **2012**, *486*, 59–67. [CrossRef]
11. Luiza-Andrade, A.; Montag, L.; Juen, L. Functional diversity in studies of aquatic macroinvertebrates community. *Science* **2017**, *111*, 1643–1656. [CrossRef]
12. Rezende, R.S.; Santos, A.M.; Henke-Oliveira, C.; Gonçalves, J.F., Jr. Effects of spatial and environmental factors on benthic a macroinvertebrate community. *Zoologia* **2014**, *31*, 426–434. [CrossRef]
13. Melo, A.S.; Froehlich, C.G. Macroinvertebrates in neotropical streams: Richness patterns along a catchment and assemblage structure between 2 seasons. *J. N. Am. Benthol. Soc.* **2001**, *20*, 1–16. [CrossRef]
14. Resh, V.H. Which group is best? Attributes of different biological assemblages used in freshwater biomonitoring programs. *Environ. Monit. Assess.* **2008**, *138*, 131–138. [CrossRef]
15. Oliveira, A.; Callisto, M. Benthic macroinvertebrates as bioindicators of water quality in an Atlantic forest fragment. *Iheringia Série Zool.* **2010**, *100*, 291–300. [CrossRef]
16. Ruiz-Picos, R.A.; Kohlmann, B.; Sedeño-Díaz, J.E.; López-López, E. Assessing ecological impairments in Neotropical rivers of Mexico: Calibration and validation of the Biomonitoring Working Party Index. *Int. J. Environ. Sci. Technol.* **2017**, *14*, 1835–1852. [CrossRef]
17. Serna, D.J.; Tamaris-Turizo, C.E.; Gutiérrez Moreno, L.C. Spatial and temporal distribution of Trichoptera (Insecta) larvae in the Manzanares river Sierra Nevada of Santa Marta (Colombia). *Rev. Biol. Trop.* **2015**, *63*, 465–477. [CrossRef]
18. Moretti, M.S.; Loyola, R.D.; Becker, B.; Callisto, M. Leaf abundance and phenolic concentrations codetermine the selection of case-building materials by Phylloicus sp. (Trichoptera, Calamoceratidae). *Hydrobiologia* **2009**, *630*, 199–206. [CrossRef]
19. Kohlmann, B.; Vásquez, D.; Arroyo, A.; Springer, M. Taxonomic and Functional Diversity of Aquatic Macroinvertebrate Assemblages and Water Quality in Rivers of the Dry Tropics of Costa Rica. *Front. Environ. Sci.* **2021**, 309. [CrossRef]
20. Colzani, E.; Siqueira, T.; Suriano, M.T.; Roque, F.O. Responses of aquatic insect functional diversity to landscape changes in Atlantic forest. *Biotropica* **2013**, *45*, 343–350. [CrossRef]
21. Motta Díaz, A.J.; Longo, M.; Aranguren-Riaño, N. Temporal variation of taxonomic diversity and functional traits of aquatic macroinvertebrates in temporary rivers on Old Providence Island, Colombia. *Actual. Biológicas* **2017**, *39*, 82–100. [CrossRef]
22. Violle, C.; Navas, M.L.; Vile, D.; Kazakou, E.; Fortunel, C.; Hummel, I.; Garnier, E. Let the concept of trait be functional! *Oikos* **2007**, *116*, 882–892. [CrossRef]

23. Mason, N.W.; Mouillot, D.; Lee, W.G.; Wilson, J.B. Functional richness, functional evenness and functional divergence: The primary components of functional diversity. *Oikos* **2005**, *111*, 112–118. [CrossRef]
24. Morrone, J.J. Biogeographic regionalization of the mexican transition zone. In *The Mexican Transition Zone*; Springer: Cham, Switzerland, 2020; pp. 103–155. [CrossRef]
25. Servicio Geológico Mexicano Conoce GeoInfoMex en 3D. Available online: https://www.gob.mx/sgm/articulos/conoce-el-sistema-de-consulta-de-informacion-geocientifica-geoinfomex?idiom=es (accessed on 18 February 2021).
26. Carabias Lillo, J.; Provencio, E.; de la Maza Elvira, J.; Ruiz Corzo, M. Programa de Manejo Reserva de la Biosfera Sierra Gorda. 1999. Available online: http://www.paot.org.mx/centro/ine-semarnat/anp/AN15.pdf (accessed on 18 September 2021).
27. Ejecutivo, P. Decreto de la Reserva de la Biosfera Sierra Gorda. 1997, pp. 1–11. Available online: http://dof.gob.mx/nota_detalle.php?codigo=4879875&fecha=19/05/1997 (accessed on 18 February 2021).
28. Sanginés, A.E. *Pobreza y Medio Ambiente en México: Teoría y Evaluación de una Política Pública*; Instituto Nacional de Ecología/Universidad Iberoamericana/Instituto Nacional de Administración: Ciudad de México, Mexico, 2003; ISBN 9688594520.
29. Instituto Nacional de Estadística, G.e I. México en Cifras. Available online: https://www.inegi.org.mx/app/areasgeograficas/?ag=22 (accessed on 25 August 2021).
30. Baird, R.B. *Standard Methods for the Examination of Water and Wastewater*, 23rd ed.; American Public Health Association: Washington, DC, USA, 2017; ISBN 9780875532875.
31. Barbour, M.T.; Stribling, J.B.; Verdonschot, P.F.M. The multihabitat approach of USEPA's rapid bioassessment protocols: Benthic macroinvertebrates. *Limnetica* **2006**. [CrossRef]
32. Cornejo, A.; Lopez, E.; Bernal, J. *Protocolo de Biomonitoreo para la Vigilancia de la Calidad del Agua en Afluentes Superficiales de Panama*; Instituto Conmemorativo Gorgas de Estudios de la Salud: Panamá, Panama, 2019; 81p, ISBN 978-9962-13-053-6.
33. Rodríguez-Romero, A.J.; Rico-Sánchez, A.E.; Mendoza-Martínez, E.; Gómez-Ruiz, A.; Sedeño-Díaz, J.E.; López-López, E. Impact of changes of land use on water quality, from tropical forest to anthropogenic occupation: A multivariate approach. *Water* **2018**, *10*, 1518. [CrossRef]
34. INEGI Instituto Nacional de Estadística, Geografía e Informática. 2020. Available online: https://www.inegi.org.mx/temas/mapadigital/ (accessed on 14 September 2021).
35. Glovis, Servicio Geológico de EE. UU. Servicio Geológico. 2021. Available online: https://glovis.usgs.gov/app (accessed on 15 July 2021).
36. Tarpley, J.D.; Schneider, S.R.; Money, R.L. Global Vegetation Indices from the NOAA-7 Meteorological Satellite. *J. Clim. Appl. Meteorol.* **1984**, *23*, 491–494. [CrossRef]
37. Dinius, S.H. Design of an Index of Water Quality. *JAWRA J. Am. Water Resour. Assoc.* **1987**, *23*, 833–843. [CrossRef]
38. Mueller, M.; Pander, J.; Geist, J. Taxonomic sufficiency in freshwater ecosystems: Effects of taxonomic resolution, functional traits, and data transformation. *Freshw. Sci.* **2013**, *32*, 762–778. [CrossRef]
39. Thorp and Covich's Freshwater Invertebrates. In *Thorp and Covich's Freshwater Invertebrates*; Academic Press: Cambridge, MA, USA, 2018. [CrossRef]
40. Thorp, J.H.; Rogers, D.C. Thorp, J.H.; Rogers, D.C. Thorp and Covich's Freshwater Invertebrates. In *Ecology and Classification of North American Freshwater Invertebrates*; Academic press: Cambridge, MA, USA, 2014; ISBN 9780080889818.
41. Merritt, R.W.; Cummins, K.W. (Eds.) *An Introduction to the Aquatic Insects of North America*; Kendall Hunt: Dubuque, IA, USA, 2008.
42. Poff, N.L.; Olden, J.D.; Vieira, N.K.; Finn, D.S.; Simmons, M.P.; Kondratieff, B.C. Functional trait niches of North American lotic insects: Traits-based ecological applications in light of phylogenetic relationships. *J. N. Am. Benthol. Soc.* **2006**, *25*, 730–755. [CrossRef]
43. Vieira, N.K.; Poff, N.L.; Carlisle, D.M.; Moulton, S.R.; Koski, M.L.; Kondratieff, B.C. A database of lotic invertebrate traits for North America. *US Geol. Surv. Data Ser.* **2006**, *187*, 1–15. [CrossRef]
44. Jesus, T.M.G.M.D. Ecological, anatomical and physiological traits of benthic macroinvertebrates: Their use on the health characterization of freshwater ecosystems. *Limnetica* **2008**, *27*, 079–092.
45. Ramírez, A.; Gutiérrez-Fonseca, P.E. Functional feeding groups of aquatic insect families in Latin America: A critical analysis and review of existing literature. *Rev. Biol. Trop.* **2014**, *62*, 155–167. [CrossRef]
46. Mondy, C.P.; Usseglio-Polatera, P. Using fuzzy-coded traits to elucidate the non-random role of anthropogenic stress in the functional homogenisation of invertebrate assemblages. *Freshw. Biol.* **2014**, *59*, 584–600. [CrossRef]
47. Villéger, S.; Grenouillet, G.; Brosse, S. Decomposing functional β-diversity reveals that low functional β-diversity is driven by low functional turnover in E uropean fish assemblages. *Glob. Ecol. Biogeogr.* **2013**, *22*, 671–681. [CrossRef]
48. Laliberté, E.; Legendre, P. A distance-based framework for measuring functional diversity from multiple traits. *Ecology* **2010**, *91*, 299–305. [CrossRef] [PubMed]
49. Villéger, S.; Mason, N.W.; Mouillot, D. New multidimensional functional diversity indices for a multifaceted framework in functional ecology. *Ecology* **2008**, *89*, 2290–2301. [CrossRef] [PubMed]
50. Sundar, S.; Heino, J.; Roque, F.D.O.; Simaika, J.P.; Melo, A.S.; Tonkin, J.D.; Silva, D.P. Conservation of freshwater macroinvertebrate biodiversity in tropical regions. *Aquat. Conserv. Mar. Freshw. Ecosyst.* **2020**, *30*, 1238–1250. [CrossRef]
51. Torres-Olvera, M.J.; Durán-Rodríguez, O.Y.; Torres-García, U.; Pineda-López, R.; Ramírez-Herrejón, J.P. Validation of an index of biological integrity based on aquatic macroinvertebrates assemblages in two subtropical basins of central Mexico. *Lat. Am. J. Aquat. Res.* **2018**, *46*, 945–960. [CrossRef]

52. Voß, K.; Schäfer, R.B. Taxonomic and functional diversity of stream invertebrates along an environmental stress gradient. *Ecol. Indic.* **2017**, *81*, 235–242. [CrossRef]
53. White, J.C.; Hill, M.J.; Bickerton, M.A.; Wood, P.J. Macroinvertebrate taxonomic and functional trait compositions within lotic habitats affected by river restoration practices. *Environ. Manag.* **2017**, *60*, 513–525. [CrossRef] [PubMed]
54. Maasri, A. A global and unified trait database for aquatic macroinvertebrates: The missing piece in a global approach. *Front. Environ. Sci.* **2019**, *7*, 65. [CrossRef]
55. Usseglio-Polatera, P.; Bournaud, M.; Richoux, P.; Tachet, H. Biomonitoring through biological traits of benthic macroinvertebrates: How to use species trait databases? In *Assessing the Ecological Integrity of Running Waters*; Springer: Dordrecht, The Netherlands, 2000; pp. 153–162. [CrossRef]
56. Tomanova, S.; Moya, N.; Oberdorff, T. Using macroinvertebrate biological traits for assessing biotic integrity of neotropical streams. *River Res. Appl.* **2008**, *24*, 1230–1239. [CrossRef]
57. Martini, S.; Larras, F.; Boyé, A.; Faure, E.; Aberle, N.; Archambault, P.; Ayata, S.D. Functional trait-based approaches as a common framework for aquatic ecologists. *Limnol. Oceanogr.* **2021**, *66*, 965–994. [CrossRef]
58. Díaz-Rojas, C.A.; Motta-Díaz, Á.J.; Aranguren-Riaño, N. Study of the taxonomic and functional diversity of the macroinvertebrate assemblages in an Andean mountain river. *Rev. Biol. Trop.* **2020**, *68*, 132–149. [CrossRef]
59. Washko, S.; Bogan, M.T. Global patterns of aquatic macroinvertebrate dispersal and functional feeding traits in aridland rock pools. *Front. Environ. Sci.* **2019**, *7*, 106. [CrossRef]
60. Theodoropoulos, C.; Vourka, A.; Stamou, A.; Rutschmann, P.; Skoulikidis, N. Response of freshwater macroinvertebrates to rainfall-induced high flows: A hydroecological approach. *Ecol. Indic.* **2017**, *73*, 432–442. [CrossRef]
61. Martínez-Trinidad, S.; Hernández Silva, G.; Ramírez Islas, M.E.; Martínez Reyes, J.; Solorio Munguía, G.; Solís Valdez, S.; García Martínez, R. Total mercury in terrestrial systems (air-soil-plant-water) at the mining region of San Joaquín, Queretaro, Mexico. *Geofísica Int.* **2013**, *52*, 43–58. [CrossRef]
62. Rico-Sánchez, A.E.; Rodríguez-Romero, A.J.; Sedeño-Díaz, J.E.; López-López, E.; Sundermann, A. Aquatic Macroinvertebrate Assemblages in Rivers Under the Influence of Mining Activities. *Sci. Rep.* **2021**, in press. [CrossRef]
63. Wang, Z.; Wang, W.; Zhang, Z.; Hou, X.; Ma, Z.; Chen, B. River-groundwater interaction affected species composition and diversity perpendicular to a regulated river in an arid riparian zone. *Glob. Ecol. Conserv.* **2021**, *27*, e01595. [CrossRef]
64. Sowa, A.; Krodkiewska, M.; Halabowski, D. How does mining salinisation gradient affect the structure and functioning of macroinvertebrate communities? *Water Air Soil Pollut.* **2020**, *231*, 1–19. [CrossRef]
65. La Jornada. 2019. Available online: https://www.jornada.com.mx/2019/05/24/estados/028n1est (accessed on 22 August 2021).
66. Williams-Subiza, E.A.; Brand, C. Functional response of benthic macroinvertebrates to fire disturbance in patagonian streams. *Hydrobiologia* **2021**, *848*, 1575–1591. [CrossRef]
67. Juvigny-Khenafou, N.P.; Piggott, J.J.; Atkinson, D.; Zhang, Y.; Macaulay, S.J.; Wu, N.; Matthaei, C.D. Impacts of multiple anthropogenic stressors on stream macroinvertebrate community composition and functional diversity. *Ecol. Evol.* **2021**, *11*, 133–152. [CrossRef] [PubMed]
68. Barnum, T.R.; Weller, D.E.; Williams, M. Urbanization reduces and homogenizes trait diversity in stream macroinvertebrate communities. *Ecol. Appl.* **2017**, *27*, 2428–2442. [CrossRef]
69. Hillebrand, H.; Bennett, D.M.; Cadotte, M.W. Consecuencias del dominio: Una revisión de los efectos de la uniformidad en los procesos de los ecosistemas locales y regionales. *Ecología* **2008**, *89*, 1510–1520. [CrossRef] [PubMed]
70. Mykrä, H.; Heino, J. Decreased habitat specialization in macroinvertebrate assemblages in anthropogenically disturbed streams. *Ecol. Complex.* **2017**, *31*, 181–188. [CrossRef]
71. Croijmans, L.; De Jong, J.F.; Prins, H.H.T. Oxygen is a better predictor of macroinvertebrate richness than temperature—a systematic review. *Environ. Res. Lett.* **2021**, *16*, 023002. [CrossRef]
72. Carrera-Villacrés, D.V.; Crisanto-Perrazo, T.; Ortega-Escobar, H.; Ramírez-García, J.; Espinosa-Victoria, D.; Ramírez-Ayala, C.; Sánchez-Bernal, E. Salinidad cuantitativa y cualitativa del sistema hidrográfico Santa María-Río Verde, México. *Tecnol. Cienc. Del Agua* **2015**, *6*, 69–83.
73. Gál, B.; Szivák, I.; Heino, J.; Schmera, D. The effect of urbanization on freshwater macroinvertebrates–knowledge gaps and future research directions. *Ecol. Indic.* **2019**, *104*, 357–364. [CrossRef]
74. Verkaik, I.; Vila-Escale, M.; Rieradevall, M.; Baxter, C.V.; Lake, P.S.; Minshall, G.W.; Prat, N. Stream macroinvertebrate community responses to fire: Are they the same in different fire-prone biogeographic regions? *Freshw. Sci.* **2015**, *34*, 1527–1541. [CrossRef]

MDPI
St. Alban-Anlage 66
4052 Basel
Switzerland
Tel. +41 61 683 77 34
Fax +41 61 302 89 18
www.mdpi.com

Diversity Editorial Office
E-mail: diversity@mdpi.com
www.mdpi.com/journal/diversity

www.ingramcontent.com/pod-product-compliance
Lightning Source LLC
LaVergne TN
LVHW070059170726
843515LV00007B/1567

* 9 7 8 3 0 3 6 5 4 9 4 1 5 *